EXTRACTION ECOLOGIES AND THE LITERATURE
OF THE LONG EXHAUSTION

Extraction Ecologies and the Literature of the Long Exhaustion

Elizabeth Carolyn Miller

PRINCETON UNIVERSITY PRESS

PRINCETON & OXFORD

Published by Princeton University Press
41 William Street, Princeton, New Jersey 08540
6 Oxford Street, Woodstock, Oxfordshire OX20 1TR

press.princeton.edu

All Rights Reserved

Library of Congress Cataloging-in-Publication Data

Names: Miller, Elizabeth Carolyn, 1974– author.
Title: Extraction ecologies and the literature of the long exhaustion /
    Elizabeth Carolyn Miller.
Description: Princeton, New Jersey : Princeton University Press, [2021] |
    Includes bibliographical references and index.
Identifiers: LCCN 2021008090 | ISBN 9780691205267 (hardback) |
    ISBN 9780691205533 (paperback) | ISBN 9780691230559 (ebook)
Subjects: LCSH: Mines and mineral resources in literature. | English fiction—
    19th century—History and criticism. | English fiction—20th century—
    History and criticism. | Industrialization in literature. |
    BISAC: LITERARY CRITICISM / Modern / 19th Century |
    TECHNOLOGY & ENGINEERING / Mining | LCGFT: Literary criticism.
Classification: LCC PR830.M56 M55 2021 | DDC 823/.809356—dc23
LC record available at https://lccn.loc.gov/2021008090

British Library Cataloging-in-Publication Data is available

Editorial: Anne Savarese and James Collier
Production Editorial: Ellen Foos
Jacket/Cover Design: Pamela Schnitter
Production: Brigid Ackerman
Publicity: Alyssa Sanford and Amy Stewart
Copyeditor: Kathleen Kageff

Cover art: Edwin Butler Bayliss (1875–1950), *Tipping the Slag*, oil on canvas, 61×92 cm.
Wolverhampton Art Gallery, Wolverhampton, UK

Publication of this book has been aided by the National Endowment for the Humanities.

NATIONAL
ENDOWMENT
FOR THE
HUMANITIES

This book has been composed in Miller

10 9 8 7 6 5 4 3 2 1

For Ambrose and Giacomo, with love

CONTENTS

ILLUSTRATIONS

ACKNOWLEDGMENTS

AS I WRITE these acknowledgments my country is in the grip of a pandemic during the waning days of a disastrous presidential administration, and even the recent past feels hopelessly remote, yet most of this book was written before the isolations of the COVID era, with the help of a network of colleagues and interlocutors who challenged and supported my thinking at every stage. This leaves me with many people to remember and thank.

Colleagues in the English Department at University of California, Davis, were my first audience for this work and fostered an intellectual home in which it could grow. Margaret Ronda read the first section of *Extraction Ecologies* I wrote back in 2014 and provided generous advice and encouragement all along the way. I owe a great debt, too, to Tobias Menely, Tiffany Werth, and Mike Ziser, who read and commented on the first draft of many sections of the book and inspired me with their own scholarship. My Victorianist colleagues, Kathleen Frederickson and Parama Roy, are a joy to work with, and their influence on my thinking is visible everywhere in this book. Fran Dolan has been a role model and a support system all in one. Other colleagues offered suggestions over email or at our faculty colloquium: thanks to Gina Bloom, Hsuan Hsu, Mark Jerng, Alessa Johns, Scott Shershow, and Matthew Vernon. Special thanks to Desirée Martín for help with the sections that touch on Mexico. John Marx and Claire Waters were supportive departmental administrators; Margaret Ferguson, David Simpson, and Scott Simmon were mentors and friends. I have learned much from conversations with Davis colleagues in environmental studies and the environmental humanities: thanks especially to Marisol de la Cadena, Claire Goldstein, Andrew Latimer, Beth Rose Middleton, Colin Murphy, Louis Warren, and Stephen Wheeler.

My work on this project was enhanced significantly by years of conversations with brilliant Davis graduate students. I owe a special debt to those who participated in an informal grad-faculty ecocritical reading group from 2016 to 2017, a rotating cast that included Kristin George Bagdanov, Rachel Dewitt, Sophia Bamert, Rebecca Hogue, Jonathan Radocay, Elizabeth Giardina, Katherine Buse, and Benjamin Blackman. Thanks, too, to Tom Hintze and Kristin George Bagdanov for organizing the Green New Deal Research Cluster. I have learned a great deal from the PhD students with whom I've worked most closely during the years I was writing this book: William Hughes, Rebecca Kling, Jessica Krzeminski, Michael Martel, Margaret Miller, Lauren Peterson, Leilani Serafin, Jennifer Tinonga-Valle, and Tobias Wilson-Bates. Finally, special thanks to the graduate research assistants who supported this project, especially Margaret Miller and Lauren Peterson, who saw me through the final hurdles of manuscript preparation, and Jennifer Tinonga-Valle, who helped navigate our library's huge and heavy volumes of the *Mining Journal*. Lauren Peterson also created the index. Research assistance from William Hughes, Jessica Krzeminski, and Rebecca Kling was also key to this book's completion.

Beyond Davis, I have many colleagues to thank for discussion and collaboration through conference panels, email exchanges, reading groups, and general intellectual camaraderie, especially Mark Allison, Sukanya Banerjee, Katharina Boehm, Siobhan Carroll, Elizabeth Chang, Devin Garofalo, Devin Griffiths, Jonathan Grossman, Daniel Hack, Taryn Hakala, Nathan Hensley, Jeffrey Insko, Benjamin Kohlmann, Deanna Kreisel, Tom Laughlin, Barbara Leckie, Casie LeGette, Margaret Linley, Kyle McAuley, Richard Menke, John MacNeill Miller, Benjamin Morgan, John Regan, Mario Ortiz Robles, Philip Steer, Michael Tondre, Lynn Voskuil, Marcus Waithe, and Daniel Williams. Allen MacDuffie and Jesse Oak Taylor provided crucial feedback on the manuscript for this book as well as the luminous example of their own brilliant work; truly, I cannot thank them enough. Robert Aguirre sent a stack of sources on Victorian Latin America, for which I am grateful. My graduate advisers, Susan David Bernstein, Caroline Levine, and Rebecca Walkowitz, have remained steadfast friends and trusted mentors for many years, and I am also deeply indebted to Florence Boos, Joseph Bristow, Linda Hughes, John Kucich, John Plotz, and Talia Schaffer. My arguments have sharpened and (hopefully) improved thanks to audiences at Rutgers University, Indiana University, Vanderbilt University, University of Colorado Boulder, Harvard University, Cambridge University, University of Warwick, Simon Fraser University, University of Southampton, Dartmouth College, and Newcastle University. I am grateful to the colleagues who arranged these talks: John Kucich, Lara Kriegel, Monique Morgan, Rachel Teukolsky, Sue Zemka, John Plotz, Aeron Hunt, Jan-Melissa Schramm, Michael Meeuwis, Ross Forman, Matt Hussey, Justine Pizzo, Alysia Garrison, and Ella Mershon. I would also like to thank Susan Hamilton and Eddy Kent, organizers of the 2017 North American Victorian Studies Association Conference, who invited me to give a keynote lecture from this project.

A number of institutions supported my work on *Extraction Ecologies*. The National Endowment for the Humanities and the John Simon

Guggenheim Memorial Foundation provided crucial fellowship support for which I am immensely grateful. (Any views, findings, conclusions, or recommendations expressed in this book do not necessarily represent those of the National Endowment for the Humanities.) I also thank the University of California, Davis, for supporting my research in many different material ways. During two periods in residence over the course of writing this book, Clare Hall, Cambridge, became a second home, and I owe special thanks to Gillian Beer for sponsoring my affiliation there. Librarians at Cambridge University Library and UC Davis's Shields Library were unfailingly resourceful, and I am also grateful to Tom Randall, trustee of Somerset Coalfield Life at the Radstock Museum, who helped with the section on *Jane Rutherford* and sent sources on the Wellsway Disaster. Anne Savarese, Ellen Foos, and others at Princeton University Press provided an ideal home for the manuscript to become a book. Parts of the William Morris section in chapter 3 originally appeared in the spring 2015 issue of *Victorian Studies* as "William Morris, Extraction Capitalism, and the Aesthetics of Surface," and some parts from chapter 1 originally appeared in the February 2020 issue of *Victorian Literature and Culture* as "Drill, Baby, Drill: Extraction Ecologies, Open Temporalities, and Reproductive Futurity in the Provincial Realist Novel."

Finally, I am grateful to my family, who made this project possible. Thanks to my parents, siblings, and in-laws: Frank, Cathy, and Cristina Miller; Sarah Miller and Jon Konrath; Stephanie Theron; Vickie Simpson; and all the Strattons. During the time I was writing *Extraction Ecologies*, we lost my beloved stepmom of thirty years, Mary Ellen Powers, and my dear grandfather James (Giacomo) Ghiardi. I remain grateful, always, for their love and care. As a child my grandfather emigrated with his family from northern Italy to the Upper Peninsula of Michigan, where a community from their home region in Piedmont had settled to work the mines of the Iron Range. Growing up, on summer and Thanksgiving trips to the UP, I saw a sublime landscape indelibly marked by mining, and I experienced a tight-knit immigrant community uprooted and replanted through the global force of extractive industry. *Extraction Ecologies* has been brewing in my imagination, at some level, ever since. Moving to California as an adult in my thirties and learning the history of the Gold Rush brought these old associations to the surface. The birth of my twins was, however, the most direct catalyst for this book. When they arrived in 2012, I found myself reflecting more and more, during wakeful nights of rocking and feeding, on what the future would hold for their generation, and I felt an urgent impulse to shift my efforts as a writer and researcher to the great collective challenge of transforming our ideas of Earth and learning to living within its bounds. I dedicate this book to Ambrose and Giacomo, in honor of their tender hopes for a better world (not to mention their joyful enthusiasm for visiting historical mining sites). To Matthew Stratton, I offer my final and deepest thanks. In a year of COVID, California wildfires, and turmoil everywhere, I could have never made it through the final hurdles of this project without his support and his love, for which I am ever grateful.

EXTRACTION ECOLOGIES AND THE LITERATURE OF THE LONG EXHAUSTION

Introduction

Come skill, and the cunning needed;
lay out, the lie of the land;
secret stories, beneath the feet,
locked up in layers, in levels below.

. . .

Unlock the store, of stories here.

MICHAEL ROSEN, "CHARMS FOR GRIME'S GRAVES" (2009)

OF ALL THE material legacies of Britain's industrial, imperial era, which will last the longest? If you ask a geologist, the answer would be mines. Jan Zalasiewicz, chair of the Anthropocene Working Group of the International Commission on Stratigraphy, has written with Colin N. Waters and Mark Williams that "the extensive exploitation of the subsurface environment" (4) that commenced with the British Industrial Revolution is an anthropogenic phenomenon with "no analogue in the Earth's 4.6 billion year history" (4). "Anthroturbation"—their term for human delving into the earth and its resulting geological transformation—"shows notable inflections" in the period following the early nineteenth-century rise of the steam engine, and while such subsurface modifications are easily neglected because they are "out of sight, out of mind," the "deep subsurface changes . . . are permanent on any kind of human timescale, and of long duration even geologically." These mines have "imprint[ed] signals on to the geological record," in other words, that will outlast almost everything (3).

The rise of industrialized mining was a geologically legible event, notable even in the context of sublimely deep timescales, but does the literature of the period attend to this unprecedented transformation taking place under its authors' feet, and if so, how? To ask these questions is to invite broader questions about the extent to which literature is embedded in natural environments and histories, and the extent to which humanist critique can take on concerns of geological scale—questions that are now being explored within and beyond the fields of nineteenth- and early twentieth-century literature.[1] To ask these questions is also to ask what industrial extraction meant, and how it transformed humans' relation to and perception of the natural world. Kenneth Pomeranz has described the period around the Industrial Revolution as the moment when correlated factors of "overseas extraction" and Britain's "epochal turn to fossil fuels" produced nothing short of a new global economy (23). Certainly, many writers and observers at the time remarked on the extraordinary new scope of underground extraction; in an 1892 account originally published in the magazine the *Graphic*, for example, Randolph Churchill reports on a treasure-hunting journey to South Africa and the colossal size of the diamond mines he saw there: "the De Beers and the Kimberley mines are probably the two biggest holes which greedy man has ever dug into the earth" (40–41). Big holes and greedy men feature frequently in mining literature, as we shall see, but the ripple effects of the global project of industrial extraction transformed literature and narrative at a far more fundamental level, and literature's mediation of extractivism reshaped form, genre, and discourse in ways that this book will describe.[2]

Extraction Ecologies sets out to show that the industrialization of underground resource extraction shaped literary form and genre in the first century of the industrial era, from the 1830s to the 1930s, just as literary form and genre contributed to new ways of imagining an extractible Earth. Industrialization was a long process that happened unevenly across the globe, and the "industrial era" is admittedly a rather imprecise and local designation, but I use the term in this book to describe the period that began in the early 1830s with the decisive shift to steam power in British manufacturing and distribution and ended in the late 1930s with the dawn of the nuclear era and the launch of the Manhattan Project.[3] With this chronology I do not intend to convey a steady, sequential parade of energy regimes, as though extracted fossil fuels were unimportant before 1830 or ceased to matter when the expansion of atomic theory gave birth to a new vision of energy as existing in all matter (not just subsurface hydrocarbons). What I do hope to capture, however, is a period when Britain came to understand itself as an empire thoroughly dependent on extraction: an extraction-based industrial society irretrievably bound up with the mining of underground material, with no viable alternative capable of preserving existing social relations.[4] Just as the rhythms of agricultural life and labor are bound up in the forms of the pastoral, I argue, the age of industrial extraction ushered in a new sense of human-natural relations, and with it a new literature.[5]

Mining has a long history, but large-scale industrial mining was a nineteenth-century phenomenon, and *Extraction Ecologies* explores the magnitude of its socio-environmental impact—an impact that extends deeply into literature and culture and deeply into the present. In this book I interpret literary form and genre as signals for habits of mind and ways of thinking about the world that have material causes as well as long-term material effects. Form and genre are important objects of environmental analysis, I argue, because they are epistemological structures that embed our most fundamental conceptual formations; what is more, they are mobile and repeatable across time and space. My aim is to show how such conceptual formations transformed under industrial extractivism, but also to express how literary form and genre produce and extend extractivism as a mode of environmental understanding because of the deep and durational qualities of discourse. In *The Ideas in Things*, Elaine Freedgood notes that "cultural knowledge is stored in a variety of institutional forms" and "is also stored at the level of the word" (23). Words, narratives, forms, and genres both preserve ways of thinking about the environment and carry them forward. Ursula Le Guin imagined fiction as a "carrier bag" for storing and sharing the story of life, prompting Donna Haraway to wonder what the "carrier bag for terraforming" might include (Haraway 121). *Extraction Ecologies* is about literary-environmental exchange, the "carrier bag for terraforming," and it rests, finally, on the idea that discourse makes environment as environment makes discourse. There is a temptation, in a project like *Extraction Ecologies*, to turn to meta-analysis focused on surface reading, text mining, and other methodological debates in literary studies, but in the following chapters I have sought instead to maintain a focus on the material impacts of extraction as mediated through literature and to avoid getting lost in the metaphorics of mining to the extent that I can. Because of the durational qualities of language, genre, and form, literature engages with environmental materiality across time, and for this reason it is a crucial archive for understanding the relation between environmental history and environmental crises today.

The urge to think now about extraction, ecology, and literature comes both from the relentless ecological calamities that surround us in our troubled present and from a recognition of the long historical roots of these calamities. Two centuries into industrial life, we find ourselves in the midst of ecological emergency, and many of the most pressing hazards associated with this crisis can be traced to the extraction-based economy that emerged with Britain's early nineteenth-century transition to steam. From metals to minerals to coal, the British imperial world saw a ramping up of extraction as the steam engine and other new technologies, including new explosives such as dynamite and TNT, contributed to a massive acceleration in extraction and the global establishment of an extractivist version of ecological imperialism.[6] The extraction boom indelibly marked the natural and social worlds of the industrial era and beyond, and this book shows how literature is bound up with industrial ecologies and the conditions of existence that govern life within them.

Extraction Ecologies

My titular phrase "extraction ecologies" is intended to suggest a tension between its two key terms. The word "extraction" is from the Latin *extrahĕre*, to draw out, and its first definition in the *Oxford English Dictionary* is "the action or process of drawing (something) out of a receptacle; the pulling or taking out (of anything) by mechanical means." "Ecology," on the other hand, was first used in 1866 by German biologist Ernst Haeckel to denote the principles of interrelationality and interdependence that characterize natural life: "By ecology, we mean the whole science of the relations of the organism to the environment including, in the broad sense, all the 'conditions of existence.' These are partly organic, partly inorganic in nature."[7] While the underlying idea of "extraction" thus presumes the ability to withdraw one component from the "receptacle" of nature, "ecology," by contrast, suggests a complex of interdependences from which no single part can be removed in isolation. The industrial era saw a pronounced tension between these two formulations of nature: just as new ecological and evolutionary theories of the natural world were coming to recognize the profound interdependence of its many parts, new industrial technologies were perfecting capacities for the removal or derangement of these parts.[8]

Human extraction of underground mineral resources has a long history, dating back to the Neolithic and even the Paleolithic eras. "Charms for Grime's Graves," the series of poetic "charms" from which I take my epigraph, was inspired by a forty-five-hundred-year-old flint mine—one of very few known to exist in Britain. The land around Grime's Graves remains, to this day, pockmarked by hollows and pits, but such early human etchings on the landscape—such stories of Earth's stores, to use the poem's alliterative language—lack the magnitude of industrial mining in terms of depth and pervasiveness. It was the water table that prevented earlier forms of mining from making an indelible stratigraphic signature of the kind Zalasiewicz, Waters, and Williams identify with the industrial era. In the struggle against groundwater, steam-powered pumps to drain the mines of water were a crucial turning point at which industrial-scale anthropogenic exploitation of the subsurface could really begin.[9] This is one reason that *Extraction Ecologies* will focus on extraction as an *activity* rather than on a particular mineral commodity such as coal, for mining of all kinds was transformed and accelerated by the technology of steam.[10]

The steam engine as a signal event in environmental history has been much discussed, but what is often unremarked is that it originally developed as a mining technology. Andreas Malm's *Fossil Capital* provides an in-depth account of how steam power came definitively to supersede water power in the 1830s English textile industry, but long before steam's capacities had developed to the point where it was able to achieve this, the earliest engines had a narrower purpose: they were built to pump water out of mines. Englishman Thomas Savery first unveiled the atmospheric steam pump in 1702, followed by Thomas Newcomen, who in 1712 "built the first really useful steam engine on the basis of Savery's patent": a pump that could "raise as much water as 5 horses" (Sieferle 129). As Matthias Dunn, a mining engineer, wrote in 1844, the steam engine was put into use "for the purpose of drawing water" in the Newcastle coalfields by 1721, and by 1769 there were at least ninety-nine "engines at work drawing water" (22, 24). At this time the engine "was imperfectly understood" and "the collieries in operation were necessarily those whose seams were lying at trifling depths from the surface, and not burthened with any considerable quantities of water" (42). The invention of the automatic centrifugal governor in 1788 was an important advance in engine technology, and in 1800, when James Watt's patent expired on his more efficient engine, "the fuel savings of his machine quickly resulted in its general success" (Sieferle 131). This was part of "a series of great and organic improvements [that] succeeded each other, not only in the erection of the various steam-engines for pumping, but in every other department of colliery engineering" (Dunn 50).[11]

By the 1840s, an integrated chain of steam-powered technologies, including everything from pumping to transport, contributed to a dramatic acceleration in coal extraction, and "the winnings of collieries, followed by the building of ships, and the extension of railways, caused an influx of that torrent of capital which has since so completely outrun all legitimate demand" (Dunn 50). The new capacity to drain mines was thus crucial to a major early nineteenth-century shift in the use of coal in Britain.[12] As E. A. Wrigley explains, "until the end of the eighteenth century coal was almost exclusively a source of *heat* energy. The principal traditional sources of *mechanical* energy, animal and human muscle, remained dominant until the early decades of the nineteenth century." The Industrial Revolution, in Wrigley's view, was "accomplished" when coal became a "convenient source" for mechanical energy (*Path* 31). With this change Britain transformed away from an organic economy and became the world's first extraction-based economy. A published letter from T. Parton of Willenhall neatly sums up this transition in the 3 April 1869 issue of the *Mining Journal*: "the Lord Chancellor now sits upon a bag of wool, but wool has long ceased to be emblematical of the staple commodity of England: he ought to sit upon a bag of coals" (238).

The inauguration of the mining press, as this quotation suggests, announced the new era of industrial extraction with periodicals such as *Quarterly Mining Review* launched in 1830 and the *Mining Journal* launched in 1835, both directed at investors, engineers, and mine owners. The *Mining Journal*, the major periodical in the field, published other works besides the journal at its office in Fleet Street, contributing to a burgeoning professional and technical print culture on extraction[13] (figure 0.1 shows an advertisement from the *Mining Journal*). Beyond such journals, literature itself was a crucial print mediator or "carrier" of extractivism, as this book will describe. Coal's rise has now been widely discussed in historical accounts of industrial Britain, but this rise was part of a larger social transformation to an extraction-based life that had cultural, aesthetic, and discursive elements as well as environmental, economic, and technological elements.[14]

It is a premise of this study that the extraction of underground mineral resources—not only coal, but gold, iron, tin, copper, silver, and more—can be conceived of as a singular activity, and that this activity of extraction was bound up with a new cluster of socio-environmental conditions: extractivism. The term "extractivism" names a complex of cultural, discursive, economic, environmental, and ideological factors related to the extraction of underground resources on a large, industrial scale. Although my use of the term focuses on the conditions that attend underground mineral resource extraction specifically, I also draw on Naomi Klein's use of "extractivism" not only "to describe economies based on removing ever more raw materials from the earth, usually for export to traditional colonial powers," but more broadly as a "resource-depleting model," a

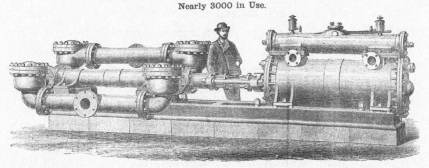

TANGYE BROTHERS AND HOLMAN,
10, LAURENCE POUNTNEY LANE, LONDON,
CORNWALL WORKS (TANGYE BROTHERS), BIRMINGHAM,
NEWCASTLE-ON-TYNE (TANGYE BROTHERS AND RAKE), OFFICES AND WAREHOUSE, ST. NICHOLAS' BUILDINGS.
SOLE MAKERS OF

THE "SPECIAL" DIRECT-ACTING STEAM PUMPING ENGINES
FOR FORCING WATER FROM MINES.
Nearly 3000 in Use.

The "SPECIAL" Direct-acting Steam Pumping Engines require no costly Engine Houses or massive foundations, no repetition of Plunger Lifts, ponderous Connecting-rods, or complication of Pitwork, and allow a clear shaft for hauling purposes.

FIGURE 0.1. Advertisement from the *Mining Journal*, 11 January 1873, 56.

"nonreciprocal, dominance-based relationship with the earth, one purely of taking . . . the opposite of stewardship, which involves taking but also taking care that regeneration and future life continue" (*This* 169).[15]

Of course, there were important differences between mining for coal and mining for gold, not least that coal was mostly mined in Britain and gold was mostly mined on the imperial frontier; I will attend to these differences with care throughout the study, but I want to emphasize here the two major similarities that yoke together these various forms of extraction as a singular activity. First, extraction of all kinds relied on the use of steam for the draining of mines, the crushing of ore, and the transport of mined commodities. Virtually every technological component of the extraction supply chain was accelerated phenomenally by steam power, and thus the accelerated extraction of coal led to more intense exploitation of all subsurface resources, and vice versa. As Rolf Peter Sieferle puts it, "The superabundance of fossil energy put metals into frenzied circulation," which "is the metabolic basis of the new scale of the pollution problem as it arose during industrialization" (137). Secondly, no matter which underground mineral resources were being mined, they were ontologically connected by their material finitude. Finitude and non-reproducibility, above all, distinguish underground resource mining as an extractive process.

Extractive industry can never benefit from regeneration or replenishment of its product but can only move on to a new vein or a new site.[16] The mood of finitude, of removing something that is irreplaceable and subject to looming environmental limits, pervades extraction ecology.

Today the term "extraction" is often used to describe other industries besides mining, industries such as fishing and forestry that likewise involve the removal of raw material from a receptacle where it is ostensibly embedded: trees from a forest or fish from the ocean, for example. These industries are also subject to limits. Old-growth trees are not capable of regeneration on human timescales, as has been brilliantly narrated in Richard Powers's recent novel *The Overstory* (2018), and worldwide fish populations have been decimated by centuries of overfishing, described movingly in W. Jeffrey Bolster's *The Mortal Sea* (2012). Forestry and fishing thus might seem to rely on the harvesting of finite resources in the same way as mining, and indeed, many now fear that soil fertility, too, could be a finite resource, subject to overextraction, such that agriculture would fit in this category as well.[17] In 1892, political economist Charles Stanton Devas worried about "exhaustive farming" as well as the "extermination or diminution of useful animals and plants" as two "injuries which the earth has received" in consequence of the Industrial Revolution (79–80), a reminder that animal species or biodiversity, like soil fertility, can similarly be considered finite resources.[18] Such losses have only accelerated since the industrial era, and indeed it is not unreasonable to say that we are now faced with apparent limits for almost every aspect of the natural world that was once considered cyclical: air, water, soil, life itself.[19] The Great Acceleration might be better termed the Great Extraction, or perhaps the Great Subtraction.[20]

Despite this current crisis of regeneration that seems to touch nearly every part of the natural world on which we depend, this study focuses on the extraction of underground mineral resources because the mining industry presents the overwhelmingly dominant example of resource finitude in the context of historical thought from the 1830s to the 1930s. Trees and fish could, after all, grow and reproduce; gold and tin could not. Regarding soil, for example, Devas affirmed that "though cultivation cannot be kept up *ad infinitum* at a very high pitch of intensity, it can be at a low pitch" (79), and as Paul Warde explains in "The Invention of Sustainability," it was understood that tree populations, properly managed, could be cultivated to maximize yield while maintaining sustainability for future populations: "the eighteenth century saw the development of 'sustained-yield' theory, the cornerstone of modern forestry" (162).[21]

Such reproductive engineering was not possible for metal and mineral resources, which were typically defined in economic terms by their special lack of regenerative capacity. As Sieferle puts it, "the subterranean forest can only be felled once" (184), and as W. Stanley Jevons memorably wrote in *The Coal Question: An Inquiry concerning the Progress of the Nation, and the Probable Exhaustion of Our Coal-Mines* (1865), "A farm, however far pushed, will under proper cultivation continue to yield for ever a constant crop. But in a mine there is no reproduction, and the produce once pushed to the utmost will soon begin to fail and sink to zero" (154–55). I will discuss this point at greater length in chapter 1, but what I want to emphasize here is that exhaustion emerged as a distinctive trajectory of extraction-based life. The emergence of a society that was economically grounded in the extraction of finite materials was understood to mean the emergence of a society that was, in a new way, unsustainable for the long run. In this sense, the nineteenth-century grappling with industrial extractivism previews the mode of living that we all experience today, a way of life that proceeds by depleting the future—in other words, the long exhaustion.[22]

The Long Exhaustion

The voice of optimism and progress—the voice that sang in the key of investment and growth—often drowns out the voice of exhaustion in nineteenth- and early twentieth-century literature, and yet industrial Britain was never without an ever-present sense that it was living on borrowed time.[23] *Extraction Ecologies* tunes into this sustained minor key, this continual note of exhaustion that pervaded literature and thinking about the environment in the aftermath of industrialism. Even in print material that was written to encourage mine speculation, where the "permanent" and "inexhaustible" resources of this or that mine were vociferously puffed, there was often rhetorical slippage acknowledging that "inexhaustible" really meant "for now." In *South African Mines* (1895–96), for example, Charles Sydney Goldmann writes of "the permanent nature of the gold-bearing deposits of the Witwatersrand to a period far beyond the life of any of those now interested" (v). A strange definition of "permanent" is at work here, where "permanent" is tied to the lifespan of current shareholders. Goldmann goes on to use the "confidence" of these shareholders as a dubious measure for the lifetime of the mine: "The confidence of capitalists in the permanency of the Witwatersrand goldfields is best illustrated by the energy with which the exploitation of its gold-bearing deposits is either being undertaken or initiated by them, at depths which have probably never previously been attempted in the history of gold mining" (vi). Deep mines are expensive to build and were attempted only where resources closer to the surface were exhausted or otherwise unavailable and where deeper resources were lucrative enough to make deep mining profitable. Deep mining is thus no evidence for the "permanency" of the goldfield, and Goldmann goes on to further qualify his definition of "permanent": "Though the majority of sceptical prognostics have been won over to acknowledge the wealth of these goldfields . . . there remain an incredulous section who would regard the forecasting of gold returns in the distant future as extremely hazardous and reckless. It may suffice, therefore, to review the past six months and anticipate only what is likely to occur in the near future" (viii).

Let us review: in a dizzying descent, Goldmann's introduction passes from the timescale of "permanent," to the timescale of the shareholder's lifetime, to the timescale of six months. He admits that one of the central questions on the minds of prospective investors must be, "what is the life of the mine?" (xv). The question haunts *South African Mines*, as it does all the technical and economic literature of extraction in this period.[24] Sometimes the answer was unintentionally comic: in his rundown of the gold mines in the Witwatersrand region, Goldmann includes an entry on the "Cornucopia Gold Mining Company, Limited." As if the discrepancy between "cornucopia" and "limited" were not jarring enough, the entry includes the crucial detail that the Cornucopia mine "has been shut down since 1891" (53).[25]

Overseas gold mines were seen as particularly volatile speculations at risk of exhaustion, and there was precedent for viewing them as such, but within Britain the more mundane prospect of coal exhaustion reared its head frequently in Parliament, in works of political economy, in a Royal Commission devoted to the question, and in the popular press. Jevons's *The Coal Question* is only the best-known and most comprehensive analysis within a complex of industrial-era discussions about coal exhaustion.[26] Discussions of metalliferous exhaustion were widespread too, as described in chapter 1. While the estimated timescales of such projected exhaustions varied, the key point to emphasize is that the timescale was understood generationally and was spoken about generationally. As Henry E. Armstrong said in a 1902 address to the British Association for the Advancement of Science, "In Great Britain we are using up our coal stores at the rate of over two hundred millions of tons per annum. Used at such a rate, the supply cannot last many generations; whence will our children derive their supplies of energy? . . . When we have squandered the wealth funded on our earth by the sun in æons past, we must fall back on the modicum we can snatch from the daily allowance the glowing orb dispenses" (825). As this suggests, the depletion of coal, the basis of industrial society, was understood to be a danger for subsequent generations in the near-to-middle-term horizon.

The predicted exhaustion of coal was particularly vertiginous to contemplate at a moment when coal's long process of formation and compression, originating with prehistoric plants, had only recently come to be widely understood.[27] How could something take so long to form and change the world so quickly, only, it seemed, to run out but a day later? Writing from the United States, but with attention to the British coal industry, P. W. Sheafer reflected in 1881, "Coal is monarch of the modern industrial world. . . . But, supreme as is this more than kingly power at the present time, comparatively brief as has been the period of its supremacy, and unlimited, in the popular apprehension, as are its apparent resources, yet already can we calculate its approximate duration and predict the end of its all-powerful but beneficent reign" (3). Sheafer expresses here the dizzying temporalities of extraction-based life, the deep timescales between the formation of coal and its extraction and use in the industrial present, and the much shorter timescales between its combustion today and its exhaustion tomorrow. His essay makes clear that Britain, who rose to industrial ascendancy on its rich resources of coal, is the nation with the most to fear from exhaustion: "There it is serious, indeed; for when Britain's coal fields are exhausted, her inherent vitality is gone, and her world-wide supremacy is on the wane. When her coal mines are abandoned as unproductive, her other industries will shrink to a minimum, and her people become familiar with the sight of idle mills, silent factories, and deserted iron works, as cold and spectral as the ruined castles that remain from feudal times" (11).[28]

Such predictions proved off the mark, of course, for as it turned out, there are far more hydrocarbon reserves underground than are at all good for us, and the globalization of extractive industry made local exhaustion less of a factor as capitalism expanded to encompass new natures. At the local level, however, mine exhaustion remains a critical factor in extraction-based life. Jessica Smith Rolston describes how in Wyoming's vast twenty-first-century coal-mining operations, the pits gradually "extend farther and farther away from the mine offices to reach the coal." Journeys from the pit to the office "take increasingly longer amount of time to complete as the mine expands" (69). In *The Road to Wigan Pier* (1937), George Orwell emphasized the same dynamic in northern British mines, depicting mineral resource extraction as an inherently centrifugal process, endlessly exhausting, requiring ever-longer travels for the miners from the pit to the coal face: "In the beginning, of course, a mine shaft is sunk somewhere near a seam of coal. But as that seam is worked out and fresh seams are followed up, the workings get further and further from the pit bottom. If it is a mile from the pit bottom to the coal face, that is probably an average distance; three miles is a fairly normal one; there are even said to be a few mines where it is as much as five miles" (22). This gradual process of exhaustion, as I discuss in chapter 2, illuminates extractivism's close relation to imperialism, since the resources of the colonial frontier are demanded as continual recompense for local exhaustion.

Exhaustion may not have played out in the way that Jevons and other contemporary observers expected, but my titular phrase "the long exhaustion" is meant to capture their correct intimation, incorrectly reasoned though it was, that extraction-based life is a future-depleting system. Like many literary authors of the era, these thinkers perceived the industrial era to be unsustainable, to be a spectacular but momentary boom entailing losses and liabilities for the generations to come. Climate change, not resource exhaustion, ultimately proved to be the most perilous environmental outcome of extraction-based life, and while hardly the only pitfall of an industrialized nature, it is now the most pressing one. Our present emergency cannot be said to have been predicted by industrial-era writers, but their sense that extraction-based life entailed a diminished future did prove to be correct.

This leads to the difficult question of what it has meant for us, as a linguistic community, to be immersed in a culture and literature so thoroughly saturated in extractivist thinking and its assumptions about the future. Have two hundred years of extractivist language and literature prepared us, in some way, for the crisis we now face? Or have they made environmental crisis seem inevitable, and thus encouraged complacency? These are questions to be pondered rather than answered, but they are the questions that motivate this book. In establishing the extent to which extractivism permeated literary form and genre in the first century of the industrial era, my goal is to show how culture, language, and discourse mediate environmental history and carry along the assumptions that emerge under one set of material-environmental conditions into the new stage that follows. Focusing on the twentieth and twenty-first centuries, Jennifer Wenzel and other scholars of petroculture have described how "narratives of limitless growth, premised upon access to cheap energy and inexhaustible resources, underwrite the predicaments of the present" (1).

What is less clear, however, is how narratives of resource exhaustion that pervaded literature before the rise of oil might also underwrite the predicaments of the present. Perhaps the gushing sense of surplus that greased the wheels of the twentieth century was in some sense a reaction to earlier narratives of exhaustion; a piece of comic verse written by Hilaire Belloc in 1928, toward the end of the period covered by this study, would suggest that the rise of oil was taken, at least by some, in that light:

Our civilization
Is built upon coal.
Let us chaunt in rotation
Our civilization
That lump of damnation
Without any soul,
Our civilization
Is built upon coal.

In a very few years,
It will float upon oil.
Then give three hearty cheers,
In a very few years
We shall mop up our tears
And have done with our toil.
In a very few years
It will float upon oil. (*Do We Agree?* 46)

For Belloc oil meant a release from the toil and tears of coal-based life, but for other thinkers in this period, oil seemed merely the next chapter in a longer process of petro-exhaustion. When Walter Darwent drilled the world's "first continually productive oil well" in Trinidad in 1866 (Hughes 2), he did not erase fears of exhaustion: as Sheafer wrote in 1881, "Partially successful experiments have been made to use petroleum as a substitute for coal to some extent. But is it not already evident, under the reckless prodigality of production, that this occult and mysterious supply of light, heat and color will be exhausted before the [coal], and can, at best, only temporarily retard the consumption of the latter?" (10). The looming specter of a long exhaustion, in other words, persisted into the oil era. With the transition to oil, as Sieferle writes, "the exhaustibility of energy resources remain[ed] a sword of Damocles hanging over the industrial system" (203). Coal was never really superseded by oil, of course; there is more coal mined today than there was before the rise of oil. What has happened is not a replacement, but rather, coal and oil have together accelerated petro-modernity. Many writers in the period under consideration here, however, understood oil as a temporary respite for coal exhaustion that would be ultimately subject to the same limits as other subterranean resources.[29]

Extractive Literature and the Literature of the Anthropocene

The literary archive from the 1830s to the 1930s bears witness to industrial extraction's transformation of the world and to the rise of what Bruce Braun has called "a 'vertical' nature" (40), stretching miles below the earth's surface. The material conditions of underground extraction are such that this transformation was difficult to perceive and comprehend, but literature is one place where we see how extractivism altered human expectations, horizons, and understandings. Literature is not merely a passive register of industrial extraction's impacts, however; it was the discursive site where this transformation was mediated. As I see it, changes in discourse and narrative operate as feedback loops whereby certain forms of environmental change or infrastructural path dependency might harden as they disseminate into the symbolic realm—or might instead be challenged in that process.[30] Organizing structures of prose narrative thus participate in a "multivalent traffic between matter and ideas," as Wenzel puts it (3). With extraction interpenetrating discourse in this way, there is a risk that the difficulty of thinking outside extractivism becomes compounded, but literature also provides forms with which to think beyond existing conditions, and such imaginative capacities, of particular concern in chapter 3, are important resources today amid our current reckoning with ecosystem collapse, how we got here, and what to do about it.

Extraction Ecologies is, then, a study of literary form and genre, but most centrally it is a study that uses literature to understand changing dimensions of the human-natural relation. It is a study of "social natures," to use Braun's phrasing, of how "practices of *representation*—deeply cultural and historical in character" are bound up with "nature's *material* transformation" (26). While the ethical stakes of my project extend beyond human communities to encompass animals, plants, and ecological relations more broadly, my focus on industrial extraction means that social natures will be the primary focus of analysis. The particular practices of representation with which I am concerned are long narrative prose works, fictional and nonfictional, the generous scope of which suits the exploration of durational, expansive topics such as time, space, and energy.[31] The prose narratives on which I focus all have some thematic relevance to extraction, but the following chapters also stretch the archive of "books about mining" beyond the obvious suspects by identifying "aboutness" in setting and worldbuilding as well as in plot. With an expanded sense of how far-reaching the impacts of industrial extraction were in this period, and how pervasive was its rewriting of the social and the natural, it now seems to me that there may be very few prose works published from the 1830s to the 1930s without *some* overt thematic interest in extraction, not to mention structuring principles rooted in extractivism. Still, all the narratives on which I focus in this study have a more or less obvious connection to the winning of underground resources, whether that connection is established through plot or setting, or, as with some texts in chapter 3, through the depiction of a post-extractive society. I have chosen to focus on extraction literature in these more overt forms to offer the most direct analysis of extractivism's impacts on literature and genre. The reverberations of industrial extraction beyond this archive of texts will, I hope, be plainer to see once we have a conceptual schema for thinking about literature and extraction ecologies—a schema that I hope to offer in this book.

A focus on extraction and literature demands a view of the natural environment as fully inclusive of the human, a perspective that recent scholarship on the Anthropocene has made increasingly familiar; my project thus rests on the premise that at least since the industrial era and probably centuries prior, there is no nature untouched by human impacts.[32] As Heidi C. M. Scott writes, the Anthropocene framing acknowledges "that today's stratigraphy is laid in the waste of industrial humans" ("Industrial" 589). Contemporary observers of the industrialization of extraction were, as we shall see, forced to much the same conclusion. Cara New Daggett calls the Victorian era the beginning of the "ideational" Anthropocene, a period of "dawning consciousness" that human-industrial impacts "might be planetary and truly catastrophic" (9). If for Charles Babbage the nineteenth-century air itself was "one vast library" of human action, for other writers the disturbed surface of the earth was the page on which the story of the human was written.[33] Troubled by thoughts of human ephemerality while searching for an African diamond mine, for example, Allan Quatermain in *King Solomon's Mines* imagines humans' lingering presence on the earth by way of our monuments: "man dies not whilst the world, at once his mother and his monument, remains" (165). This conception of the earth as a "monument" of the human, a bearer of the signature and memory of the human, anticipates the Anthropocene imagination where we understand Earth to be indelibly marked by anthropogenic impacts, where we "imagine a world in which an alien geologist from the future detects in the strata of the ground evidence of the presence of humans long after we have gone extinct" (Bubandt G135). "Truly the universe is full of ghosts," as Allan Quatermain reflects (166).

Extraction is, of course, in large part responsible for this anthropogenic signature, not only from the atmospheric residue of fossil fuel combustion, but also from other extractive pollutants including the radioactive deposits of nuclear weapons that originate in uranium mining (the signature that the Anthropocene Working Group has currently settled on as the "golden spike" marking the new era).[34] Fossil fuel extraction and nuclear weapons persist, too, in contemporary forms of capitalism and militarism. A focus on extraction will thus convey, I hope, that the Anthropocene concept does not entail "a turn away from the critique of sociopolitical power relations," but rather is a tool to help widen "the focus of sociopolitical critique," to see human power relations and struggles within a larger context of "geophysical actors" and earth systems (Davies 62). This is the larger environmental-material context within which any solutions will also need to work.

Scholars of the Anthropocene have sometimes dated its origin to the invention of the steam engine, beginning with Paul Crutzen and Eugene Stoermer's first coining of the term in 2000, but that narrative is now contested. Some argue that technologies such as agriculture led to humans' irreversible impact on earth systems, as extensive tracts of land were repurposed to grow grain, and others say that it was not a particular technology that spawned the blight that surrounds us but rather a set of social and economic relations such as capitalism, colonialism, or the plantation system.[35] Tobias Menely and Jesse Oak Taylor have discussed the Anthropocene Working Group's debate over the timing of the human signature on the stratigraphic record and the consequences of this debate for the humanities, showing how "geologists give narrative shape to history" when they select this or that boundary event as definitional (3). Scholars of nineteenth- and early twentieth-century literature have a stake in these debates, given the primacy of our era in the arrival of fossil capitalism, but considerations of scale and acceleration should caution us against any easy link between the invention of steam power and the more than 400 parts per million of CO_2 that hang heavy in our atmosphere today. As David Wallace-Wells writes, "many perceive global warming as a sort of moral and economic debt, accumulated since the beginning of the Industrial

Revolution and now come due," but "more than half of the carbon exhaled into the atmosphere by the burning of fossil fuels has been emitted in just the past three decades" (4). Here it may be helpful to think of steam power less as a material trace than a form, a form that has been expanding and accelerating since its inception, subject to various historical encouragements and, occasionally, historical checks. Carbon dioxide from coal burned during the Industrial Revolution still floats in our atmosphere, but it is the broader complex of extraction-based life and the forms and practices that support it that are responsible for our current impasse.[36]

Wherever the steam engine fits into the story of the Anthropocene, and however it contributed to the rise of fossil capitalism—that ever-accelerating juggernaut of waste and productivity powered by the stored solar energy of long-dead lifeforms—none deny that the birth of steam was one of the signal events in environmental history, nor that it happened in coal-rich Britain.[37] But despite the prominent role of mining in the environmental and social history of Britain and its empire, and despite the recent flourishing of work on literature and the Anthropocene, our critical understanding of British literature has been inadequately attentive to the epistemology of extractivism. Amitav Ghosh argues that art and literature since the Enlightenment have developed "modes of concealment" that prevent us from recognizing the environmental catastrophes of modernity, but my study is premised on the idea that extraction does play a crucial structural role in the literature, albeit one that we have failed to observe (11).[38] If literary criticism has, in the main, tended to overlook how language and literature are shaped by the natural world and its transformations, my book contributes to the work of addressing this oversight. But *Extraction Ecologies* also suggests that we find a particularly influential vision of the natural world in the literature of Britain's industrialized empire. First to transition to fossil-fueled industry, Britain was the first extraction-based society, and the literature of the British imperial world is thus in the remarkable position of originating the literature of fossil capitalism and industrial extractivism. In this role it reckons with a new vision of civilization where humans now depend on finite, nonrenewable stores of earthly resources that are incapable of replenishment through seasonal rebirth, and the threatening horizon of exhaustion works its way into narrative form. Themes of degeneration and decline have long been recognized as preoccupations of modern literature, but we have yet to connect this literary turn with the descent down the mine shaft that was a base structure of modern life.

Organization and Chapter Overview

To uncover extraction's multifaceted role in the literature of industrial and imperial Britain, I have organized *Extraction Ecologies* conceptually, with three long chapters broadly devoted to three central categories (time, space, and energy) and three corresponding literary genres (provincial realism, adventure literature, and speculative fiction). Each chapter, after presenting its overall argument, includes five subsections each focused on a particular text. For readers looking to the book with an eye for the individual case, the subsections are listed in the table of contents. Orders of time (when things happen and in what order), space (where things happen and how they move), and energy (how things happen and from what cause) transform with the rise of extraction-based life, and thus each chapter traces industrial extraction's shadow and formation in one major conceptual domain. This is admittedly an unusual structure, but it allows me to make a case for extraction ecologies as a feature of this era's litera-ture by drawing together multiple textual examples for each major point, foregrounding the project's broad conceptual interventions and its claims about genre while still allowing for close literary analysis. The argument of *Extraction Ecologies* is not one that can be proven through long read-ings of a few texts; it seeks instead to showcase a pattern or trend beyond the individual case. Genre, as a category of analysis, offers something like a middle ground between close and distant reading, allowing us to see larger patterns without detaching us from the singularity and nuance of individual texts. The mobility and repeatability of literary form and genre across time also get to the problem of historicity at the heart of this study: environmental history and environmental knowledge require a long-term view, and literary genre and form carry ideas across historical periods in ways that transcend individual texts. To plumb the literary archive of the past is to find discursive and conceptual formations that have remained with us, to our detriment, as well as formations that have been left to the wayside and are worth revisiting today.

My first chapter, "Drill, Baby, Drill: Extraction Ecologies, Futurity, and the Provincial Realist Novel," demonstrates how the provincial real-ist novel incorporated exhaustion as a temporal structure to depict the new horizons of human life under extractivism. Provincial realism's long-standing reliance on the marriage plot and the inheritance plot, on pro-viding closure via social reproduction, transforms against the backdrop of extractivism to withhold the promise of reproductive futurity. As the steam engine and other industrial technologies were transforming the scale and impacts of mining in the backwaters of global empire, discourse around exhaustion, futurity, and decline reached a new stage as well, transforming the endings, trajectories, and temporalities of the provincial realist novel. All the novels discussed in chapter 1 take place in settings of extraction or exhausted extraction—sacrifice zones—and all explore the temporal structure of an extraction-based present claimed at the expense of future generations.

The chapter's first major subsection focuses on Joseph Conrad's *Nostromo* (1904), a novel that gathers a large cast of characters around an out-of-the-way silver mine in the fictional South American country of Costaguana, interweaving the story of the mine with three broken and infertile marriage plots and revealing how exhaustion's temporal features pervade the trajectory of the provincial realist novel. Next I turn to George Eliot's *The Mill on the Floss* (1860) and its key setting, the Red Deeps— an exhausted ironstone quarry where Maggie Tulliver and Philip Wakem enter a forbidden engagement that will never be consummated in marriage, just one of the novel's failures of futurity. The third major subsection considers Fanny Mayne's *Jane Rutherford: or, The Miners' Strike* (1854), a lesser-known novel that treats the conditions of working-class family life in a mining community, toggling between a strike story and a marriage plot to underscore the forms of social reproduction demanded of workers within extraction economies. Charles Dickens's *Hard Times* (1854), discussed next, is also set in a coal-mining district, but here the long-awaited marriage between workers Stephen Blackpool and Rachael never happens because of Stephen's tragic fall into an exhausted coal pit. The last major subsection of this chapter focuses on D. H. Lawrence's *Sons and Lovers* (1913), a novel that links its mine-ridden landscape with Paul Morel's difficult sexual maturity, transforming the provincial bildungsroman to conceive of individual human development in the context of extractivism's socio-environmental entanglements.

Chapter 2, "Down and Out: Adventure Narrative, Extraction, and the Resource Frontier," turns from the temporal to the spatial imaginary and from realism to adventure writing, arguing that industrial-era adventure literature exhibits a newly energized orientation toward the horizon of the resource frontier, stimulated by the constant search for new lodes that defines the extractivist age. Focusing on adventure narratives that take place in Latin America and Africa, I show that they are premised on a collapse of the vertical and the horizontal, where a journey across the earth becomes the necessary complement to downward delving into the earth. Jason Moore's *Capitalism in the Web of Life* has helped us understand the appropriation of "cheap nature" as part of the historical tendency of capitalism, and that a restless global reach toward the frontier must accompany any notion of "free" nature. Imperial adventure narrative is a genre full of treasure hunting on the frontier, one that was born in the context of the mineral resource scrambles that dominated geopolitics in the industrial era, from the Mexican mining boom to the Californian and Australian gold rushes to the South African Mineral Revolution.

The first major subsection of chapter 2 focuses on Mary Seacole's *Wonderful Adventures of Mrs Seacole in Many Lands* (1857), a memoir that foregrounds the epistemological challenges of frontier space as it details Seacole's supporting role in one of the great extractive dramas of the era: she ran a hotel in Panama catering to miners heading to and from the California Gold Rush and tried her own hand at gold mining in several failed schemes. I turn next to Robert Louis Stevenson's *Treasure Island* (1883), a fictional adventure romance that shares Seacole's Spanish Caribbean setting and similarly foregrounds in its narrative forms the limited perspective from which extractive imperialism precedes. The third major subsection focuses on H. Rider Haggard's *Montezuma's Daughter* (1893), a historical adventure novel about the Spanish quest for gold in the Americas, which strives to justify Britain's extractive ascendancy in Latin America after the decline of Spanish and Portuguese rule. Turning next to adventure narratives set in Africa, I show how Haggard's *King Solomon's Mines* (1885) offers a vision of colonial extraction and mineral wealth waiting to be won on a rich frontier in a narrative structure that codifies the extractivist worldview. Finally, I look to Joseph Conrad's *Heart of Darkness* (1899), a novel about fossil ivory in the era of fossil capital, which merits inclusion for the iconic manner in which it folds the imperial extraction plot into its experimental narrative forms.

My third and final chapter, "Worldbuilding Meets Terraforming: Energy, Extraction, and Speculative Fiction," addresses the energy imaginary within the industrial extraction boom and how this imaginary shaped the political and social projections of speculative literature. Speculative genres such as hollow earth fiction, utopian fiction, and fantasy fiction burgeoned alongside industrial extraction, and my chapter focuses on the ruminations on energy and exhaustion that grounded these literary speculations. Extractive energy supplied the material conditions from which speculative fiction takes flight, but these worldbuilding genres also offer imaginative resources for envisioning energy beyond extractivism, even as they narrate, through their secondary worlds, energy's determinative role in culture, environment, and society.

The first major subsection of this chapter focuses on Edward Bulwer Lytton's *The Coming Race* (1871), a hollow earth novel that begins when the protagonist is exploring an underground mine and falls into the world of the Vril-ya, a subterranean civilization built around a mysterious energy source, vril. Next I turn to Rokeya Sakhawat Hossain's "Sultana's Dream" (1905), a feminist energy utopia originally published in the *Indian Ladies' Magazine* that depicts a world fueled by extraction-less solar power and utterly transformed gender relations. William Morris's utopia *News from Nowhere* (1890), the subject of my third major subsection, likewise imagines a social evolution away from extractive energy, with capitalism and the human-environment relation, rather than gender, depicted here as the primary social vectors of extractivism. From Morris I turn to H. G. Wells's *The Time Machine* (1895), which, like *The Coming Race*, features a subterranean society, in this case inhabited by the evolutionary victims of extraction ecology. Finally, J.R.R. Tolkien's *The Hobbit* (1937) depicts a quest for underground treasure that brings to speculative fiction's subsurface settings and chthonic character types the new energy agencies of the early nuclear age. *Extraction Ecologies* then offers a brief conclusion, reflecting on the question of how extractive literature of the past can helpfully intersect with environmental politics and thought today.

Sacrifice Zones and the Settings of Extractivism

As the above chapter summaries suggest, the narratives on which this volume focuses vary significantly in terms of reputation, regard, and canonicity. We begin with *Nostromo*, perhaps Conrad's most complex and difficult novel, and end with *The Hobbit*, a fantasy novel written for children; along the way, we analyze underdiscussed writings by women of color (Seacole and Rokeya Sakhawat Hossain) alongside popular romances by imperialist writers like Haggard. What all these various works share are features of setting expressed through narrative form: they are set in spaces of extraction or exhaustion, or in a post-extractive future, and such settings foreground especially clearly the extractivist elements of the works' formal and generic structures. The three primary genres with which I am concerned—provincial realism, adventure narrative, and speculative fiction—are all expressly setting dependent. Provincial realism draws its sense of place from its out-of-the-way-ness; adventure narrative features a journey into the frontier, or sometimes beyond the frontier; and speculative fiction's imaginative worldbuilding creates new settings in an alternative reality. Mining communities, resource frontiers, imaginary worlds with new energy formations: the extractivist currents of these narrative genres are particularly noticeable because of these settings, even though extractivist forms can also be said to pervade industrial-era literature more generally.

The key role of setting in *Extraction Ecologies* suggests the logic through which I have chosen my central texts and why I have approached them conceptually via genre. Extractivism produced new genres and transformed old genres as literature intersected with industrialism and its impacts on the natural world. Elizabeth Chang, in her recent study of plants and the global nineteenth century, has argued, relatedly, that the landscape of empire "was becoming increasingly nontransparent in its infrastructure" (*Novel* 18), which necessitated the rise of detective fiction, a genre where the setting steps forward from a stable narrative background to become interpretable. Setting is, for obvious reasons, a primary focus of much ecocritical work, and recent ecocriticism has challenged us to theorize setting more robustly.[39] Still, my ambition for this book is that the overall argument will prove portable beyond novels with explicitly extractive settings. In the following chapters I aim to expand notions of what qualifies as an extractive setting and to test the flexibility of that category within industrial-era literature. Even novels that might not initially seem to be about extraction, such as *The Mill on the Floss* or *News from Nowhere*, emerge as extractive literature when placed in the context of environmental history and considered from the standpoint of genre. By drawing together works that are quite obviously about mines and underground treasure, such as *Nostromo* and *King Solomon's Mines*, with other less obvious examples, I hope to illustrate the breadth of extractive literature as a category.

Setting references time as well as place, and insofar as we are still living in the world that industrial extraction created, these settings of the past continue to persist. Some would argue that a book about industrial-era extraction ecologies is necessarily presentist since it attends to, and is designed to attend to, the environmental crises of today, especially global warming and its roots in the coal-fired capitalism of the British Empire. Debates about presentism and strategic presentism have now occupied literary studies for some time, but what I aim to practice in this book is, rather, a methodology capable of working on multiple timescales.[40] Thinking about the literature of underground resource extraction in the first century of the industrial era, we can imagine at least four temporal frames in which to position these texts: a deep timescale in which coal, diamonds, and other extractible commodities took form over long stretches of geological change; a fragile nineteenth- and early twentieth-century present in which such commodities were understood to be abundant, though physically resistant and labor intensive to acquire; an imagined future of depletion in which, it was thought, extractable commodities would eventually be exhausted; or the actual future we live in now, an era of anthropogenic climate change and other toxic remainders that can be linked back to the historical rise of large-scale extractive industry. Reading extraction-based literature with an eye for all these temporal registers, the following chapters ask whether intimations of our present exist in literatures of the past, whether intimations of future decarbonization exist in literature of the past or our readings of it today, and whether the environmental imagination of the past can reveal possible futures, roads not taken, that we can learn from in our present impasse. Ultimately, we can make sense of form and setting only in durational terms, as products of history, and neither form nor setting can truly be said to mean anything outside of history, and yet "historicism" is often discussed as though it were a more temporally static method than it actually is. What I aim to practice in this book is a heterotemporal historicism that is sensitive to the multiple, nested time lines of environmental change and environmental devastation across this long era of exhaustion in which we yet remain.

CHAPTER ONE

Drill, Baby, Drill

EXTRACTION ECOLOGIES, FUTURITY, AND
THE PROVINCIAL REALIST NOVEL

"The conditions of production are at the same time the conditions of reproduction."

KARL MARX, *CAPITAL*, VOLUME 1 (1867), 711

"What elegant historian would neglect a striking opportunity for pointing out that his heroes did not foresee the history of the world, or even their own actions? . . . Here is a mine of truth, which, however vigorously it may be worked, is likely to outlast our coal."

GEORGE ELIOT, *MIDDLEMARCH* (1871), 61

A NITROGLYCERINE ADVERTISEMENT in the 4 January 1868 issue of the *Mining Journal* boasts, "The EXPLOSIVE FORCE of this BLASTING OIL IS TEN TIMES that of GUNPOWDER, and the ECONOMY and SAVING in TIME, LABOUR, and COST in removing granite and hard rock, in sinking shafts, driving tunnels, and opening forward in close ends is immense" (13) (see figure 1.1). Mining, drilling, and extraction accelerated exponentially under the global force of industrialism, transforming the face of the earth, but also, as this advertisement suggests, the face of the clock. The regime of industrial extraction was variously said to speed up time or to save time, to make the future from the past or the present from the future. Indeed, as this chapter will discuss, extractivism occasioned a complex chronological alchemy, where metals, minerals, and coal proved capable of changing the experience of time and transforming ideas of futurity.[1] This new sense of time was born of the perception that future human life

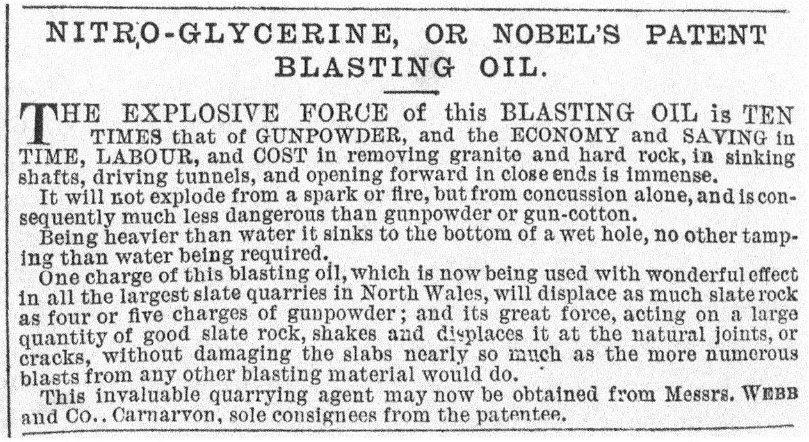

NITRO-GLYCERINE, OR NOBEL'S PATENT BLASTING OIL.

THE EXPLOSIVE FORCE of this BLASTING OIL is TEN TIMES that of GUNPOWDER, and the ECONOMY and SAVING in TIME, LABOUR, and COST in removing granite and hard rock, in sinking shafts, driving tunnels, and opening forward in close ends is immense.

It will not explode from a spark or fire, but from concussion alone, and is consequently much less dangerous than gunpowder or gun-cotton.

Being heavier than water it sinks to the bottom of a wet hole, no other tamping than water being required.

One charge of this blasting oil, which is now being used with wonderful effect in all the largest slate quarries in North Wales, will displace as much slate rock as four or five charges of gunpowder; and its great force, acting on a large quantity of good slate rock, shakes and displaces it at the natural joints, or cracks, without damaging the slabs nearly so much as the more numerous blasts from any other blasting material would do.

This invaluable quarrying agent may now be obtained from Messrs. WEBB and Co.. Carnarvon, sole consignees from the patentee.

FIGURE 1.1. Advertisement from the *Mining Journal*, 4 January 1868, 13.

could not continue to be sustained by the stock of nonrenewable resources that powered industrial society. In the above epigraph from *Middlemarch*, Eliot refers to the failures of human foresight as "a mine of truth . . . likely to outlast our coal" (61); the use of this metaphor in a landmark novel like *Middlemarch* attests to the role of literature in mediating extractivism's new horizons of exhausted and uncertain futurity.

This chapter focuses on provincial realist novels set in places of mineral extraction: the sacrifice zones of Britain and its empire, where industrial terraforming pulverized the land, casting up at least as much waste and damage as it did treasure. Understanding provincial realism as the genre that best theorizes "quasi-removal" (Plotz, "Provincial Novel" 369) allows us to see the significance of these texts within imperial, capitalist, and ecological systems, for the regions depicted in these novels provided the mineral material to power British growth.[2] As discussed in the introduction, all forms of industrial extraction surged with the steam engine, and extraction begot extraction. Coal-fired engines were required to drain mines of all kinds and were used for boring and for processing ore; steam-powered railways transported mined materials on an industrial scale. But while the project of industrial extraction was a major source of wealth in this era of global mineral rushes, it was also understood to be a temporary bonanza that would not last. Geologists and political economists varied in their estimates of when coal exhaustion and other kinds of mineral exhaustion would occur, but the timescale was short enough to be understood generationally. This was a new view of the issue, as Louis Simonin explains in his

1868 text *Mines and Miners*: "The duration of the coal-fields, which was formerly estimated by geologists at thousands of years . . . will, perhaps, not exceed a few hundred years" (263). This widespread conclusion led to a new understanding of industrial life in the present as occurring at the expense of future human generations—a situation familiar enough today, though for different reasons. A question this chapter will begin to frame, one that hangs over the book as a whole, is whether we have been acclimated to the horrors of climate change and the potentially uninhabitable future it portends by two centuries of extraction-based life premised on a model of depletion to which we have become accustomed.[3]

Britain's national discourse on mineral exhaustion cannot be said to have begun in the nineteenth century, but it seems to have taken form around the time of the invention of the steam engine and took some years to enter widely into discourse. One of the earliest writers to address the subject was John Williams, in *The Natural History of the Mineral Kingdom* (first published in 1789), who expressed "not the smallest doubt that the generality of the inhabitants of Great Britain believe that our coal mines are inexhaustible" (159). While Williams felt "a strong reluctance against sounding the alarm" (160), he believed that "silence would be unpardonable" (161), for the question of exhaustion has "never been considered in this light, nor has any person, public or private, to the best of my knowledge, ever taken the trouble to make a fair representation of this matter" (170–71), despite the fact that "the commerce, wealth, importance, glory, and happiness of Great Britain will decay and gradually dwindle away to nothing, in proportion as our coal and other mines fail" (172–73).

The idea that modern existence was now fully reliant on coal, and thus reliant on the depletion of the coal supply available for future generations, made for a profound shift in ideas of time and futurity—a shift that occurred with the rise of industrial extraction. Indeed, a British government cabinet report from 1903 made the case that the problem went beyond coal and was symptomatic of the whole extractive industry: "coal is an exhaustible product of the earth which cannot be replaced by labour. . . . If it is true of coal, it is true of iron ore, which is believed to be much nearer exhaustion than coal. . . . It must also be true of . . . the entire metal industries—lead, copper, brass, tin, also of all bye-products" (*Fiscal Problem* 24). Seen in this light, the basis of industrial society appeared to be so much sand slipping through the hourglass, and political economists pointed, meanwhile, to sharply increasing rates of coal consumption. Faced with such circumstances, Simonin wrote, "the question naturally arises as to the time when the coal-fields will be exhausted, and as to the nature of the fuel which will replace coal after its total exhaustion—a double problem such as has never been presented until now during the history of the world" (262). This new horizon of exhaustion was difficult to conceptualize, but the provincial realist novel proved to be one genre where mineral exhaustion's "slow causality," to use Tina Young Choi and Barbara Leckie's phrase, could be imaginatively worked out.[4] Rather than pointing definitively toward a future end point and collapse, provincial novels set in extraction zones often posit instead an unknown future, an undead future, a future of diminishing returns that recedes as we move toward it. This new future changed the character of the present, for it meant a daily existence bereft of the cyclical comforts of a continuing natural system.

The temporal structures of provincial realist novels set in extraction landscapes convey a growing sense that an extraction-based society is no longer tethered to the seasonal rhythms of the living earth, and they convey a new conception of futurity imbued with the realization that Britain and its empire are now reliant on an industrial system powered by a nonrenewable, diminishing stock of resources. They challenge novelistic forms premised on the human life cycle, like the bildungsroman, or the seasons of the year, like the pastoral, and move instead toward an exhausted future that will not have grown from the past but will have been drained by it. As the steam engine and other industrial technologies were transforming the scale of resource extraction, achieving new dimensions of depth and new heights of production, cultural discourse around exhaustion, supply, futurity, and decline reached a new stage as well and leached into the endings, trajectories, and temporalities of that most characteristic literary form of the industrial era: the realist novel.

REPRODUCTIVE FUTURITY AND EXTRACTION CAPITALISM

After an overview of the major terms of my argument, focusing on reproductive futurity and extractive exhaustion, this chapter will take up five provincial realist novels set in landscapes of extraction: Joseph Conrad's *Nostromo* (1904), George Eliot's *The Mill on the Floss* (1860), Fanny Mayne's *Jane Rutherford: or, The Miners' Strike* (1854), Charles Dickens's *Hard Times* (1854), and D. H. Lawrence's *Sons and Lovers* (1913). All five depart from the conventions of the marriage plot—or at least from the assurances of reproductive futurity that typically accompany this plot—and their collective deviation from chrononormativity, to use Elizabeth Freeman's word, reflects the new understanding, which accompanied the rise of fossil-fueled capitalism, of an extraction-based life claimed at the expense of future generations.[5] Echoing extractivism's scrambling of temporal norms, I have structured my reading of these texts against chrononormativity and have arranged the novels that serve as my case studies not chronologically but according to their primary mineral resource: first the novels of silver and ironstone, then, a more common formation, three novels of the coal-fields.

Let me clarify, first, that my use of the "'no future' paradigm," as Macarena Gómez-Barris has called it (144), draws on the work of postcolonial and feminist critics focused on environmental justice rather than the work of Lee Edelman, whose *No Future: Queer Theory and the Death Drive* (2004) is better known in literary studies. Edelman's critique of reproductive futurity made a key intervention into US queer theory but does not easily scale out to an analysis of global or transspecies justice, nor translate readily into such domains as Indigenous or environmental justice critique, which are often animated by a sense of ethical obligation to future life— human and nonhuman alike. As Neel Ahuja has argued, the climate crisis asks us to think "more broadly about reproduction than Edelman does, recognizing that bodies and atmospheres reproduce through complex forms of socio-ecological entanglement" (368).[6] Viewed in this light, all humans are engaged in reproductive processes, whether they realize it or not, as well as activities that interfere with reproductive processes. Indeed, given ongoing rates of species extinction and climate-related threats to precarious human communities, we are now witnessing what Ahuja describes as a "staggering scale of 'reproductive failure,' human and nonhuman" (370). I seek to address reproductive futurity from this broad environmental standpoint, even as I look to marriage and reproduction plots as the provincial novel's formal conventions for imagining the future of earthly life and its continuance.[7]

Central to my argument is a cross-historical parallel between the "no future" paradigm of the nineteenth and early twentieth century (resource exhaustion) and the "no future" paradigm of today (climate change). Resource exhaustion did not turn out to be the fatal flaw of fossil-fueled industrialism that many nineteenth- and early-twentieth-century thinkers predicted, but it did occasion early reflection on the unsustainability of this mode of ecological relations. Gómez-Barris locates extraction capitalism at the heart of what she calls the "'no future' paradigm" (144), and in her recent analysis of South American extraction regimes and the social movements opposed to them, she contrasts the "'no future' model that is extractive capitalism" (34) with the growing movement toward "transgenerational stewardship" (48), visible, for example, in recent South American legal frameworks that grant "rights to future generations" (27). Gómez-Barris situates her argument squarely within queer feminist thought yet maintains that any "critique of reproductive futures has to be balanced against the historical weight" of eugenics and anti-Indigenous policies, which have sought to fix Indigenous peoples in the past and deny their claims on the future (145). Similarly, Maristella Svampa describes how resistance to extraction regimes in South America has built on a "strengthening of ancestral struggles for land by indigenous and campesino movements," movements that encompass "the defense of the common, biodiversity, and the environment" (68). The framework of ancestral and multigenerational collectivity has in this instance enabled a powerful critique of extraction capitalism and its inequities across generations. Thea Riofrancos, too, has discussed the success of the anti-extraction movement in Ecuador in achieving a legal basis for an expanded notion of territory "as a space of cultural and ecological reproduction" ("*Extractivismo*" 287).

Ecofeminist critics focused on North America have similarly emphasized reproductive futurity writ large—a global, transspecies affair—as a key component of environmental justice. Indeed, given that the human communities most threatened by climate change are communities of color, many recognize an urgent need to insist on this point in the interest of racial justice as well as transspecies justice. Naomi Klein, drawing on the work of Indigenous feminists, has posited an environmental "right to regenerate" as "the very antithesis of extractivism, which is based on the premise that life can be drained indefinitely" (*This* 442). Despite significant differences, recent work by Donna Haraway can also be situated alongside this argument; Haraway's call to "Make Kin Not Babies" (5–6) asserts the need for a reduction in human reproduction, but Haraway makes this call in the name of humans and other species who face an uncertain future in these "times of burning and extraction called the Anthropocene" (90). Earth's growing human numbers, she says, "cannot be borne without immense damage to human and nonhuman beings" (208). Her work thus shares a goal of ensuring a future for ongoing generations of earthly creatures, though she insists on the ethical gains of "staying with the trouble" and "learning to be truly present" (1).

In my view, learning to be present will also require attending to the past, for the present is long, extending backward into the historical circumstances that produced it and the narrative and discursive formations with which we confront it. The complex, many-sided environmental crisis we face today, whose scope we are still struggling to grasp, presents pressing hazards such as climate change and air, water, and soil pollution that have material and formal connections to the extraction ecologies that emerged with the Industrial Revolution. The geographies of present-day extraction capitalism are also an extension of the past: today's large-scale extraction projects are often relegated to Indigenous lands or so-called developing nations, making large profits for international corporations and little for the regions being exploited.[8] Attending to our present crisis and its uneven burdens also means attending to the conditions that produced this situation, conditions that are recorded, mediated, and, at times, reproduced within the archive of literature.

The provincial realist novels under discussion here all appeared in an era of accelerating imperial extraction; while most are set in England, one is set in South America, for the provincial backwater and the extraterritorial backwater were both targets for sacrifice and exploitation.[9] (My focus turns more squarely to the colonial frontier in chapter 2.) That these novels share features of setting as well as temporality, or, more specifically, that provincial realist novels set in extraction zones defy conventional novelistic temporalities and futurities, suggests that they express what M. M. Bakhtin has called a "chronotope," the "intrinsic connectedness of temporal and spatial relationships that are artistically expressed in literature" (84). While the concept of the chronotope is key to all the extraction literature I will discuss in this study, my understanding of the particular chronotope of the provincial novel is more finely tuned than Bakhtin's and more historically situated. Bakhtin reads the provincial novel as a "family-labor, agricultural or craft-work idyll" that conveys "the uninterrupted, age-old link between the life of generations and a strictly delimited locale" and where "the rhythm of human life is in harmony with the rhythm of nature" (229). If we look more narrowly at provincial novels set in landscapes of extraction during Britain's industrial era, however, we instead find a chronotope that works precisely against the formulation of the provincial novel that Bakhtin elaborates here. The concept of the chronotope is crucial, however, in helping us to see the connections between provincial space and exhaustion temporalities in all these novels, for in addition to shared features of setting, the five novels I discuss here all share a resistance to typical plots of continuity and reproductive futurity such as the marriage plot and the inheritance plot. By grouping the five works together I hope to demonstrate how extraction ecologies intersected with the provincial realist novel and its rendering of time and futurity during the industrial extraction boom, formalizing the environmental severance across generations that was inherent to extraction-based life.

EXTRACTION ECOLOGIES AND
EXHAUSTION TEMPORALITIES

Literature of the industrial era found new ways to express through form and genre the transformation to an extraction-based society where the living and nonliving earth were brought together in a new relation. Recent work by critics such as Jason Moore has demonstrated the centrality of such human-nonhuman relations to the history of capitalism: Moore argues that "capitalism depends on a repertoire of strategies for *appropriating* the unpaid work/energy of humans and the rest of nature outside the commodity system" (54), including "the congealed work of extremely ancient life (fossil fuels)" (220). Elizabeth Povinelli has argued, relatedly, that "key to the massive expansion of capital was the discovery of a force of life in dead matter, or life in the remainders of life: namely, in coal and petroleum" (167). The long-standing enlightenment project of externalizing nature, of separating "nature" and "society" in binary opposition, proves untenable, both critics suggest, when one considers the work done by this "force of life in dead matter." The nature/society binary has, moreover, prevented us from reckoning with capitalism's dependence on the work of what Moore calls "cheap nature," the "process of getting extra-human natures—and humans too—to work for very low outlays of money and energy" (304). Binary categories of human and nonhuman, living and nonliving, Povinelli argues, have served the same end. It would seem to confirm such reasoning that while studies of nineteenth-century industrialization have been quite attentive to mining—arguably the signature form of industrial labor—and its social history, we have not tended to consider the miners' plight within a larger ecology of mining, one that includes humans and nature bound together.

Industrialized mining was an occupation of everyday environmental danger, struck through with toxicity, but it also produced long-term environmental and epistemological change. To examine extraction in this light will, I hope, contribute to an urgent project with which our field is now engaged: the project of reconceptualizing the industrial era by way of the fossil fuel economy it engendered, for this economy has proven obstinate in its staying power and difficult to move beyond. That the nineteenth century saw the moment of our entanglement in this particularly insidious form of path dependency is now evident in W. Stanley Jevons's 1865 declaration: "Day by day it becomes more obvious that the Coal we happily possess in excellent quality and abundance is the Mainspring of Modern Material Civilization. . . . Coal, therefore, commands this age—the Age of Coal" (vii–viii). Andreas Malm's *Fossil Capital* explores in detail the historic shift from water power to steam power in the 1830s English textile industry as the moment when, "at a certain stage in the historical development of capital, fossil fuels become a necessary material substratum for the production of surplus-value" (288). In literary studies, Jesse Oak Taylor's *The Sky of Our Manufacture: London Fog in British Fiction* has helped us think about the effect of coal burning on the atmosphere of London and the climate imaginary of British literature, as well as about our own methods as critics "rethinking history in a time of climate change . . . [where] artifacts take on new agency in distant times and places" (69), while Allen MacDuffie's *Victorian Literature, Energy, and the Ecological Imagination* has traced "the contours of our ongoing energy crisis" in Victorian narratives that both map and mystify the changing energy regime they depict (12).

My goal is to reframe this ongoing discussion around extraction itself, and to position coal within a broader network of industrial extraction— a global infrastructure of labor, capital, and material devoted to the unearthing of buried treasure. Extraction, I will suggest, possessed social and aesthetic forms of its own, and during this period of rapid environmental change, literature transformed to convey how extraction is bound up in industrial ecologies and in the conditions of existence that govern modern life. Literature captured a new vision of civilization in which humans were now dependent on finite, nonliving, nonrenewable stores of earthly resources. Preindustrial trajectories of closure, completion, and revival gave way to a horizon of exhaustion, a transformation expressed in novelistic form.

Numerous social scientists were thinking through the implications of extraction-based life in the era when these novels appeared, and just as they were grappling with the prospect of future exhaustion, provincial realist novels were fictionalizing the consequences for a society built on a finite supply of exhaustible material. Together these discourses convey extraction ecologies' temporal structure of exhaustion: one that defies trajectories of renewal and rebirth as well as of progress and fulfillment. Jevons's *The Coal Question: An Inquiry concerning the Progress of the Nation, and the Probable Exhaustion of Our Coal-Mines* (1865) is one such text that attempts to account for the social and temporal conditions of extraction-based life. Influential and widely circulating, *The Coal Question* is a long rumination on exhaustion as a material threshold and its remarkably complex timescale. The study was taking part in what was, by 1865, a decades-long debate about the threat of coal exhaustion in England.[10] Prior to writing it, Jevons spent five years in Australia at the height of the gold rush, and if he saw firsthand how gold mining can produce the accelerated pace of a "rush," he was at pains a few years later to warn England that its ever-intensifying rush for coal would ultimately run up against a terminal pressure in the form of exhaustion.[11] Bringing his economic and mathematical chops to bear on the geological field, Jevons took factors such as rising coal consumption and population increase into his projections, resulting in nearer-term expectations of British coal exhaustion. He also treated mining as a special industry, as I note in the introduction, by virtue of its nonrenewability: "A farm, however far pushed, will under proper cultivation continue to yield for ever a constant crop. But in a mine there is no reproduction, and the produce once pushed to the utmost will soon begin to fail and sink to zero" (154–55).

Jevons's emphasis on the lack of reproductive capacity within extraction ecologies was typical of industrial-era political economists and other observers of coal. John Holland's *The History and Description of Fossil Fuel* (1841) emphasizes that because coal is "incapable of reproduction or increase," all established consensuses about "free trade . . . do not legitimately apply" (439). Holland adds that coal, tin, lead, and other extracted commodities "differ so essentially from other articles produced by English industry" (442) because of their "prospective exhaustion, at some remote period" and the "undoubted fact, that our mines are not inexhaustible" (454–55). Simonin's *Mines and Miners* (1868) similarly stresses the difference between timber and fossil fuel, noting that the "management of collieries . . . is far more interesting than that of forests and coppices, for the coal when removed does not grow again" (123).[12] The extraction economy thus implied a new relation to futurity, and Jevons accordingly insists in *The Coal Question* that Britain's coal reserves be measured in time rather than volume. In the graphs that appear opposite the title page of Jevons's book, centuries are the primary unit of measurement, going up to the year 2000 (see figure 1.2). Glossing the bottom of the graph, Jevons writes, "supposed future consumption of Coal at same rate of progress showing the impossibility of a long continuance of that progress."[13]

Such Victorian hand-wringing about coal exhaustion may seem tragically misguided from the perspective of today, at a historical moment of reckoning with the geological fact that there is far too much coal in

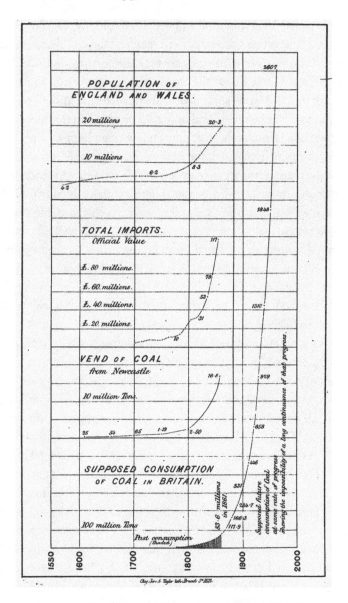

FIGURE 1.2. Graph from W. Stanley Jevons, *The Coal Question:*
An Inquiry concerning the Progress of the Nation, and the
Probable Exhaustion of Our Coal-Mines (London: Macmillan, 1865),
inside front cover.

the ground for our own good, far too many fossil fuels buried beneath
us for our atmosphere to bear. Indeed, from a policy point of view, our
best course of action to prevent the worst-case scenario would be precisely to leave the coal, oil, and gas in the ground, not for the use of future

generations but for their survival. This represents, in some ways, a clear distinction between the earlier extraction debate and our own, but there are continuities as well. If Jevons and others worried about a finite store of coal and other extracted materials, we now worry about a finitude of sinks in which to discard their waste products—not just the CO_2 emitted by fossil fuel combustion, but also the heavy metals of e-waste and other toxic remainders of extraction. Industrial-era exhaustion debates demand reconsideration now because they show that the transition to extraction-based life was understood *at the time* to entail a depleted earth for future generations; the transition required, in other words, the accommodation of a dynamic that is all too familiar today.

Even so, many social scientists of the era recognized that exhaustion correlates not only with supply but also with numerous social and political factors. Despite Jevons's ostensible focus on the time frame of exhaustion, for example, he strains to identify the moment at which such a conclusive end point might occur: he admits that mines might always be dug deeper and are constrained only by costs and risks from doing so; that other parts of the world, beyond England, might contain rich coal stores; and that new mining technologies could lead to new profits even in abandoned sites. Exhaustion was inevitable, but also overdetermined: it was a temporal threshold determined by economics (that is, the viability of profit in a particular mine) and demographics (that is, the rate of consumption, which depends on population) as well as geology and capacity.[14]

Decades prior to Jevons, Alexander von Humboldt's *Political Essay on the Kingdom of New Spain* (1808–11) also addressed the topic of resource exhaustion in its survey of Mexico's mining potential for European investors, but here Humboldt relied on an imperialist notion of an expandable resource frontier to offset the finitude of underground resources. In this way his essay confronts the strangely undead timescales of extraction capitalism, arguing that while in "a great number of works of Political Economy . . . it is affirmed that the mines of America are partly exhausted, and partly too deep to be worked any longer" (*Selections* 186), the vast frontier works against this tendency: "The abundance of silver is in general such in the chain of the Andes, that when we reflect on the number of mineral depositories which remain untouched, or which have been very superficially wrought, we are tempted to believe, that the Europeans have as yet scarcely begun to enjoy the inexhaustible fund of wealth contained in the New World" (185). This enticing vision of "inexhaustible" wealth was repeated frequently in the print culture surrounding the early years of British investment in Latin America, and as Barbara A. Tenenbaum and James N. McElveen write, "Thanks to the writings of Alexander von Humboldt . . . many in Britain . . . were eager to invest in Mexico, believing that the mines so lovingly described in his works could be brought to their former state of bonanza with the infusion of British capital and know-how" (52).[15]

It was not only Humboldt who produced this effect, but those who took his essay and rehearsed its argument to audiences of speculators, often for their own benefit. An author writing as G. H. echoed Humboldt in *The American Mines* (1834), stating that "so far from the mineral riches of America being exhausted . . . the richest districts both of Spanish and Portuguese America have never yet been subjected to the operations of the miner" (15), and that "Spanish America . . . must eventually prove an inexhaustible mart for British industry" (13). In 1824, Humboldt's essay was edited and published in England by John Taylor, mining engineer and treasurer to the Geological Society, whose gloss on the essay had the explicit goal of encouraging British investment in Mexican mine schemes: "if the skill and experience in mining which we possess, and the use of our engines, could ever be applied to the mines of Mexico, the result would be that of extraordinary profit" (vi). Taylor was himself heavily involved in such schemes and was key to the 1824 formation of the Real del Monte Company (an English mining company in Mexico).[16] Later in the century he founded another firm, Taylor Brothers, heavily involved in South African mining (Romero 73).

If imperial expansion appeared, at this early point in the century, to offer deliverance from the temporal threat of exhaustion—a subject I will take up at greater length in the next chapter—the Latin American mining bust of the 1820s also complicated this supposed escape hatch. It was a banner event in the history of mine exhaustion: British firms and investors had thought that abandoned mines left by the Spanish and Portuguese could be brought back to life with British industrial technologies, but the going proved much harder than expected, and the failures of these mining projects were such that they ultimately "led to the London market crash toward the end of 1825" (Heath 264). Accounts such as Humboldt's, according to Sharae Deckard, "almost single-handedly produced a British investment boom in silver mining," but when the "boom went bust . . . Humboldt was widely blamed for exaggerating Mexico's production capability" (38). Despite fiascos like this that kept mine exhaustion at the forefront of public discourse, the chroniclers of industrial extraction did not envision exhaustion as a discrete end point.[17] Extractive exhaustion instead appears in the literature to resemble that "untravelled world" of Tennyson's "Ulysses," "whose margin fades / For ever and for ever when I move" (142). Exhaustion loomed; mines could be brought back from the dead, but mines could also die again. The provincial realist novel incorporated this exhaustive timescale into its plots, forms, and conventions, upending the genre's foundational means of providing closure and continuity.

"Mine-Ridden": Nostromo

A sense of undeadness hovers over the San Tomé silver mine in Joseph Conrad's *Nostromo* (1904), a novel that unites provincial realism with aspects of the colonial adventure story (the subject of chapter 2), just as it brings together the temporality of extractive exhaustion with a spatial conception of the expanding resource frontier.[18] How can *Nostromo*, which takes place in South America, be read as provincial realism? The question is one of genre and theme as well as setting. John Plotz has described the experience of provincial life in the Victorian provincial novel as one of "semi-detachment," "far away from the seemingly inescapable centrality of the metropolis, yet still connected to it" (*Semi-detached* 108); this rhymes with Nathan Hensley and Philip Steer's recent description of Costaguana, the fictional South American setting of *Nostromo*, as a "place [that] has yielded its natural resources for the benefit of those elsewhere" ("Signatures" 74). Within Costaguana, much of the novel's action takes place in the town of Sulaco, which, the narrator says, "had never been commercially anything more important than a coasting port" (5). But ports turn out to be key arenas of action within extraction regimes, since they connect remote mines to the centers of finance—in this case, San Francisco, a city bred of the California Gold Rush. If sleepy, windless Sulaco had once been "an inviolable sanctuary from the temptations of a trading world" (5), the San Tomé mine, along with the steamships and steam railways that attend it, is changing all that. Money to develop the mine has come from San Francisco, but the mine owner, engineers, and railway bigwigs are all from England. The mine is "the biggest thing in Sulaco, and even in the whole Republic" (31), yet Sulaco remains semi-detached from the centers of Anglo-American power and "dwarfed almost to nothingness" when considered in light of the global portfolio and ambitions of its primary investor, Holroyd (58).

The combustion of coal pervades industrial-era novels set in London, as Jesse Oak Taylor and Allen MacDuffie have shown, but we must go outside the city, to provincial and overseas settings like Costaguana, to find accounts of resource extraction rather than consumption in the landscape of fiction.[19] This chapter's focus on provincial realism goes to some extent against the grain of recent urgings that ecocriticism turn to speculative and nonrealist genres as a corrective to an earlier overemphasis on realism, but as I hope to show, the temporal structure of the provincial realist novel—which is heavily reliant on the marriage plot, the inheritance plot, and the ideal of social reproduction—intersects with extraction temporalities in remarkable ways.[20] My argument thus focuses on provincial realism set in extractive zones amid the industrial mining boom. I take the term "extractive zone" from Gómez-Barris, whose book of that name points to the "successive march of colonial and neocolonial actors operating in relation to South America as if it were an extractible continent" (xvi). Industrial Britain played a starring role in this particular parade; in nineteenth- and early twentieth-century South America, "much of the trade was British trade. A majority of the merchants were British merchants. And the foreign capital investment was almost wholly British" (Humphreys 151).

Nostromo narrates this history of British engagement just as it anticipates the increasing dominance of US investment in the period after World War I, but while there has been a good deal of recent work on *Nostromo* that discusses the novel in terms of financialization, institutionalization, and the place of both within the imperial project of global capitalism, few critics have discussed at any length the "material interests" at the base of such developments: the extraction of silver from the San Tomé mine.[21] Cannon Schmitt describes *Nostromo* as a "documentation of the arrival" of finance capitalism and informal imperialism from Britain into Latin America, and he interprets the novel's difficult style and narrative complexity as a formal analogue for "the pervasive effects [of] the culture of investment" ("Rumor," 194). Benjamin Kohlmann likewise reads the novel as an account of the rise of a new imperialism rooted in "globalized speculation" (454), and Regina Martin as a parable for the rise of absentee capitalism and the concomitant demise of an earlier form of capitalism based in the family-owned firm. All three critics have brought critical finance studies to bear on the novel in important ways, but I would suggest that extractivism, not just finance capitalism, must be central to our understanding of the novel.

With its large cast of characters, *Nostromo*'s pivotal node is not, in fact, the eponymous sailor Nostromo, but rather the San Tomé mine itself, whose abandoned status causes its unwilling recipient, Gould Sr., to be haunted by dreams of the undead. After being granted the unproductive mine from the Costaguana government as a means of getting more taxes from him, Gould Sr. "became at once mine-ridden," and he "began to dream of vampires" (43). Obsessed with the mine, Gould Sr. begins to "suffer from fever, from liver pains, and mostly from a worrying inability to think about anything else" (44). This is not the novel's only portent that the subsurface treasures of Costaguana are somehow cursed: the narrator opens the first chapter by retelling a local legend about "heaps of shining gold" that lay buried on the nearby Azuera Peninsula, beneath a barren patch of land "blighted by a curse" where not "a single blade of grass will grow." Locals believe the wasteland is haunted by the ghosts of two gringo treasure hunters, whose "souls cannot tear themselves away from . . . the discovered treasure" (5–6).

The association between treasure and blight extends to the San Tomé mine as well. Gould Sr.'s son Charles, who will inherit the mine, grows up being "told repeatedly that [his] future is blighted because of the possession of a silver mine" (44). His father warns him that the mine is a poisoned chalice, an evil inheritance, and of the "fatal consequences attaching to the possession of that mine" and "the apparently eternal character of that curse" (44). Such a morbid fixation has the inevitable result that mines come to acquire "a dramatic interest" for young Charles: "Abandoned workings had for him a strong fascination. Their desolation appealed to him like the sight of human misery, whose causes are varied and profound. They might have been worthless, but also they might have been misunderstood" (45). At school in England, "the young man in Europe grew more and more interested in that thing which could provoke such a tumult" in his father (45). That Charles's feelings for the mine and commitment to resurrecting it stem from his father's fatal obsession—an obsession that ultimately kills Gould Sr.—suggests from the beginning of the novel how mining and extraction are tied up with reproduction, generation, and failed inheritance. Charles determines to bring the mine back to life, both to avenge his father's death and in explicit transgression of his father's "solemn wishes" (50), for he imagines that a productive mine might bring stability to Costaguana as well as to his patrimony.

The history of the mine would suggest otherwise. Under Spanish colonialism, the San Tomé mine had been worked "mostly by means of lashes on the backs of slaves," but "with this primitive method it had ceased to make a profitable return" (40), thus it was deemed exhausted. An English company, the narrator tells us, obtained rights to the mine after the fall of Spanish imperialism, but a revolt of native miners prevented the mine's development under this firm. San Tomé is henceforth neglected to the point where it "was no longer an abandoned mine; it was a wild, inaccessible, and rocky gorge" (42). Believing it exhausted and worthless, the Costaguana government then has an "ingenious insight into the various uses a silver mine can be put to, apart from the sordid process of extracting the metal from under the ground." San Tomé is off-loaded to Gould Sr. in the form of a "perpetual concession" (41), one that hounds him to the grave.

Contrary to all expectations, however, the mine comes to give again, under the right pressures. After studying mine engineering in Europe, Charles Gould, the "Englishman of Sulaco" (37) manages to revive San Tomé into a "fabulously rich mine" (31). He accomplishes this with the help of new machinery sent by his San Francisco investor (61), and the help of six hundred Indigenous mine workers, who receive, as a perk of their position, protection from the police and from forced enlistment in the army of the republic (73). The mine promises "work and safety" (76) to the Indigenous Costaguanans while they, in the narrator's ironic phrasing, remain "waiting for the future" (66). Like mining, however, politics in Costaguana is "a brazen-faced scramble for a constantly diminishing quantity of booty" (86), and ultimately the San Tomé mine generates not just wealth, but war, exposing the connection between extraction regimes and warfare now conveyed with the term "conflict mineral."[22] In the course of *Nostromo*'s convoluted narrative, the dictator who was financed and propped up by Gould and his mine falls to a revolutionary dictator and his army of "negro Liberals" (286)—who seek to control the mine themselves—which leads to yet another revolution, this one engineered by Gould's allies and funded, again, by the mine.[23] The plot, though fictional, conveys how Britain's informal empire in Latin America contributed to political upheaval by way of extraction and investment.[24]

Interwoven with this history of the mine is the story of a marriage and its own downward trajectory; the romance between Charles and Emilia Gould, "the only English woman in Sulaco" (25), is from the beginning a romance of extraction. Emilia is described as "the first, and perhaps the only person" to understand Charles's "secret mood . . . towards the world of material things" (45–46). His enthusiasm for the mine becomes hers: "His faith in the mine was contagious" (57). Gould is a third-generation Costaguanan, but he "looked like a new arrival" (37); his family origins are English, and the Gould family "always went to England for their education and for their wives" (36), evincing their concern with extraction of another sort.[25] Charles meets Emilia in Europe, and during their courtship they visit a marble quarry in Italy, "where the work resembled mining in so far that it also was the tearing of the raw material of treasure from the earth"

(46). When Charles proposes, Emilia has "the physical experience of the earth falling away from under her" (48). This is a romance told in rock, mineral, and earth.

After they wed, Charles's resuscitation of the mine is described as "in essence the history of [Mrs. Gould's] married life" (51). She "had seen it all from the beginning; the clearing of the wilderness, the making of the road, the cutting of new paths up the cliff face of San Tomé" (80). At first she "talk[s] of the mine by the hour with her husband with unwearied interest and satisfaction" (53), but by the middle of the novel there is "a wall of silver-bricks, erected by the silent work of evil spirits, between her and her husband" (160). The marriage has proven to be a disappointment in its temporal features: "It had come into her mind that for life to be large and full, it must contain the care of the past and of the future in every passing moment of the present. Our daily work must be done to the glory of the dead, and for the good of those who come after" (373). For Mrs. Gould, this means children, and the children never come, which is in some obscure way because of the mine: "she saw clearly the San Tomé mine possessing, consuming, burning up the life of the last of the Costaguana Goulds; mastering the energetic spirit of the son as it had mastered the lamentable weakness of the father. A terrible success for the last of the Goulds. The last! She had hoped for a long, long time, that perhaps—But no! There were to be no more" (373).

Such passages suggest how exhaustion's temporal features pervade the provincial realist marriage plot, denying its normative resolution and defying conventional novelistic means of ensuring futurity. The only birth scene in the novel, indeed, is the mine's first parturition of silver, a delivery at which Emilia serves as midwife: "she had seen the first spungy lump of silver yielded to the hazards of the world by the dark depths of the Gould Concession; she had laid her unmercenary hands, with an eagerness that made them tremble, upon the first silver ingot turned out still warm from the mould" (80). Such ironic framing of mine production as a substitute form of reproduction actually highlights the gulf between them—that mining operates on a timescale of exhaustion rather than renewal.[26] Still, precisely because of the threat of exhaustion, extraction does beget more extraction, and at the end of *Nostromo* the San Tomé mine has become San Tomé Consolidated Mines, producing not just silver but also gold, copper, lead, and cobalt (361).

If *Nostromo*'s extraction plot leaches into the Gould's marriage plot, dismantling its narrative capacity to ensure regenerative futurity, we see a similar effect with two other thwarted marriage plots in the novel:

between Martin Decoud and Antonia Avellanos and between Nostromo and Giselle Viola. Neither of these marriages comes off at all owing to the premature deaths of Decoud and Nostromo, deaths that are directly traceable to the influence of the San Tomé silver. Martin Decoud and Antonia Avellanos, to take one example, are both of the Costaguana settler elite— educated in Europe, politically aligned with the Goulds, and possessing skin of a "healthy creole white" (111). Their relationship appears doomed from the beginning, as Decoud suggests in a letter to his sister: "[Antonia] is more to me than his precious mine [is] to that sentimental Englishman," but "there was for us no coming together, no future" (173). In an effort to keep the San Tomé silver out of the hands of the revolutionary forces, Decoud joins with Nostromo in a desperate plan to spirit a load out of the country via a small boat, but as Decoud himself says, "so much treasure is very much like a deadly disease" (191). He ends up stranded on the small island where he and Nostromo bury the treasure, and he drowns himself before Nostromo comes back for him, weighing his body down with bars of San Tomé silver. The solitude and waiting, the narrator explains, cause Decoud to lose "the sustaining illusion of an independent existence as against the whole scheme of things of which we form a helpless part" (357). Before leaving him, Nostromo told Decoud to try to imagine time from the treasure's point of view: "Time is on its side. . . . Silver is an incorruptible metal that can be trusted to keep its value forever" (216). But such mineral timescales merely encourage Decoud's reflections on his own insignificance. Mineral timescales seem to dictate Antonia's love, too—for she never marries after Decoud's death and instead memorializes him in a marble medallion in the Sulaco Cathedral.

Nostromo's engagement to Giselle Viola similarly ends in silver-streaked tragedy. Our hero is secretly engaged to Giselle, unbeknownst to her sister, Linda, to whom he is publicly engaged. Even after the revolution is put down, Nostromo never tells Gould that he and Decoud successfully buried the silver on the island, and his secret relationship with Giselle mirrors his secret hoard of silver, from which he determines to "grow rich very slowly" (360) to avoid exciting suspicion. Linda anguishes over their long engagement, telling Nostromo, "I shall go grey, I fear, before the ring is on my finger" (392). Giselle, meanwhile, plans to elope with her sister's fiancé, and her description in the novel identifies her with precious metals: "coppery glints rippled to and fro on the wealth of her gold hair" (383); her voice is like "the tinkling of a silver bell" (383). The triangulated engagement plot ends abruptly when Nostromo is accidentally shot by Giselle and Linda's father while prowling about the island one night for more of his buried silver. When Giselle goes to Emilia Gould for comfort, the latter feels only "cynical bitterness" and responds, "console yourself . . . very soon he would have forgotten you for his treasure" (402). Thus Nostromo dies unmarried, heirless, with no provisions for the future. On his deathbed, a man asks, "have you any dispositions to make?" Nostromo dies without answering (402).

All these broken and infertile marriage and inheritance plots point toward an exhausted futurity, a demise captured in many aspects of the narrative. The section of the novel where Decoud and Nostromo are on their mission to save the silver, for example, is suffused with dramatic irony, in a positive symphony of unevenly distributed knowledge, with the effect of casting general confusion over novelistic certainties of every sort.[27] Never has extraction capitalism been so accurately depicted as an affair of profound ignorance and hedged bets. Among the novel's large cast of characters, confusion and misinformation circulate as to the whereabouts of Decoud, Nostromo, and six months' worth of silver from the mine. Many think the men have drowned and the silver is at the bottom of the sea; readers have enough information to know this is not the case, but Conrad still leaves us in extended suspense as Nostromo and Decoud are absent from the narrative for a long portion of part 3. When Nostromo reappears, his first words are, "I am not dead yet" (296), which he repeats just a few pages later, "I am not dead yet" (312). His disappearance and reappearance suggest how, in this ostensibly realist novel, narrative temporalities of life and death become as ghostly and untethered to natural cycles as the temporalities of the mine.[28]

By the end of the narrative, Nostromo, like Gould, has become peculiarly invested in the San Tomé silver, so much so that the silver appears to have leached into him: "his life had become bound up with it" (305), and he "had welded that vein of silver into his life" (377). As Nostromo himself confesses to Doctor Monygham, "there is something in a treasure that fastens upon a man's mind. . . . He will never forget it till he is dead—and even then—Doctor, did you ever hear of the miserable gringos on Azuera, that cannot die? . . . There is no getting away from a treasure that once fastens upon your mind" (331). Amalgamated with the silver, Nostromo, wraithlike, can no longer be narrated through realist trajectories of novelistic closure, a point captured in the novel's ghostly closing line: "away to the bright line of the horizon, overhung by a big white cloud shining like a mass of solid silver, the genius of the magnificent [Nostromo] dominated the dark Gulf containing his conquests of treasure and love" (405). Nostromo is dead by this time, but the idea that his "genius" haunts the place where his treasure is buried suggests the ghostly horizon of unfixed limits that attends extraction ecologies.

The mine, too, operates on a timescale at odds with regenerative cycles, and its future, the novel suggests, will be as complexly drawn out as its past. San Tomé is under constant threat of detonation because Gould has loaded it with "enough dynamite . . . to send it down crashing into the valley . . . to send half of Sulaco into the air" (148). This is his way of protecting the mine from "the approach of an armed force" (286). Convinced that such an explosion would destroy the mine for good, Gould asks a leader of the revolutionary opposition, "once dead, where was the power capable of resuscitating such an enterprise in all its vigour and wealth out of the ashes and ruin of destruction?" and "where was the skill and capital abroad that would condescend to touch such an ill-omened corpse?" (288). But the dynamite is never lit, and San Tomé is never exploded. A literary explosion, in Sue Zemka's phrase, is a "trope of the portentous moment" (176), a gathering up or concentration of meaning into one fateful instant. In other writings, like "Youth" (1898) and *The Secret Agent* (1907), Conrad relies on such punctualist explosions, and in *Lord Jim* (1900) he "build[s] a novel around an instant," as Zemka argues, with Jim's jump as "the pivot on which everything turns" (174). But if such texts show "the rise of the moment's symbolic duties" in modern literature (64), *Nostromo* is temporally tuned instead to the slow violence and slow causality of environmental catastrophe. Even without the dynamite, the expiration of San Tomé is well underway and has been for many years. Already, a wild ravine of sublime beauty has been transformed by the workings of the mine: "The waterfall existed no longer. The tree-ferns that had luxuriated in its spray had dried around the dried-up pool, and the high ravine was only a big trench half filled up with the refuse of excavations and tailings" (79). "Mrs. Gould," the narrator goes on, "had seen it all from the beginning" (80). The mountain is dying, and its fortune is told in the novel's marriage and inheritance plots—stories of exhausted futurity.

"The Red Deeps, Where the Buried Joy Seemed Still to Hover": The Mill on the Floss

The nonproductive and unfulfilled marriage plots of *Nostromo* mirror those of other provincial realist novels that foreground extraction ecologies, including George Eliot's *The Mill on the Floss* (1860). Eliot's second novel is set at a water-powered mill during the 1830s at a historical moment of energy transition from water power to coal-fired steam power;

moreover, it features a thwarted romance between Maggie Tulliver and Philip Wakem, whose courtship unfolds in an abandoned ironstone quarry called the Red Deeps. Here Eliot, like Conrad, adapts the provincial realist novel's long-standing focus on social renewal by way of marriage, reproduction, and inheritance to the extraction-based society of industrial Britain, which is undergirded by a temporal structure of exhaustion rather than seasonal renewal. "The completion of the harvest had been arrested," Eliot's narrator tells us just before the novel concludes in a catastrophic flood (507).[29] The flood kills Maggie and her brother and with them the Tulliver line whose generational drama had heretofore been the focus of the book.

Like *Nostromo* and *Jane Rutherford* (which I will discuss next), *The Mill on the Floss* takes place in the recent past.[30] Devin Griffiths has described the historical novel as it emerged in the nineteenth century as "a new technology for writing about past events and thinking about their complex relations to present experience" (*Age of Analogy* 2), one that represents "a new sense of everyday life as being *lived in history*" (26).[31] Such a perspective is apparent in *The Mill on the Floss* not only from its 1830s setting, but from the narrator's transtemporal perspective and archival ambitions.[32] We know from Eliot's notebooks that the novel's provincial setting reflects her detailed study of real-world locations, down to the likelihood of flood conditions on the River Trent and the dates of recent inundations.[33] These specificities are important because they show how Eliot is using the temporal technology of the historical novel to reflect on what ecologists call the "shifting baseline," the way that the human perspective on environmental norms can change profoundly even in a generation or two.[34] England's environment transformed greatly in the thirty years from when Eliot's novel is set (1830-ish) to when it was published (1860), years that saw a momentous transition toward industrialism and extraction-based life. If many of the novel's characters seem ignorant of the magnitude of the changes taking place around them, the narrator explains that "the present time," which is actually circa 1830, "was like the level plain where men lose their belief in volcanoes and earthquakes, thinking to-morrow will be as yesterday, and the giant forces that used to shake the earth are for ever laid to sleep" (156). As is clear with the flood at the novel's end, however, such forces of natural change have been awake the whole time.[35]

Eliot's fictional account of the major transitions that took place in 1830s industry is of new interest today because of the world-historical significance of the energy shift her novel documents—a shift that set the course for anthropogenic climate change. As Malm explains, it was during the period from 1825 to 1848 that a "decisive shift to steam occurred" in British industry (63); in this "new type of economy that Britain had developed, coal was the lifeblood of all manufacturing" (228). The relation between this historical novel's past and our present is particularly striking if we imagine the fossil economy, as Malm does, "as a train put at a point in the past on the current perilous track" (13). Eliot's novel tells the story of how the train switched tracks. At one point, one character asks another about his "intentions concerning steam" (459); the divide between such intentions and their effects has never been more visible than it is today.

Discussion of the possible conversion of Dorlcote Mill from water power to steam power courses through the novel, along with a general sense that the world is speeding up on the back of coal-fired capitalism.[36] As Maggie's Uncle Deane puts it, "It's this steam, you see, that has made the difference: it drives on every wheel double pace, and the wheel of fortune along with 'em" (403). If the wheel of the water mill previously defined the rhythm of life on the Floss, time's wheel is quickening its pace under a new energy regime. Deane explains to Maggie's brother Tom, "the world goes on at a smarter pace now than it did when I was a young fellow." Back then, "the looms went slowish, and fashions didn't alter quite so fast" (403). He connects steam's accelerated temporality to the rise of global trade and increased production: "Trade, sir, opens a man's eyes. . . . It's a fine thing . . . to further the exchange of commodities" (403–4). Dorlcote Mill is, Tom tells Deane, "a very good investment, especially if steam were applied" (406), and in the same conversation, Deane sends Tom on a business trip to Newcastle, center of the coal trade (403). Fossil capitalism is driving the accelerated pace of progress in the novel, transforming the characters' environment and their experience of time and history. Reflecting on the generational span of time antecedent to the novel's 1830 setting, the narrator notes the temporal difference between the older economy and the newer one, contrasting "the industrious men of business of a former generation, who made their fortunes slowly" with "these days of rapid moneygetting" (159).[37]

Moneygetting has accelerated because of extraction and global trade, as the opening lines of the novel convey: "A wide plain, where the broadening Floss hurries on between its green banks to the sea, and the loving tide, rushing to meet it, checks its passage with an impetuous embrace. On this mighty tide the black ships—laden with the fresh-scented fir planks, with rounded sacks of oil bearing seed, or with the dark glitter of coal—are borne along to the town of St Ogg's" (51). The novel's first sentence

("A wide plain . . .") is not, in fact, a full sentence at all, since it is missing a predicate for "wide plain." The plain is a subject without a verb, raising the question of what kind of action we can expect from an environmental agent—a question that will be answered in the novel's final, tumultuous flood, where the water takes Maggie's life and Dorlcote Mill along with it. These opening lines also paint a picture of multiple energy forms: the currents of water that move the ship, the ship itself laden with the "dark glitter of coal" (51). We see, too, evidence of global trade: the oil seed comes from Russia, as we later learn (156), and the coal is likely from northern England. Extraction and trade are bound up with each other, and social life (or at least "good" social life) now depends on extracted commodities: "good society," the narrator says, "is of very expensive production; requiring nothing less than a wide and arduous national life condensed in unfragrant deafening factories, cramping itself in mines, sweating at furnaces, grinding, hammering, weaving under more or less oppression of carbonic acid."[38] Such extraction and industry are, in the narrator's ironic phrasing, "necessary for the maintenance of good society and light irony" (312)— qualities embodied in the person of Stephen Guest, whose own perfume bears no resemblance to those "unfragrant" factories, for his "diamond ring, attar of roses, and air of nonchalant leisure . . . are the graceful and odiferous result of the largest oil-mill and the most extensive wharf in St Ogg's" (373). Phrased this way, the maintenance of a Stephen Guest hardly seems worth all the extraction and labor it requires.

While *The Mill on the Floss* depicts the rise of steam and the related expansion of global trade, much of the plot concerns a legal dispute about local water rights. Maggie's father, Mr. Tulliver, depends on the flow of the river to power Dorlcote Mill, a mill that has been in his family "a hundred year and better" (191). At the outset of the novel, Tulliver has successfully fought off a neighbor's attempt to dam the river, but he soon engages in a new legal entanglement against Mr. Pivart, a farmer establishing an irrigation scheme further up the river.[39] Tulliver eventually loses this case, to the ruin of his family. His legal woes point to a fundamental aspect of the broader energy transition happening at the time the novel is set: the affordances of water are unsuited to privatization and to capitalization on a large scale, whereas coal, in its precombusted form, is more easily privatized. Water, Tulliver says, is "a very particular thing—you can't pick it up with a pitchfork" (191). Malm writes similarly that water "respect[s] no deeds or titles, bow[s] to no monetary transactions; it continue[s] on its course, unmoved by conceptions of private property" (117). Even if one owns the land on which a stream of water flows, the water also belongs to others upstream or downstream, which is why the laws around water's use were a matter of dispute at the time Eliot's novel is set. Not until coal is burned does it belong to everyone, and become everyone's problem.

Water power is also subject to fluctuations based on the weather and the season. Early in the novel, when Maggie goes inside the mill, she hears "the unresting motion of the great stones," seemingly ceaseless, yet she also senses "the presence of an uncontrollable force" (72). Water power entails a human harnessing of the river, but dry weather as well as wet weather and storms can impact its capacity, and as Jean-Claude Debeir, Jean-Paul Deléage, and Daniel Hémery explain, abundant rain, drought, and freeze can all immobilize a water wheel (75). Malm writes, "traditionally, weak streams during dry summers were no more aberrant or maddening than the fact that grain could not be harvested in midwinter or ploughed in a thunderstorm," but such "indulgence toward erratic rivers" waned with the rise of global capitalism (166). With the option of steam, the temporal ebb and flow of water power became newly intolerable for 1830s manufacturers, even though water power remained an accessible and inexpensive source of energy in England.

Eliot's novel links water power closely to the temporal rhythm of the seasons by situating the devastating flood at the end of the novel—the flood that destroys the mill and kills Maggie and Tom—in the "second week of September . . . about the equinox" (507–8). A long-held belief in the so-called equinoctial storm, which held that "a severe storm is due at or near the date of the equinox," was gradually debunked with the improvement of meteorological science in the late nineteenth century ("Equinoctial Storm"), but in Eliot's novel, the equinoctial storm serves as a climatic and climactic event that ties together weather, water, and seasonal time, binding the temporality of water power to astronomical rhythms. Dorlcote Mill's water-powered business is acknowledged by Deane to be "a good one," but he considers it a worthy speculation in that it "might be increased by the addition of steam-power" (270). Steam offers an escape from the rhythm of the seasons. As Dickens writes in *Hard Times*, a novel that I discuss later in the chapter, "no temperature" affects steam engines, which go "up and down at the same rate, in hot weather and cold, wet weather and dry, fair weather and foul" (104).

Eliot could not have known the accretional effects of coal combustion across time, which would ultimately destabilize the seasons altogether, but her novel does suggest that humans are ill equipped to understand the longer temporal arcs of the energy systems we use because of our short lifespan and transient memories. Maggie and Tom are unprepared for the catastrophic flood at the end of the novel, the flood caused by the equinoctial storm, in part because of a lapse of human memory across generations: "the rains on this lower course of the river had been incessant, so that the old men had shaken their heads and talked of sixty years ago, when the same sort of weather, happening about the equinox, brought on the great floods, which swept the bridge away, and reduced the town to great misery. But the younger generation, who had seen several small floods, thought lightly of these sombre recollections and forebodings" (508). Here we see the insufficiency of individual human understanding in the face of the longer time line of the floods and the protracted intervals between them. Eliot's notes on flooding in her research for the novel circle around this point by documenting accounts of various floods and how they compare to the memories and records of the local communities (Eliot *Writer's*, 36–38).

The tragedy at the novel's end thus represents a failure of temporal perspective, one that the narrative form of *The Mill on the Floss* attempts to correct. At the novel's conclusion, the narrator is poised between the future and the past, reading them in relation to one another, and calls for a reader with a similarly dialectical temporal orientation. Surveying the land that the waters once engulfed from the standpoint of five years into the future, the narrator observes, "Nature repairs her ravages—repairs them with her sunshine, and with human labour. The desolation wrought by that flood, had left little visible trace on the face of the earth, five years after." But the narrator goes on to restate the point more precisely and more pessimistically: "Nature repairs her ravages—*but not all*. The uptorn trees are not rooted again; the parted hills are left scarred; if there is a new growth, the trees are not the same as the old, and the hills underneath their green vesture bear the marks of the past rending. *To the eyes that have dwelt on the past, there is no thorough repair*" (522, my emphasis). Here Eliot describes something like the shifting baseline discussed by Jeffrey Bolster and others, where "each generation imagine[s] that what it saw first was normal, and that subsequent declines were aberrant. But no generation imagined how profound the changes had been prior to their own careers" (Bolster 10). The point is not simply that there is no normal in the natural world, but that the human tendency to assert a normal has the effect of masking the profound reach of human impacts into the natural world.

These profound impacts are especially visible in one of Eliot's settings, the Red Deeps. If extraction shapes *The Mill on the Floss*'s temporal forms by way of coal and steam, it also shapes its landscape in the abandoned quarry known as the Red Deeps. This is a key setting in the novel and a place haunted by exhaustion, "broken into very capricious hollows and mounds by the working of an exhausted stone-quarry, so long exhausted that both mounds and hollows were now clothed with brambles and trees" (316). Eliot does not specify the type of stone that was quarried here but was most likely thinking of ironstone, given the redness of the dirt and the fact that Lincolnshire, where the novel is set, was the richest site in England for ironstone mining. Ironstone was typically mined from open pits or quarries, especially in the area around Scunthorpe and Frodingham, near the River Trent (model for the fictional Floss); it was then shipped up the Trent to the ironworks and blast furnaces of Yorkshire.[40] In Eliot's novel, the telltale red dirt at the Red Deeps is, then, both a reminder of past extraction and the clue that leads Tom to discover his sister's secret rendezvous in the exhausted quarry with the hunchback Philip Wakem—in one of two thwarted marriage plots that shape the novel's trajectory.[41]

In the Red Deeps, Maggie and Philip meet secretly for a year before they are found out. Tom eventually puts two and two together after hearing his mother "scold Maggie for walking in the Red Deeps when the ground was wet, and bringing home shoes clogged with red soil" (355) and his aunt describe Philip Wakem as always "a-scrambling out o' the trees and brambles at the Red Deeps" (354). When Tom confronts Maggie, her blush "displays on her body" the "red landscape" she has been visiting (Buckland 235), suggesting that ironstone has permeated Maggie as the silver did Nostromo. The termination of her meetings with Philip leaves Maggie with a sense that "her future . . . was likely to be worse than her past" (384) and leaves Philip with "nothing but the past to live upon" and dismay that "the future will never join on to the past again" (448). These lines suggest how the Red Deeps, a site of exhaustion, occasions a rupture of temporal continuity in the text, like a narrative without an ending: "That book never will be closed," Maggie says of her time with Philip in the Red Deeps (449). Just as the landscape continues to show its past life of extraction on its surface, Maggie's yearlong idyll with Philip seems always present and never over. When she meets Philip later she feels herself back in the past: "in Maggie's mind the first scenes of love and parting were more present than the actual moment," and suddenly, again, "she was looking at Philip in the Red Deeps" (449).

That Eliot locates this time-rending romance in an exhausted quarry is crucial. Earlier in the novel, describing Tom's poor education, the narrator asks, "how should [his teacher] be expected to know that education was a delicate and difficult business? any more than an animal endowed with a power of boring a hole through a rock should be expected to have wide views of excavation" (202). Humans may have the capacity to dig, but that does not mean we have any sense of the effects such activities will have in the wide view or the long view. Perhaps for this reason, the Red Deeps feels haunted, like other sites of extractive exhaustion in industrial-era literature: "In her childish days Maggie held this place . . . in very great awe, and needed all her confidence in Tom's bravery to reconcile her to an excursion thither" (316). Later, the place attracts her by its aberrance from the rest of the landscape and by its privacy: "now it had the charm for her which any broken ground, any mimic rock and ravine, have for the eyes that rest habitually on the level." Here she can "sit on a grassy hollow under the shadow of a branching ash, stooping aslant from the steep above her . . . sure of being unseen" (316–17).

Maggie and Philip's relationship grows in this exhausted quarry but lacks an apparent future because of Philip's deformity and because their fathers hate each other: "Any prospect of love and marriage between her and Philip [was] put out of the question by the relation of the two families" (396). Indeed, "the very name of Wakem made her father angry, and she had once heard him say, that if that crook-backed son lived to inherit his father's ill-gotten gains, there would be a curse upon him" (219). Still, their futureless relationship thrives in secret in the abandoned quarry, at odds with chrononormativity and repronormativity, which Elizabeth Freeman has described as the "interlocking temporal schemes necessary for genealogies of descent" (xxii). This is not to say that their relationship exhibits the liberatory quality that Freeman ascribes to queer time, the release from a timescale oriented toward "maximum productivity" (3). Rather, their relationship operates on what I would call an extractive trajectory, diverging from the provincial novel's temporal structures of renewal and rebirth.[42] Maggie tells Philip, the first time they meet, that they must not do so again, but he begs, "let me see you here sometimes" (321); Maggie fears from that moment that the agreement "would act as a spiritual blight" (321). The association of extractive spaces with blight, which we also saw in *Nostromo*, conveys a sense of ruin for the future.

In rhythm with the exhaustion that haunts the Red Deeps, Maggie and Philip's relationship is denied the possibility of futurity and consummation but is also denied an ending or closure and seems to live on after Maggie's dramatic death. After Maggie is swept away in the climactic flood, Philip remains to haunt the Red Deeps, his grief persisting in an undead future that extends beyond the novel's penultimate line: "His great companionship was among the trees of the Red Deeps, where the buried joy seemed still to hover—like a revisiting spirit" (518).[43] Like *Nostromo, The Mill on the Floss* ends with a dead protagonist, a haunted vision of extraction's remains, and marriage plots left unfulfilled. Jed Esty says that Eliot's novel marks the early edge of a shift in narratives of "provincial growth" to "assimilate an increasingly global logic of economic transformation into the very texture of the protagonist's emergence" (53), and Ian Duncan argues that it shows woman's "banishment from the historical avenues of *Bildung*" (*Human Forms* 177). I want to stress, however, how much its "antidevelopmental conclusion" (61), to use Esty's phrase, mirrors other narratives of resource extraction from this period.

The revised marriage plot that we find in provincial realist novels set in extraction zones conveys how human life and reproduction are bound up in the ecologies and economies of extraction. Philip's companionship with the trees that emerge from the "buried joy" of the Red Deeps emblematizes the operations of extraction ecologies as I have been describing them: the dynamic bundling of human, animal, and plant life with extracted commodities and the nonliving earth; the undead futures of extraction and exhaustion. Years ago, in *The Country and the City*, Raymond Williams highlighted marriage as a symptomatic terrain in the postpastoral novel of environmental displacement: "One of the most immediate effects of mobility, within a structure itself changing, is the difficult nature of the marriage choice. This situation keeps recurring in terms which are at once personal and social" (210). His impulse to think of marriage in socio-environmental terms anticipates those critics who are now rereading literary history with an eye to the ways that human choice is coproduced by natural circumstance. Benjamin Morgan, for example, argues against "a conventional literary historicism that attributes causal primacy to human culture" without attention to "entanglements of human action with geological and climatological events . . . as motive forces of history" ("After" 2). Such an argument conveys in wider terms the operations by which extraction ecologies can be said to leach into the provincial realist marriage plot in the age of industrial extraction and revise its horizon of time and futurity.[44]

"To Teem with Life": Jane Rutherford: or, The Miners' Strike

The mysterious seepage of the nonliving earth into human lives and human futures is at the very heart of *Jane Rutherford: or, The Miners' Strike* (1854), a novel by Fanny Mayne that unites the provincial realist marriage plot with an account of extractive labor. Centering on a coal-mining village, the novel, unlike *Nostromo* or *The Mill on the Floss*, closely examines labor conditions within extractive zones and the social forms that attend the necessary reproduction of workers in extraction econo-mies. This is a subject more famously taken up thirty years later in Émile Zola's *Germinal* (1885), a novel that is similarly focused on labor, repro-duction, and strikes in a coal-mining community, but is far more radical in its politics.[45] *Jane Rutherford*'s politics are not radical; Mayne's narrator is the sort who will pause in the middle of the story to dispute the labor the-ory of value: "Socialists and Communists may boast as they will about the omnipotence and all-sufficiency of labor-capital, but disconnect it from the ordinary operations of trade and commerce, allow it to remain inactive or unemployed, and where is its value? It possesses literally none" (159–60). As a contemporary review in *Tait's Edinburgh Magazine* flatly put it, the novel "is a domestic story, intended to illustrate the evils of strikes" (Review of *Jane Rutherford*). Despite this seemingly crude political moral, however, the *Tait's* reviewer still finds the novel of interest because the writer seems "at home among" the miners, "intimate with their circum-stances and condition," and it is this admixture of working-class extractive labor and domestic life that makes the novel worthy of closer attention.[46]

Mayne was not herself a member of the working classes, but her brother Charles Otway Mayne was the vicar of Midsomer Norton, a town in the heart of the Somerset coalfield where the novel is set. He officiated at the burial of the miners who were killed in the Wellsway Disaster on 8 November 1839—a disaster that inspired the plot of *Jane Rutherford*.[47] Of the twelve miners who died in the Wellsway Disaster, seven were age sev-enteen or under, and of those, five were age thirteen or under. The young-est, Amos Dando, was twelve.[48] Perhaps it was the youth of the victims in the context of an extraction economy expressly dependent on young work-ers that led Mayne to produce a novel with notable narrative resistance to repronormativity.

Before its 1854 volume publication, *Jane Rutherford* was initially seri-alized in Fanny Mayne's periodical *True Briton* from June to October 1853. The paper was evangelical and populist in tone and staunchly abolitionist, but not antiestablishment.[49] As the editor states in a column on the front page of the 1 August 1853 issue, the paper aims to be "rational, elevating, and useful in its literature, without exhibiting party spirit," and it directs itself at "the *people*—the *masses*," who, the editor hopes, will "look upon us as their friend and fellow-labourer." The journal catered to readers in mining districts with numerous articles such as "Interesting Facts Related

THE RUN AND READ LIBRARY.

JANE *393*

RUTHERFORD:

THE MINERS' STRIKE.

BY A FRIEND OF THE PEOPLE.

LONDON:
CLARKE, BEETON, & CO., 148, FLEET STREET.

ONE SHILLING AND SIXPENCE.

The tale is the best of the kind in tone and treatment we have met with.' WEEKLY TIMES.

'It is a broad and well-charged picture of life among the lowly.' CHRISTIAN WITNESS.

FIGURE 1.3. Cover of *Jane Rutherford: or, The Miners'
Strike* (London: Clarke, Beeton, 1854).

to Minerals" (7 October 1852), "The Ventilation of Mines" (16 December 1852), "Coal and the Collieries" (6 October 1853), and a series titled "The Gold Hunt" (20 January 1853–21 April 1853), the last installment of which ended with a melancholy reflection on exhaustion: "The gold hunt, and all other earthly pursuits of man, will soon come to an end" (part 3, 676). Following its serialization, *Jane Rutherford* was published in volume form as part of the Run and Read Library, a series of books intended as travel reading for "Railway, Road, and River" (see an image of the cover in figure 1.3). The novel presents a case for reading the railway novel as extractivist literature, for *Jane Rutherford* not only is designed for reading on steam-based transit; it tells the story of the labor entailed in powering such transit.[50]

Though serialized in 1853, *Jane Rutherford* is set in 1844 in the immediate wake of the 1842 Mines and Collieries Act, the key provisions of which were intended to mitigate the cruelties of industrial-era mining by prohibiting women and children under ten from working underground.[51] The novel depicts life in the Somerset coalfield, which supplies the wealthy neighboring city of Bath; this Somerset mirrors Conrad's Costaguana as a place that generates extracted resources for the wealth of others, elsewhere. In the opening sentence of the novel, the narrator looks at the "dim and dingy" scene from atop a hill and reflects, "There are many places and things more pleasant to look at from a distance than near," including the "dirty and smoky collieries with which [this] part of England abounds" (9), a provincial region "defiled by coal and chimneys, and smoke and steam" (11). To present its account of mining labor as accurate, *Jane Rutherford* is preceded by "A Miner's Preface," signed by Peter Richards, "a Newcastle miner" who asserts that the novel "is all so true about mining . . . how the pits be worked, and what the different miners have got to do." He guesses that the author of this anonymously published novel "must have been a miner himself" (iii). The inclusion of this preface—and that significant word "himself"—suggests that Mayne was anxious about establishing her authority on the subject of mining, unsurprisingly given her class position and given that women were legally barred from underground mine labor after the 1842 act. She makes a claim for authority in the middle of the novel when the narrator breaks off to ask, "Have any of my readers ever been down a coal-mine? . . . The humble writer of this narrative—which probably never would have been written but for this circumstance—descended one fine summer's day into a coal-pit. . . . All this has to be endured if you would wish to see what colliers have to undergo" (97–98).

While the human drama of the novel takes place almost entirely above ground, it is thoroughly interwoven with those coal deposits below, as the narrator explains in the novel's opening pages: "where coal abounds, there also factories abound . . . caus[ing] the centre of labor, and the surrounding villages and hamlets, to teem with life" (10). Human life "teems" in conjunction with extraction economies despite the ecological risks and environmental toxicities they entail. Such risks and toxicities are everywhere in *Jane Rutherford*. The narrator details how the heroine's embittered father, Jonathan Rutherford, has experienced "the ordinary mischances of his calling": "Once he was in bed for ten weeks, from hurts received in a serious explosion of fire-damp. At another time, from the compound fracture of his thigh. . . . One spring he got the miner's inflammation of the lungs, and was ever after 'touched in the wind'" (49).

Cumulative risks are likewise visible in the distinct physical type that the miners present: "At first sight you know he is a miner. A sort of earthen, yellow, mole-like cast of countenance, with sunken peering eyes, a great development of bone and muscle, especially about the shoulders and hips, and legs so small that they seem scarcely to have enlarged or elongated since they left the cradle" (14). Like goldfish that grow to the size of their bowl, Mayne's miners take on the form of the mine, a result of having begun the work as young boys. In *Child Workers and Industrial Health in Britain*, Peter Kirby has described how nineteenth-century observers reckoned that "the physical shape of miners was fundamentally different from that of other workers," just as Mayne's novel suggests. The observation, Kirby says, led to a belief "that colliery communities were populated by a discrete shorter genotype which was sustained by the high degree of occupation succession between male members of coal-mining families in isolated mining settlements" (117). The prevalence of such beliefs may explain *Jane Rutherford*'s interest in occupational intermarriage, emblematized in the romance between Jane Rutherford, daughter of a miner, and William Norman, who works in a paper mill; for the reproduction of miners was a key component of the extraction economy in Britain.[52]

That mining communities were thought of as a race apart is well established in the primary and secondary literature of the period. In *The History and Description of Fossil Fuel*, Holland writes that underground labor differs "so much from the scene of man's ordinary daylight avocations above ground, as often to impress very distinct traits upon those who are wholly brought up therein" (287). Anne McClintock discusses the use of the phrase "a race apart" in journalism and reports about miners, and in her analysis of Arthur Munby's photographs of the pit-brow women, who did surface work at the mines but were no less sooty for not working underground, she argues that mine workers were "deeply implicated in the emerging discourse on racial degeneration" (115). Covered in coal dust, the mine workers' whiteness was compromised, and "the relationship between industrial capitalism and imperialism made itself constantly felt in analogies to slavery" (116).

For British mining communities, however, there were also positive aspects to this sense of difference. Many critics have discussed the unique bonds of solidarity that developed among miners because of the isolated nature of their work and their close reliance on one another for safety; some argue that it was these bonds that made the miners' unions such a force in the labor movement.[53] Miners were often figured as the cornerstone of the working-class movement in Britain, and in fact "miners were the only group of workers Friedrich Engels mentioned by name during his 1883 speech at Marx's graveside" (Rolston 35). As early as 1825, in the wake of the repeal of the Combination Acts, the Colliers of the United Association of Durham and Northumberland were publishing their grievances in a work titled *A Voice from the Coal Mines*, and in 1841, when Martin Jude formed the Miners' Association of Great Britain, it "quickly grew to 100,000 members" (Webb 39). From such beginnings would come coal strikes and general strikes of national and international impact, such as the strike of 1912, which saw "nearly a million coal miners walk off the job in protest" and led to minimum wage legislation for coal miners (Bob Johnson 36). Even beyond miners' success at collective organizing, however, feelings of belonging developed through aspects of mining culture such as pit languages, including "pitmatic" in the northeast of England. Such "pit talk," as Natalie Braber describes, "emphasised and strengthened the comradeship . . . amongst mining people" (244).[54]

The envisioning of mining communities as separate worlds helps us understand one of the peculiar subplots of *Jane Rutherford*: the backstory of Jane's parents' courtship, which is told in a long flashback and provides ostensible explanation for why Jane does not fit in with the colliery community where she has grown up. Mayne dwells insistently on the physical differences between Jane and her father: "What a contrast was there between that father and daughter! . . . Jonathan was below, rather than above, the middle height—not that his body was short, but his legs were small and dwindled. Unlike his daughter—who had that peculiar erect and elastic gait . . . she inherited from her mother" (27–28). Jane's mother dies before the novel begins, but the narrator moves back in time to explain that Mrs. Rutherford had married Jane's coal-miner father not from love but from a strange sense of fate, perhaps "mesmerism, or some other mysterious agency" (38–39). The marriage was not a happy one. Jonathan brutally beat his wife, and one particular beating brought on Jane's premature birth (41). (The depiction of spousal abuse with the protagonist in utero anticipates the later colliery novel *Sons and Lovers*.) Over time, Mrs. Rutherford is worn down by sorrow and eventually fails to resemble the daughter who takes after her: "no one who had known Mrs. Rutherford only as a married woman could trace her in the bright complexion and animated countenance of her daughter" (45). Being a miner is hard on the body, but in the case of Mrs. Rutherford, being married to a miner is also hard on the body. Ultimately she dies from consumption, comforted only by her daughter and her close friendship with William Norman's mother.

Returning from the flashback, the narrator apologizes for the "length-ened digression" but trusts "that our readers will pardon it on the plea of its having made them better acquainted with the history of our heroine and her suitor" (67). But if the backstory cements the idea of miners as a separate physical type and explains Jane's bond to William Norman and why they might want to marry, it also offers a rationale for why the narra-tive takes so long to marry them off. *Jane Rutherford*, as a novel, exhibits pronounced ambivalence about marriage and procreation, and the story of Jane's mother is one instance of this. How much comfort can we take from the novel's assurances of futurity if those assurances are made by means of a marriage that could resemble Jane's mother's?[55] Ultimately Jane and William's marriage will conclude Mayne's novel, but the event is serially delayed by a host of plot complications: his enlistment, her emigration and return six years later, false reports of William's death in Africa, and so on. These deferrals of the conclusion resemble the open time frame of exhaus-tion as described by Jevons and others, the threshold that recedes as we move toward it. At one point, Jane is heading off to the United States and William to the Cape of Good Hope, and the sudden expansion of the novel's geographic scope threatens to excise the two altogether from the delimited purview of this provincial novel: "And now, leaving our hero-ine and her lover both on the broad waters of the Atlantic, though bound on different errands, with different prospects and to different quarters of the globe, let us return and see . . . the events that occurred to those who were left behind" (253). When the two characters do return, and the marriage does finally happen, it is an older, childless marriage that will produce no new workers. William is disabled by his war injuries, and in this sense his marriage to Jane resembles that long-deferred union of Jane Eyre and Rochester; but in this case, crucially, there is no child.

While *Jane Rutherford*'s anti-repronormative outcome mirrors that of *Nostromo* and *The Mill on the Floss*, in a novel focused on working-class life in an extractive economy the outcome also serves to stave off tragedy, given the number of child deaths that happen in the course of the story. In the novel's climactic accident at Downton Wood Pits—an accident that is modeled on the 1839 Wellsway Disaster—a miner named Abraham Pearce and his two sons, aged twelve and ten, are descending the shaft with other miners, and when the rope snaps they all die. Earlier in the novel, we are introduced to the Pearce family and the two eldest boys who work in the pit, "one, aged twelve, as a carrying-boy; and the other, only ten years old, as a trapper" (96–97). (Carrying boys helped transport the coal, and trap-pers, the youngest mine workers, operated the ventilation doors.) Both are old enough to work in the pit even after the 1842 act raised the minimum age, but we learn that before the act had passed the "eldest boy was employed there when aged only six; and though it went to his parents' hearts to hear and see him cry when the hour came, yet they felt themselves compelled by poverty and hard times to insist on it" (101).

Mayne depicts the deaths of Abraham Pearce and his sons from the perspective of the miners at the top of the shaft, watching from above: "There they go, hanging to the chain like bees, each one seated on a cross stick, passed through a link in the chain. . . . There they go, in a moment they are out of sight of the men above ground, when Eliezer cries out, 'The Lord ha' mercy on their zoulz, for they are all dead men. Ztop the wheel!' But it is too late, [*sic*] The rope has snapt, and all who hang in are launched into eternity!" (257). This fictional account mirrors the 1839 accident at Radstock Wellsway Pit, when seven boys and five men fell 756 feet to their deaths while being lowered down the mining shaft, victims of a broken rope. To better establish the generational quality of the risks of mining labor, however, *Jane Rutherford* also flashes back to an earlier accident in Abraham Pearce's own boyhood, when he was left in the pit for five days but survived; that moment is illustrated in figure 1.4, with the caption "Mrs. Pearce reproaching the colliers for leaving her son in the pit." The Mrs. Pearce pictured here, in flashback, is Abraham Pearce's mother, but the novel also depicts his wife, the second Mrs. Pearce, mourning later for her husband and sons who have died in this latest pit accident. The recurring grief of the first Mrs. Pearce and the second Mrs. Pearce conveys the cycle of human reproduction and maternal anguish on which the exhaustion of the mine depends.[56]

The Pearce family endures all manner of child death, in fact, for this mining family has already lost two children earlier in the novel even before the accident in the shaft. Lucy Pearce, aged five, drowns in the river, and the Pearce baby dies shortly thereafter. A double funeral is held for the two small children (232). Soon after, the Pearce brothers die in the pit with their father, and their death is so horrific it leaves no bodies to be buried: "The bodies were so crushed, so mangled by the depth of the fall, that the parts could not be separated or recognised. Eleven sacks were sent up filled with the remains of humanity, but these were all that were seen by the sorrowing survivors" (259). In the Wellsway Disaster, too, only one of the twelve mangled bodies could be identified. What prevents Mayne's fictionalization of this disaster from serving as a straightforward exposé of the dangers of mining labor, however, is that she makes Jonathan Rutherford, Jane's disgruntled father, responsible for the accident, attributing the

FIGURE 1.4. "Mrs. Pearce reproaching the colliers for leaving her son in the pit."
Illustration from *True Briton* 2, n.s., no. 1 (4 August 1853): 1.

broken rope to sabotage rather than faulty equipment or unsafe conditions. Sabotage was also the ruling in the Wellsway Disaster, despite the absence of a suspect or motive and a lack of evidence.[57] In effect, Mayne takes the mystery of the still-unsolved Wellsway Disaster and fills in the narrative gaps with a perpetrator *who is himself a miner*, mirroring common anti-regulation rhetoric that protected owners from liability by blaming workers for accidents.[58]

Despite this turn to sensational sabotage at the novel's climax, the multiple mining accidents in *Jane Rutherford* reflect a hard truth about life in a mining community: the death or injury of workers, including young workers, is an omnipresent danger that hangs always over the social forms of extractivism. As Humphrey Jennings puts it in *Pandæmonium 1660–1886*, "In the fantastic symphony of the Industrial Revolution . . . the great and horrible pit disasters return . . . like reminders from the unconscious (as in dreams) of this work that goes on, out of sight, night and day" (133). "The Dark Shadow of the Miner's Life" is the aptly named title for the section on mining accidents in Sidney Webb's book *The Story of the Durham Miners* (1921).[59] Such accidents, as we have seen, often involved boys as well as men, and fathers and sons frequently died together. In the Haswell Colliery Disaster of September 1844, which came hard on the heels of the Great Strike that same year, ninety-five underground workers were killed, including thirty-six between the ages of ten and eighteen. Among the dead were many family groups: brothers, fathers and sons, grandfathers and grandsons dying together.[60] The explosion resulted in calls for "a special scientific investigation," and Charles Lyell and Michael Faraday were among those enlisted to prepare a report (Webb 49). Similarly, in the 1862 Hartley Colliery Disaster, 204 men and boys were trapped underground and eventually suffocated, and as with the Haswell Disaster, about one-third of the victims were under age nineteen.[61] Many newspaper accounts of the Hartley Colliery Disaster emphasized how entire male lines of families were extinguished and how mothers persisted in grief afterward, as in the *Leeds Intelligencer*: "One poor woman has her husband and six sons in the pit, besides a boy whom they had adopted" ("Dreadful Colliery Accident"). The *Times* reported, "families are lying in groups; children in the arms of their fathers; brothers with brothers" (qtd. in Thesing and Wojtasik 36). As the monument to this tragedy records, with its list of victims from age ten to seventy-one, such events could destroy generations of a mining family—brothers, fathers, sons—in a moment (see figure 1.5). In *Jane Rutherford*, a similar monument is erected in the churchyard to the twelve victims of the mining accident that kills the Pearce boys (261), a sign of the omnipresence of this threat for mining communities.[62]

The death of the child miners in *Jane Rutherford*, not to mention in the Hartley Colliery Disaster, the Wellsway Disaster, and other mining accidents, anticipates a key feature of Anthropocene temporality that this chapter explores, one that Rob Nixon eloquently sums up in the idea of "borrowed time": "In this interregnum between energy regimes, we are living on borrowed time—borrowed from the past and from the future" (69). The death of child workers in a coal mine conveys this sense of a depleted future—a future we are not just borrowing from, but taking from— that shines through in all these novels' accounts of the social ecology of

ERECTED

TO THE MEMORY OF THE 204 MINERS, WHO LOST THEIR
LIVES IN HARTLEY PIT, BY THE FATAL CATASTROPHE
OF THE ENGINE BEAM BREAKING, 16ᵗʰ JANUARY 1862.

J. AMOUR.	ACED 43	C. WANLESS.	ACED 20
R. AMOUR.	ACED 11	T. WANLESS.	ACED 19
J. TERNENT.	ACED 44	J. WANLESS.	ACED 14
C. TERNENT.	ACED 15	W. JACK.	ACED 24
W. PAPE.	ACED 11	W. CLEDSON.	ACED 71
T. SHARP.	ACED 18	W. CLEDSON.	ACED 43
H. SHARP.	ACED 14	C. CLEDSON.	ACED 41
A. ELLIOTT.	ACED 20	T. CLEDSON.	ACED 36
C. SHARP.	ACED 19	T. CLEDSON.	ACED 16
C. SHARP.	ACED 15	W. LIDDLE.	ACED 10
J. SHARP.	ACED 13	W. LIDDLE.	ACED 17
J. BEWICK.	ACED 31	J. LIDDLE.	ACED 13
J. BEWICK.	ACED 32	J. LIDDLE.	ACED 17
R. BEWICK.	ACED 30	T. LIDDLE.	ACED 13
T. ROBINSON.	ACED 12	C. LIDDLE.	ACED 15
T. DAWSON.	ACED 49	J. LIDDLE.	ACED 11
J. DAWSON.	ACED 12	T. LIDDLE.	ACED 14
A. RICHARDSON.	ACED 22	T. LIDDLE.	ACED 15
J. JOHNSON.	ACED 11	T. LAWS.	ACED
R. JOHNSON.	ACED 12	C. LAWS.	ACED 23
T. COAL.	ACED 37	W. LOUCE.	ACED 30
T. CHAMBERS.	ACED 55	J. LONG.	ACED 18
C. CHAMBERS.	ACED 19	P. LONG.	ACED 12
J. HUMBLE.	ACED 17	M. MURRAY.	ACED 24
W. DON	ACED	I. MULLER	ACED
			ACED 2

FIGURE 1.5. Monument in St. Alban's parish churchyard,
Earsdon, Tyne and Wear, to victims of the Hartley Colliery
Disaster. https://commons.wikimedia.org/wiki/File:
Hartley_Pit_Disaster_Monument.JPG.

extraction. Thwarted reproduction across the novels in this chapter is sometimes enigmatically or allegorically connected to extraction, as with the Goulds in *Nostromo*, and sometimes quite directly connected to extraction, as with the death of the Pearce children in *Jane Rutherford*; but what is remarkable is that the trope remains durable across all these provincial realist novels, all set in extractive zones though different in many other respects. Gómez-Barris has pointed out how "the lesser-known story of mining is its dependence on the labor of women and children" (114), not only in the mines, but in the communities that sustain the mines. A poem titled "The Trapper's Petition" that ran in the 28 July 1853 issue of *True Briton*, along with chapter 15 of *Jane Rutherford*, gave voice to such an imaginary child working as a trapper in "that dark and dismal hole . . . buried amidst the blackest coal" (898). The exploitation of child mine workers emblematizes, here and elsewhere, an extractive economy that drains future life to power the present, and the nonreproductive marriage plot of *Jane Rutherford* provides no such symbolic continuity for extractivism.

"Country of the Old Pits": Hard Times

"Now, what I want is, Facts. Teach these boys and girls nothing but Facts. Facts alone are wanted in life. Plant nothing else, and root out everything else" (3). The famous opening lines of Charles Dickens's *Hard Times* (1854), spoken by Thomas Gradgrind, satirize utilitarian doctrine but also associate it with a postagricultural turn of mind: "plant nothing else," "root out everything else."[63] Such language appears everywhere in the novel, and it speaks to Britain's transformation from a society based in agriculture to one based in extraction, a transformation that is all too evident in the smoke-filled skies of Coketown, where much of the novel is set. *Hard Times* is divided into three sections, "Sowing," "Reaping," and "Garnering," but despite this agricultural structure, which hearkens back to the pastoral mode, we see little actual agriculture in the novel, and both its urban and its rural setting are devoted to coal-fired industrial life.

Hard Times's urban setting has been much discussed and has often been considered the only setting that matters in the novel, but Dickens is careful to situate Coketown within a broader provincial landscape of extraction that also figures importantly as setting.[64] He offers a systematic view of the countryside as place of extraction connected to the city as place of consumption and combustion, as we see, for example, in the description of Mr. Bounderby's exurban domicile: "Mr. Bounderby had taken possession of a house and grounds, about fifteen miles from the town, and accessible within a mile or two by a railway striding on many arches over a wild country, undermined by deserted coal shafts, and spotted at night by fires and black shapes of stationary engines at pits' mouths" (153–54). A number of financial and material connections bind Bounderby's pastoral retreat to the mills of Coketown: Bounderby makes his wealth off the back of the factories in the city but lives in the country; the country is crisscrossed by steam-powered railways that link to the cities via trains that come "shrieking and rattling over the long line of arches that bestrode the wild country of past and present coal-pits" (178); and the phrase "coal-pits past and present" is repeated several times in the novel (178, 189), reminding us that Coketown consumes that which is hauled up from the neighboring subsurface, exhausting the pits one by one. The city burns coal that is dug up from the pits of the countryside, some of which are now deserted, and the smoke in the city where this coal burns is one of Dickens's most persistent figurations across the novel—his use of anthropomorphism here "verg[ing] on the toxic sublime," as Jesse Oak Taylor puts it (*Sky* 47).

At times the poetics of air pollution in *Hard Times* suggest the greenhouse gas effect. Dickens writes that in Coketown, "Nature was as strongly bricked out as killing airs and gases were bricked in" (60).[65] He later adds, even more evocatively, "The air that would be healthful to the earth . . . tear[s] it when caged up" (201). While the greenhouse gas effect was not fully understood at this time, such passages suggest how Dickens's novel is confronting pollution via coal combustion and the question of what happens when "killing airs" are hemmed in.[66] This figuration of a human-created atmosphere helps us see the connection between urban and rural in the novel, for the coal goes from underground in the country to aboveground in the air of the city, and the narrator offers a perspective that identifies Coketown as part of a wider provincial surround: "Seen from a distance . . . Coketown lay shrouded in a haze of its own. . . . You only knew that the town was there, because you knew there could have been no such sulky blotch upon the prospect without a town. A blur of soot and smoke, now confusedly tending this way, now that way . . . a dense, formless jumble" (103). It is the smoke that makes the town, and the town seems to blow this way and that—expanding and contracting, without definite volume, inseparable from its context: "Coketown cast ashes not only on its own head but on the neighbourhood's too" (235).[67] Despite its formlessness, Coketown's smoke is yet, somehow, permanent: "interminable serpents of smoke trailed themselves *for ever and ever, and never got uncoiled*" (23, my emphasis). *Hard Times* imagines no temporal horizon in which this smoke will ever dissipate.

In one of the later chapters of the novel, Sissy and Rachael "[help] themselves out of the smoke" by taking a train to the rural area outside Coketown, but "the green landscape was blotted here and there with heaps of coal" (235). The countryside is no mere escape from the bad air of the town, but rather a place with its own hazards that are closely connected to the smoke of Coketown: the deserted works and abandoned mine shafts in that "country of the old pits" (236).[68] It is a landscape of extraction and exhaustion, pocked with the "wreck of bricks and beams overgrown with grass, marking the site of deserted works" (235–36). Disused coal shafts were indeed a common hazard in many parts of England, and the *Mining Journal* reported in 1868 that the South Staffordshire district, about one hundred miles south of Preston, where *Hard Times* is supposed to be set, encompassed "544 collieries, many of which are not . . . in operation," resulting in thousands of deserted pit shafts that had to be covered or closed ("Old Pit Shafts"). In *Hard Times*, Sissy and Rachael are careful to avoid the "mounds where the grass was rank and high," for "dismal stories were told in that country of the old pits hidden beneath such indications" (236). But even as they gingerly make their way through the landscape, Stephen Blackpool has already taken the fall that will become yet another of these "dismal stories" of death by descent.

Long before Stephen's dramatic death, *Hard Times* has already made every effort to place us in not only an extractive landscape, but an extractive community and an extraction-based society. The names of the locales, such as Coketown and Stone Lodge, refer to mined and quarried materials. At Stone Lodge, the little Gradgrinds—even their surname suggests the grinding rasp of mineral processing—have a "little metallurgical cabinet, and a little mineralogical cabinet; and the specimens were all arranged and labelled, and the bits of stone and ore looked as though they might have been broken from the parent substances by those tremendously hard instruments their own names" (14). Dickens's language labors in this passage to express the difficult, grinding work of metallurgical and mineralogical extraction and processing: "the greedy little Gradgrinds grasped at more than this, what was it, for good gracious sake, that the greedy little Gradgrinds grasped at!" (14). Such velar repetition of the hard "g" sound produces the same choking effect as the name of the children's schoolmaster, M'Choakumchild. The hard, grinding edges of all these characters are not just utilitarian—they are extractive.

Indeed, *Hard Times*'s characters often seem themselves to be produced by industrial extraction. William Cohen, in a discussion of Dickens and sexuality, writes that "in Dickens, the interior of the person is reached and reshaped by means of the bodily openings through which the material world enters it—by means, that is to say, of sensory keyholes" (17); I want to suggest that their interiors are reached and reshaped by industrial toxicities as well as other kinds of material or sensory inputs.[69] Far from being absent of interiority, Dickens's characters' interiority *is* an expression of the smoke and soot that enter their bodies, and of the larger extractive environment they inhabit. Mr. Bounderby, for example, is described as having "a metallic laugh" and a "brassy" voice (17). Sissy Jupe, we are told, will be "reclaimed and formed" (47). Many critics of the novel have commented on the book's "metallic note," as Elizabeth Starr puts it (317), but they have not tended to connect such language to the actual processes of extraction that constitute the landscape and milieu of the novel. Surprisingly, it has been more common to read the metallic figuration as a metaphorical account of literature, language, or writing in the industrial era, rather than a trace of the expanding material and ideational impacts of extractivism—suggesting how literary studies has tended to overlook the question of environmental disruption and how it is chronicled in literature.[70] To me, rather than pointing recursively back to language and literature, the "metallic note" of *Hard Times* points outward to the ecologies and economies of extraction that shape characters, books, and ideas in a broader framework of industrial-era meaning making.[71]

In an extraction-based society, people themselves are metallurgical, imbued with subsurface industry. Mrs. Gradgrind is described, for example, as being without fancy, "and truly it is probable she was as free from any alloy of that nature as any human being . . . ever was" (20). "Fancy" itself is represented here as a metal—not the organic counterpoint to metallic fact that we might expect, but rather an "alloy," a metal produced by the amalgamation of two or more metals. *Hard Times* is often read as staging a battle between fact and fancy, but here, fancy is as much the product of extraction as fact. Such depictions suggest characters whose bodies and souls are pervaded by the material outputs of industrial extractivism. The novel's smoke is perhaps its most obvious figuration of such pervasion, but smoke is hardly the only industrial pollutant in the novel: Coketown's river "ran purple with ill-smelling dye" (23), for example, and its "waste-yard" is full of "the litter of barrels and old iron" (66).

Extraction ecologies have transfused not only the novel's settings, but its marriage plots too, echoing the provincial realist novels discussed earlier in the chapter. The two characters on which *Hard Times*'s bifurcated plot focuses—Louisa Bounderby (née Gradgrind) and Stephen Blackpool—both endure unhappy marriages and broken romances, and despite the class difference between them, in both cases the extractive environment they inhabit militates against reproductive futurity. The Bounderby courtship is the novel's first attempt at a romance plot, and when Mr. Bounderby proposes to Louisa, he is fifty and she is twenty, so the temporal features of their union are distorted from the start (92). Young Louisa is described as "a fire with nothing to burn" (16): all spark, no fuel. Bounderby begins to visit Stone Lodge "as [her] accepted wooer," and "love was made on these occasions in the form of bracelets; and, on all occasions during the period of betrothal, took on a manufacturing aspect" (100). What kind of manufacturing? "Jewellery was made," for one thing. The use of passive voice in these passages—"love was made," "jewelry was made"—ironically emphasizes the invisible labor and work of nature that goes into the production of the bracelets and jewelry, from the jeweler to the jewels themselves, which are extracted and processed by human labor and formed in their rawest state by the work of nature. Humans are bound together in the reproductive structure of marriage through the symbolic work of extracted material: "with this ring, I thee wed."

The trajectory of Louisa's plot after her marriage, in the final third of the book, is inexorably downward, a figuration that most obviously suggests the trope of the "fallen woman," but that connects as well to significant subsurface sites in the novel such as mines, shafts, and pits. Mrs. Sparsit, hoping to see Louisa fall, "erected in her mind a mighty Staircase, with a dark pit of shame and ruin at the bottom" (182), not unlike those abandoned pits that pockmark the landscape around the Bounderby home. Sparsit "waited for [Louisa's] last fall" to the "black gulf at the bottom" (186), "step by step, onward and downward, towards some end" (153). For readers who have still failed to catch the narrative drift, chapter 11 is titled "Lower and Lower," and chapter 12 is simply titled "Down." (This anticipates book 3 of *The Mill on the Floss*, titled "The Downfall.")

The Bounderby marriage is a disaster from which Louisa eventually escapes, falling not into the murky pit of Sparsit's imagination, but into "an insensible heap" at her father's feet (197). (The effect on her father is profound, as he expresses through extractive metaphor: "The ground on which I stand has ceased to be solid under my feet" [199].) But Louisa remains alone, and childless, after the demise of her marriage. Many critics have read this fate as a kind of punishment: Starr notes that Louisa often appears in the criticism as "a stunted character who ends badly" (319); Patricia Johnson says that she is "denied the only satisfactions . . . that her society allows women" (135); and Anne Humpherys writes that her "failure to remarry after Bounderby's death is a kind of death" (178). But in the wider landscape of the novel, Louisa's "punishment" is not self-evident. As her own mother tells her when Louisa becomes engaged, "if your head begins to split as soon as you are married, which was the case with mine, I cannot consider that you are to be envied" (96). Mrs. Gradgrind's recurring line in the novel is to tell her children, "you're enough to make one regret ever having had a family at all" (20). Surely the work of reproducing Gradgrindism is enough to bring anyone a lifetime of regret.[72]

Stephen Blackpool's tragic marriage plots are similarly downward pointing in their trajectory, and similarly unproductive. His legal wife is a drunken dissipate who, Stephen says, "went bad—soon" (70). She is never given a name. They were married eighteen years before the action of the novel (83), but by now she is, as the narrator says, "a creature so foul to look at, in her tatters, stains, and splashes, but so much fouler than that in her moral infamy, that it was a shameful thing even to see her" (64). *Hard Times* is certainly, on one level, a plea for divorce law reform, but I want to suggest that its tragic romances are environmental as well as social, an aspect of the novel that becomes clearer when we read it next to other provincial realist novels set in extraction zones. Fallenness in *Hard Times* is as much a condition of living near abandoned pit shafts as it is a moral, gendered condition. Married when Stephen was twenty-one and his wife twenty, the Blackpools are a case study against early working-class marriage, but it is not overfertility that dooms their union, as conservative moralizers of the day were fond of foreboding, but something like its opposite—desolation and blight.

Although married to his wife, Stephen loves Rachael, a fellow worker, and "he knew very well that if he were free to ask her, she would take him" (78). From the moment Rachael appears in the novel, the narrator emphasizes her age and her reproductive shelf life: "in the brightness of a lamp" we see her "quiet oval face . . . not a face in its first bloom; she was a woman five-and-thirty years of age" (62). "Bloom," a word that describes a stage in a flower's reproductive cycle, suggests how the language for female appearance is closely tied to designators of sexual maturity. Later we learn that Rachael was "young when [she and Stephen] were first brought together in these circumstances," but she is "mature now . . . soon to grow old," having loved Stephen for many years without hope for a future together: "He thought of Rachael . . . of the number of girls and women she had seen marry, how many homes with children in them she had seen grow up around her, how she had contentedly pursued her own lone, quiet path—for him—and how he had sometimes seen a shade of melancholy on her blessed face, that smote him with remorse and despair" (78). Reflecting on these regrets, he has "an unwholesome sense of growing larger, of being placed in some new and diseased relation toward the objects among which he passed" in Coketown (79). The toxicities of his environment are bound up, in other words, with the futureless romance he shares with Rachael. The novel invites us to wish for the death of Mrs. Blackpool before it is too late—a death so close we can taste it, in the sickbed scene with the poison (84)—but instead presents us with Stephen's death at the bottom of an abandoned coal shaft, an extractive exhaustion if there ever was one. What is more, we endure a long, drawn-out wait to find out that this is, indeed, the end Stephen will face—a wait that hearkens back to *Nostromo* and Decoud's long absence from the narrative in *Nostromo*.

"Another night. Another day and night. No Stephen Blackpool. Where was the man, and why did he not come back?" (235). Four chapters from the end of the book, we are waiting, wondering if Blackpool has been murdered, perhaps by Tom Gradgrind, while the characters in the novel wait and wonder if he is really a thief as he is accused of being. We wait a long time. That the novel was originally published in weekly serial intervals in Dickens's *Household Words* would have reinforced this sense of drawn-out waiting that is pronounced even in the novel's volume form. Indeed, critics have long described *Hard Times* as being too drawn out, or too belated, and of ending with a general sense of exhaustion. Emily Heady writes that "*Hard Times* is replete with stories that only half end," and that its "domestic plots . . . fizzle out before they even reach a climax," leading to "readers' disappointment at these sorts of irresolutions" (134). Jesse Rosenthal argues that *Hard Times* "is not one of Dickens's finest productions of plot," in part because Stephen's narrative "carries the final third of the novel—carries the novel, that is, to its conclusion. This conclusion is repeatedly delayed" (30).[73] Priti Joshi approaches *Hard Times* from the subgenre of the industrial novel, reasoning that Dickens was "[tardy] to the scene" and that this generic tardiness is crucial to the novel's account of industrial life (222): "by the end of the '40s, the market for industrial fiction was saturated, and the topic seemed more or less exhausted" (225). It is generic exhaustion, then, in Joshi's reading, that accounts for the novel's labored ending, "the dying gasps of the industrial novel" (236), but as the setting of the abandoned coal pits makes clear, Dickens wants us to think of fuel exhaustion and ecological exhaustion as well.[74]

When Sissy and Rachael happen upon "the brink of a black ragged chasm hidden by . . . thick grass," they immediately realize what lies below: "He's down there! Down there!" (236). The shaft, it turns out, is a famous one, Old Hell Shaft, "with a curse upon it" according to a local pitman and "worthy of its bad name to the last" (240). When Stephen is retrieved, and before he dies, he explains what haunts the abandoned pit, what force from the past is powerful enough to exert such dangerous pressure on the present: "I ha' fell in the pit, my dear, as have cost, wi'in the knowledge o' old fok now livin', hundreds and hundreds o' men's lives—fathers, sons, brothers, dear to thousands an' thousands . . . I ha' read on 't in the public petition, as onny one may read, fro' the men that works in pits, in which they ha' pray'n an' pray'n the lawmakers for Christ's sake not to let their work be murder to 'em" (241). If the plot has been moving in a downward trajectory for some time, it finally lands on the cursed bottom of Old Hell Shaft, the site of the deaths of hundreds of laboring miners.

Along with these hauntings from the past, the novel also offers a glimpse "into futurity" in its final chapter—into events that are supposed to have happened after the end of the narrative. Rachael wears black until the end of her days and is "a woman working, ever working, but content to do it, and preferring to do it as her natural lot" (263). We are teased with a false vision of reproductive futurity for Louisa, "herself again a wife— a mother—lovingly watchful of her children," until the narrator walks it back: "such a thing was never to be" (263). This imagining of a future that "was never to be" makes us remember the fictionality of all visions of the future, for the future, as Lee Edelman puts it, is "continually deferred by time itself," is "*always* / A day / Away" (30). And yet barring a massive asteroid collision or similar cataclysmic event, there will be a future for some of the earth's inhabitants, and that future, their present, will be marked by the old pits of the past. Dickens asserts the ethical claim of that future in the final line of *Hard Times*, a voice from the dawn of the Anthropocene: "Dear reader! It rests with you and me whether, in our two fields of action, similar things shall be or not. Let them be! We shall sit with lighter bosoms on the hearth, to see the ashes of our fires turn grey and cold" (264). Moving from the fictional mode to the imperative mood, he seeks to cross the line from discourse to action and from present to future, before our fires "turn grey and cold."

"The Habit of the Mine": Sons and Lovers

Hard Times concludes after Stephen Blackpool falls down Old Hell Shaft, while D. H. Lawrence's *Sons and Lovers* (1913) opens with the line "'The Bottoms' succeeded to 'Hell Row'" (9). A shared extractive language unites the two novels, and a shared system of extractive meaning, symbol, and form. "Hell Row" is a block of miners' cottages that "was burned down" after having acquired "an evil reputation" (9). A mythic figuration of mines as hellish underworlds is evident just from the place names in these two ostensibly realist novels, but in the opening paragraph of *Sons and Lovers*, Lawrence emphasizes the kind of changes that have taken place in the mining industry over the span of sixty years, the same amount of time that separates *Hard Times* (1854) and *Sons and Lovers* (1913): "some sixty years ago, a sudden change took place. The gin-pits were elbowed aside by the large mines of the financiers. The coal and iron field of Nottinghamshire and Derbyshire was discovered. Carston, Waite and Co. appeared. Amid tremendous excitement, Lord Palmerston formally opened the company's first mine at Spinney Park, on the edge of Sherwood Forest" (9). The mythic age of Robin Hood is passed, but so is the era of donkey-powered "gin-pits" and of a natural world that was "scarcely soiled by these small mines" (9).[75] Large-scale industrial extraction now dominates this corner of Nottinghamshire, and it is the context of Paul Morel's life, as it was for Lawrence's. Robert Butler has called Lawrence "the first great artist in any medium . . . to describe the world of coal mining from the point of view of an insider" (62), for his father was a miner, and so were three of his uncles. His uncle James died in Brinsley pit on 24 February 1880, trapped and suffocated when a roof collapsed underground (Worthen 42); the death inspired Lawrence's short story "Odour of Chrysanthemums" (1911).

Lawrence's family was the source for the Morel family in *Sons and Lovers*, and because the Morels live about ten miles outside of Nottingham, we see the same rural extraction–urban consumption dynamic here that is also at work in *Hard Times*. Lawrence's provincial extraction setting is reminiscent of Dickens's description of the rural areas around Coketown, yet, sixty years on from Dickens, the memory of a preindustrial landscape has faded, and the pits have become part of the natural world. Walking outside one day, Gertrude Morel says to her son Paul, "the world is . . . wonderfully beautiful," and Paul replies, "so's the pit. . . . Look how it heaps together, like something alive, almost" (152). The provincial landscape that Paul inhabits is a palimpsest of historical extraction projects, as we see when Paul and his friends climb to the top of a nearby hill and "at their feet fell the precipice where the limestone was quarried away" (206), but the emphasis of the novel is on the profound shifts occasioned by large-scale coal mining. And while the provincial landscape is marked by such large-scale extraction, the nearby city is marked by large-scale consumption: coming up to Nottingham one day through the fields, Walter Morel, Paul's coal-miner father, sees "the town spread upwards before them, smoking vaguely in the midday glare . . . with spires and factory bulks and chimneys" (29).

The smoke moves upward, but extraction moves downward, signaling the impossibility of transcendence in the miner's plot.[76] Indeed, Walter Morel has become so accustomed to life underground that it has become normal and natural for him, just as the pits have become part of the countryside. He "prefer[s] to keep the blinds down and the candle lit, even when it was daylight. It was the habit of the mine" (38). This is a story of energy path dependency as social form, told at the level of individual adaptation. In 1913, when *Sons and Lovers* was released, Nottinghamshire was on the heels of a "rapid expansion of coal mining" from 1881 to 1911 (Wrigley, *Path* 40–41). Walter's life was thus a common one, and his adaptation to the pit is such that "it scarcely seemed hard to him . . . [to] go down. He was so used to it, it came simply and naturally" (38). Like the mushrooms that he enjoys hunting, Walter Morel, whose surname also alludes to a mushroom, is habituated to dark and dank places; he is what grows in extractive ecologies. In this sense the novel could be said to anticipate Anna Lowenhaupt Tsing's *The Mushroom at the End of the World*, a book that takes the mushroom as a figure for that which continues to grow in disturbance-based ecologies, that which "emerges in damaged landscapes" (A. L. Tsing 18). Morel's children, too, like to hunt for mushrooms and for the blackberries that grow in the "old quarries," abandoned places of mineral exploitation (93). As this suggests, in this novel there is no "natural" apart from the extractive landscape.

If industrial extractivism has brought a new version of the natural world, it has brought a new relation to reproductive futurity as well. No longer is reproduction the end point toward which the novel works, the end point that ensures continuity and that is never achieved in the novels I have discussed so far. I read Paul Morel's struggles with sexual maturation as connected to the extractive landscape he inhabits, but it is crucial to note that *Sons and Lovers*'s skepticism toward reproduction is apparent even before his birth. As the novel opens, Mrs. Morel is pregnant, "but she felt wretched with the coming child. The world seemed a dreary place. . . . And the children! She could not afford to have this third. She did not want it. . . . This coming child was too much for her" (13). Several of the novels that appear in this chapter express ambivalence about reproduction in the context of an extractive economy, where children born into a mining community are the fuel to keep the industry burning. But none expresses that ambivalence so directly as *Sons and Lovers* does via Gertrude Morel—arguably the first working-class woman character, at least among those I have discussed, drawn with any effort at psychological realism. Her desire not to have children is certainly not predictive of her feelings for the children once they arrive but stems rather from her desire not to proliferate the conditions of life that she experiences, conditions where, in the novel's extractive phrasing, "the prospect of her life made her feel as if she were buried alive" (14). Her husband, a miner, faces the daily risk of actually *being* buried alive, yet the feeling of claustrophobic suffocation extends to Gertrude, and indeed the novel describes it as a more general affective disposition of the mining community. In this instance it is Paul's life, taking form inside Gertrude, that occasions her feelings of claustrophobia, just as her intense maternal love will later threaten to suffocate and stifle Paul.

Like Walter, who has adapted to life in the pits, Gertrude appears in the novel in terms of the extractive economy she inhabits and upholds. From a higher social position than her husband, her hair is "as bright as copper and gold, as red as burnt copper, and it has gold threads" (16). Her hair may resemble extracted mineral wealth, but she is curiously ignorant of the work of the miners beneath her feet. When she meets Walter, he tells her he has been working in the coal mines since age ten, and "a new tract of life suddenly opened before her. She realised the life of the miners, hundreds of them toiling below the earth and coming up at evening. He seemed to her noble. He risked his life daily" (19). Within months of her marriage, however, "something in her proud, honorable soul had crystalised out hard as rock" (21). She is now brought directly into the extractive economy, and her labor is meant to produce the next generation of miners, but she comes to devote herself to thwarting such replication and to keeping her children out of the pits. The task requires her to be the "motor force" for her children's upward mobility (44).

The difficulties of Walter Morel's work in the mines are alluded to but rarely depicted directly in the novel, and Walter himself is made responsible for those difficulties, in a manner reminiscent of *Jane Rutherford*.[77] As a worker he is accident prone and cantankerous, "a heedless man, careless of danger" (108). Although "very steady at work, his wages fell off" because "authority was hateful to him" (25); thus he is assigned to those parts of the mine closest to exhaustion: "he came gradually to have worse and worse stalls, where the coal was thin, and hard to get, and unprofitable" (26). The narrator, who is learned in the ways of coal mining and translates for ignorant readers, explains that the miners work in groups of two or three and are "given a certain length along a seam of coal, which they are to mine forward to a certain distance" (26). If their "stall was a good one . . . they got a hundred or two tons of coal out, and made good money. If their stall was a poor one, they might work just as hard, and earn very little. Morel, for thirty years of his life, never had a good stall" (26). His stall is not only poor; it is also far from the shaft—about a mile and half, requiring him to travel far underground in getting to and from the coal face (41). While underground he depends on his watch for the time, hidden as he is from the movements of the sun, and mine time moves slowly: "at two o'clock he looked at his watch . . . again at half past two" (41). News of the weather

sometimes travels from up top: "it's rainin'" (42). On the particular after-
noon described here, as Walter eyes his watch and learns that it is raining,
Gertrude is giving birth to Paul. Walter is in the mine, out of touch with
natural cycles of daylight and weather, while the reproductive cycles of
the surface world continue overhead. When Walter comes home from
the mine, exhausted, "the fact that his wife was ill, that he had another
boy, was nothing to him at that moment. He was too tired, he wanted his
dinner" (43).[78]

Paul's birth is narrated, then, against the backdrop of the mine and
the mine's peculiarly antithetical relation to reproduction and natural
cycles.[79] This relates, too, to the novel's unusual mode of telling the story
of Paul's life, for while *Sons and Lovers* is a bildungsroman, it is also, as
numerous critics have discussed, conspicuously eccentric to the genre in the
lengths to which it goes to embed its protagonist in a social context that
precedes him.[80] The action of the novel begins long before Paul's birth, with
accounts of his mother's youth, his parents' courtship, his brother William's
childhood, and then, especially, his mother's pregnancy with Paul. In the
chapter where he is born, titled "The Birth of Paul, and Another Battle,"
Paul is not named in the text—apart from the chapter title—for many
pages. He is described only as "the baby" and "this last baby" (49), and
when he knits his brows and lifts his fingers, Mrs. Morel "felt, in some far
inner place of her soul, that she and her husband were guilty" for bringing him into existence (50). All this happens before Paul is even named as
an individual. When his mother finally determines, "I will call him Paul,"
immediately "a fine shadow was flung over the deep green meadow, dark-
ening all" (51). A pall attends the birth of Paul, in other words, like the
smoky skies of the mining country that surrounds him. The scene of Paul's
naming encapsulates a more general sense in the novel that the Morel
family is part of a toxic ecosystem, one that the family members take into
their bodies unknowingly and unavoidably. Such "trans-corporeality," to
use Stacy Alaimo's term (2), is evident, for example, in Paul's chronic bron-
chitis (90) and his mother's tumor (415). Before his mother dies, "her body
was wasted to a fragment of ash" (436), figuratively recalling the ash tree
that grows in front of Paul's childhood home. Paul listens to her "terrible,
long-drawn breaths" on her deathbed, as they alternate with "the hooters
of the iron works" and "the blowers of the collieries" (441). The Morel family is deeply embedded in the ashes and fires of coal combustion; it is the
air that they breathe, the rhythm of their lives.

Sons and Lovers conceives of human development, as this suggests,
in the context of endless social and natural entanglements, and as a

bildungsroman it blurs the boundaries of where Paul's life begins and where it ends. This is a departure for a genre that emerged in the philosophical context of Enlightenment liberalism, but I would suggest that the novel's atypical generic qualities are an effect of its extractive setting.[81] It is not until the fourth chapter, titled "The Young Life of Paul," that the novel really begins to focus on its central character, and this adaptation of the bildungsroman prepares us to see how Paul enters into a community and an ecology that shape the conditions of his life. Paul emerges from within the extraction industry that is his family's and community's livelihood. His childhood is dominated, too, by the toxic conflict between his parents, and the conflict between them is produced, in large part, by Walter's work in the mine—work that leaves him exhausted, underpaid, and thirsty: "A man gets that caked up wi' th' dust . . . that clogged up, down a coal mine, he *needs* a drink" (47). One toxic atmosphere begets another. The parents fight, and "the children breathed the air that was poisoned" (56). While there is no one moment when the Morel marriage collapses, it is always in the process of expiring, in an exhaustive trajectory that looks much like the marriage in *Nostromo*: "There were many, many stages in the ebbing of [Gertrude's] love for [Walter], but it was always ebbing" (62).

It is Gertrude's great wish that her children not be like their father, and thus the novel is peculiarly committed to an anti-reproductive vision of family life. Although his children turn from him, Walter's desire to reproduce himself in them persists: when William, the eldest boy, takes his first job at the Co-op, his father objects, "Put 'im i' th' pit wi' me, an' 'e'll earn a easy ten shillin' a wik." Gertrude replies, "He is *not* going in the pit . . . and there's an end of it. . . . If your mother put you in the pit at twelve, it's no reason why I should do the same with my lad" (70). Walter retorts that he was put in the pit "a sight afore that"—he was actually ten—and indeed, he has been in the pit so long that his body is marked by "blue scars, like tattoo marks, where the coal-dust remained under the skin" (235). Gertrude wants to see herself, but not her miner husband, reproduced in her children; in William, for example, she "wanted to see her life's fruition, that was all" (77).[82]

Paul is so committed to his mother's vision of eradicating the paternal influence that he prays for his father's death: "let him be killed at pit" (85). And yet ironically, while Paul manages to avoid working in the coal mine, his work will mirror his father's in unexpected ways. When he takes a position as clerk for a manufacturer of surgical appliances in Nottingham, the factory is described in terms that liken it to his father's mine, suggesting that the coal pit extends into all areas of industry and indeed all areas of life in an extraction-based society. The building is "like a well" and "like a pit" (118), and is arranged around a "wide shaft" with a "big, oblong hole in the ceiling" (128). Much like the mine, "it was always night on the ground floor" (128). To catch the early train to Nottingham to get to work, Paul must wake early and walk through "the valley . . . full of a lustrous dark haze . . . in which the steam from Minton pit melted swiftly" (127). He has avoided going down the pit, but his clerk job requires a ride on a steam train and a walk through a valley polluted by smoke from the mine. This "black valley space" is regularly "violated by a great train" (140), and the sexual imagery here indicates the novel's desire to connect the story of Paul's sexual coming of age with the extraction ecology that surrounds him. Paul cannot help reproducing the extractive dynamics that determine his birth and upbringing; the novels anti-reproductive energies, then, must be directed toward Paul's own sexual life.

Like the other provincial novels set in extraction zones that are discussed in this chapter, *Sons and Lovers*'s marriage plots are peculiarly frustrated. Paul's older brother, William, dies of pneumonia on the verge of his marriage, and shortly after, Paul contracts pneumonia too. His mother nurses him back to health, and from then on, "Mrs Morel's life now rooted itself in Paul" (171). Many critics have discussed the close bond, nigh incestuous, between Paul and his mother. At age fourteen, his greatest hope is that when his father dies he might "have a cottage with his mother, paint and go out as he liked, and live happily after" (114). The Oedipal plot of the book—Paul's inability to break from his love for his mother to form bonds with other women—has been central to the novel's critical reception, but when we place the novel in a longer trajectory of provincial realist novels set in extractive zones, we can see Paul's love for his mother as a species of impossible romance, not altogether unlike Maggie's relationship with Philip Wakem. It is a passion without a future, and in this case, it is also a reproductive impulse that has turned backward, to its origins, rather than outward: "It was like a circle where life turned back on itself, and got no further. She bore him, loved him, kept him, and his love turned back into her, so that he could not be free to go forward with his own life, really love another woman" (389). Paul feels that the place of his mother is the "one place in the world that stood solid and did not melt into unreality" (261), and at the end of the novel, with his mother's death, his sense of meaning and reality turn out to be bound up in her. His relationship with his mother is not the only frustrated and futureless romance in the book, but it is the one that reveals most, I think, about the novel's environmental imaginary. Gertrude is a figure for mother earth: the grounding for everything, the medium for existence—a medium that is taken for granted even as it enables life to continue. "She was the only thing that held him up, himself, amid all this" (464). In an extractive ecology, the earth mother is consumed but not replenished, drained to exhaustion. Paul's Oedipal issues thus point to a broader set of problems with the conditions of life and conditions of need that operate in extractive relations with the earth.

Paul's relationship with Miriam Leivers is similarly permeated with the novel's extractive setting. Mid-narrative, Paul becomes friends with the Leivers family, one of the few rural families in *Sons and Lovers* who live by agriculture rather than extraction. Paul relishes his time on the farm with the Leivers brothers, who seem "so cut off from the world," like "'les derniers fils d'une race épuisée'" (180), the last sons of an exhausted race. Paul's failed romance with Miriam compounds this figuration of the Leiverses as a dwindling family line, one tied to the agricultural life of the past. Although she has grown up on the farm, Miriam is prudish to the extreme about "the continual business of birth and of begetting which goes on upon every farm. . . . Her blood was chastened almost to disgust of the faintest suggestion of such intercourse" (198). Paul's romance with Miriam is thus in no way a return to pastoral nature or agricultural rhythms of life, nor is it a departure from the industrialized life he was born into in the mining district. In fact, their relationship is represented as out of sync with reproductive seasonal rhythms: "Spring was the worst time. He was changeable and intense and cruel. So he decided to stay away from her" (231). Miriam and Paul are both out of step with normative sexual time, "late in coming to maturity, and psychical ripeness was much behind even the physical" (198).[83] Paul's love for Miriam is also mysteriously connected with the pits that he passes on his way to and from the Leiverses' farm: he "never went past the pits at night, by the lighted lamp-house, the tall black head-stocks and lines of trucks, past the fans spinning slowly like shadows, without the feeling of Miriam returning to him keen and almost unbearable" (230).

Miriam's feelings for Paul are similarly unbearable. She is "full of twisted feeling," her "soul coiled into knots of shame" (208). When they eventually consummate their relationship, for Miriam, "it [is] a sacrifice, in which she felt something of horror" (330). The next time, again, "she lay as if she had given herself up to sacrifice" (333). In such passages, Miriam is the zone of sacrifice, and it is Paul who takes something irreplaceable, but elsewhere in the text it is Miriam who is the extractive force in the relationship.[84] Paul's mother fears that Miriam will extract the life from her son, will exhaust him: "She wants to draw him out and absorb him till there is nothing left of him, even for himself. . . . She will suck him up" (230). Paul himself feels that Miriam "wanted to draw all of him into her" (232), and he tells her, "you absorb, absorb, as if you must fill yourself up with love, because you've got a shortage" (258). Later, in a letter that attempts to break things off, Paul tells Miriam that if he continues with her, "I should die of exhaustion" (294). When he finally does end the relationship for good, he says, "I don't want to marry. I don't want ever to marry. And if we're not going to marry it's no good going on" (339). During this conversation, "he dug viciously at the earth" (339–40), and, in a repetition with slight variation, "dug at the earth viciously" (340). He "hacked at the earth" and "sat flinging lumps of earth" (343). Even at its end, their relationship is figured via extractive penetration of the earth.

Eventually Paul forms a romantic relation with another woman, Clara Dawes, but she is already married and separated from her husband, making her to some extent unavailable for a future with Paul. Clara disrupts the novel's apparent progression toward a marriage plot not only by her marital status, but also by her feminist views, which challenge the essential patriarchy of the capitalist family structure in which women provide unremunerated reproductive labor that industry relies on for workers. In a debate with Clara and the Leiverses about women's wages and equal pay, Paul argues that "a woman [is] only an accessory in the labour market," "a transitory thing, supporting herself alone for a year or two" (273). Clara argues to the contrary that "so much work should have so much pay, man or woman," and notes "the number of women who supported father, mother, sisters, etc." (273). She is a character who disturbs the logic of the marriage plot by denying the premise on which it is ideologically based: that of the nuclear family and of woman's essentially reproductive rather than productive role within that unit. Clara is strong and can "run like an Amazon" (290), but she drudges at home with her mother making lace, and when Paul asks her if she likes the work, she says, "What can a woman do?" (303). Paul sees her "stranded there, among the refuse that life has thrown away" (304). If Clara constitutes a feminist case against the marriage plot as a narrative horizon—a position not overtly inhabited by any of the earlier novels I have discussed—her presence nevertheless fails to stimulate alternative means of imagining futurity in an extraction-based society. When she asks Paul his plans, he says, "Don't ask me anything about the future. . . . I don't know anything" (397).

Echoing Maggie and Philip's illicit meetings in *The Mill on the Floss*, albeit less chastely, Clara and Paul's relationship is consummated on a riverbank of red clay, and Clara's "shoes and skirt bottoms [are] covered with red earth" (354). The two novels are set in neighboring regions of England, about sixty miles apart, both rich in ironstone. Stanton Ironworks Company, which operated quarries and blast furnaces and was a major employer for more than a century, is directly named in *Sons and Lovers* (200). The same red dirt thus clings to Clara's shoes as to Maggie's—the mark of outdoor romance in a region of extraction, outdoor romance that is peculiarly absent of a future and resistant to repronormativity. As their two pairs of besmirched shoes suggest, desire, affection, structures of feeling between people and for the land are all generated in the context of the extractivism that dominates the novels' settings. Wiedenfeld has suggested that "sex is so central to Lawrence's metaphysics" because "sex destroys the illusion of detached autonomy" (308); Lawrence's rendering of outdoor sex, I would suggest, more particularly destroys the illusion of detached autonomy from webs of environmental relatedness.

The role of iron-rich dirt in both novels' erotic imaginary relates to the broader insistence of *Sons and Lovers* on human relations as synecdoche for ecological relations writ large. In Paul's relationship with Clara, he experiences outdoor sex as an encounter with "strange, wild life," with "the tremendous living flood which carried them always": "it was all so much bigger than themselves" (398). If their relations expose, momentarily, a broader network of ecological relations of which human connections are just one part, they also expose the extractive medium of these ecological relations. The ecology that "carried them always" is an extractive ecology, and the moments in the novel where ecological awareness pokes through the veneer of everyday consciousness are also moments that expose a relation to the natural world profoundly shaped by the extractivism that surrounds them. When Paul and Clara walk pass the colliery, she opines, "What a pity there is a coal-pit here where it is so pretty," and Paul replies, "I am so used to it, I should miss it. . . . I like the pits here and there" (364).

How do you end a novel without a future? The ending of *Sons and Lovers* is famously ambiguous. Like the other novels discussed in this chapter, it refuses to provide a sense of continuity and futurity for the provincial, extractive life it depicts. After his mother's death, Paul moves to Nottingham, his relationship with Clara now over, and searches internally for a new future: "'you can go on with your painting,' said the will in him. 'Or else you can beget children'" (456) Dissatisfied with both options, he tells Miriam soon after, again, that he has no desire to marry (462). At the novel's end, Paul looks out at the nighttime lights, "beyond the town the country, little smouldering spots for more towns—the sea—the night—and on and on! And he had no place in it . . . on every side the immense dark silence seemed pressing him, so tiny a speck, into extinction" (464). As with *The Time Machine*, discussed in chapter 3, extinction is the ultimate temporal horizon of the conclusion of *Sons and Lovers*: not marriage, not reproduction, not inheritance, nor any other conventionally novelistic means of indicating cyclical renewal. This is fitting for a novel set in the Nottinghamshire coal pits, which are slowly being exhausted, as we learn early on when Walter, as noted previously, comes "gradually to have worse and worse stalls, where the coal was thin" (26). At the novel's conclusion, Paul looks out to darkness on one side, the "glowing town" on the other, and turns and walks "towards the city's gold phosphorescence" (464). Phosphorescence is, like fluorescence, a so-called cold light that does not produce heat, but phosphorescent materials emit light for longer and burn out slowly rather than rapidly.[85] Wrapped up in Paul's famous and ambiguous turn in this final line toward "gold phosphorescence," then, is the novel's extractive timescale: one that burns out slowly, almost imperceptibly, but is always burning out.

CONCLUSION

These five novels set in extraction zones, *Nostromo*, *The Mill on the Floss*, *Jane Rutherford*, *Hard Times*, and *Sons and Lovers*, all convey how provincial realism set in the context of the industrial extraction boom adapted the temporal features of the novel to account for the new horizon of exhaustion that accompanied the shift to extraction-based life. In a 1902 discussion of coal exhaustion, Henry E. Armstrong complained—prefiguring Amitav Ghosh's argument in *The Great Derangement*—that "a scuttle full of coal excites no emotions in the literary mind; it should be one to call up harrowing visions, as well as a vista of memories extending far back into the ages of time—for in no other stone can we find a more wonderful sermon" (825). These novels show that literature did take up harrowing visions of extraction and exhaustion, working subtly through generic form to orient readers toward a new future, one untethered to past cycles and drained by an extraction-based present. Novelistic renderings of an extractivist chronotope, these works express how human lives are interpenetrated with the nonliving earth, and their temporal forms correspond with a future-depleting extractive system.

If extraction's future is distressingly unmoored from the cyclicity of the past and the bonanza of the present, these five novels do envision what it might look like via glimpses of post-extractive settings. In *Jane Rutherford*, a nearby hillside has been blazing intermittently for thirty years, burning with the combustible refuse of the coal pit, and the characters have become thoroughly inured to the sight: "Why it is only the hill on fire, as it has been for the last thirty years; and, as far as I know, it will continue for the next thirty" (164). In *The Mill on the Floss*, Dorlcote Mill and the Guest & Co. oil mill hum along noisily while the Red Deeps, exhausted, has retired into a "garment of Silence" (316). In *Nostromo*, "the jungle-grown solitude of the gorge" has been filled with the "refuse of excavations and tailings" (79). Such sites are conventionally called sacrifice zones, yet all these novels belie the logic that the term "sacrifice" implies, which is also the logic of extraction: the idea that one part can be forfeited without affecting the whole. Together these novels convey the extractivist chronotope as expressed by the provincial realist novel: one where extraction ecologies leach into human marriage, growth, and reproduction, transforming horizons of time and futurity as well as landscapes and environments. They point ahead to the unintended consequences of extraction-based life—to the dynamic, open webs of causation in which we are still enmeshed today.

CHAPTER TWO

Down and Out

ADVENTURE NARRATIVE, EXTRACTION,
AND THE RESOURCE FRONTIER

*"There seems, to you and me, no reason why the electric forces of the
earth should not carry whatever there is of gold within it at once to the
mountain tops, so that kings and people might know that all the gold they
could get was there; and without any trouble of digging, or anxiety, or
chance, or waste of time, cut it away.... But Nature does not manage
it so. She puts it in little fissures in the earth, nobody knows where: you
may dig long and find none; you must dig painfully to find any."*

JOHN RUSKIN, *SESAME AND LILIES* (1865), 34

*"In reviewing the history of diamond mining in South Africa ... it is
easy to look back and criticise, but for the pioneers of the industry there
was absolutely no previous experience to be guided by. It was a perpetual
groping in the dark, and a necessary waiting for events, and the wonder
is that with all the drawbacks of the situation so much should have been
accomplished."*

REUNERT'S *DIAMOND MINES OF SOUTH AFRICA* (1892), 24

THE PREVIOUS CHAPTER ended in a discussion of sacrifice zones and
their novelistic depiction in provincial realism. These novels undercut the
logic of sacrifice through their ecological plots, I argue, narrating how one
place cannot be drained for the sake of another without damage to a larger
interconnected whole; in *Jane Rutherford*, for example, Bath's guzzling
of Downton Wood's coal rips the social fabric of the greater community.
But what if Downton Wood were not next door to Bath, but on the other

side of the planet? Does the literary ethic of ecological interdependence between sites of resource extraction and sites of consumption still hold when the extraction site is far away on the colonial frontier? *Nostromo* touches on this question, as we saw in the previous chapter, but since the owner of the San Tomé mine actually lives in Costaguana, the spatial effect of planetary distance is not fully engaged.

In contrast, consider a novel like Anthony Trollope's *John Caldigate* (1879), where the eponymous protagonist must make his fortune in the remote gold mines of Australia after accruing gambling debts at university that force him to sell his landed inheritance at home in Cambridgeshire. Arriving in the deep interior of New South Wales, Caldigate encounters "those hideous signs of mining operations which make a country rich in metals look as though the devil had walked over it, dragging behind him an enormous rake" (74). He sees "the bare trunks of a few dead and head-less trees, the ghosts of the forest which had occupied the place six or seven years previously." These ghost trees are "signs of mining vitality" (72), for everything green and growing has been sacrificed to the life of the mine: "There was not a blade of grass to be seen. As far as the eye could reach there stood those ghost-like skeletons of trees in all spots where the soil had not been turned up; but on none of them was there a leaf left" (74). Here the colonial mine is a site of extraction on the most temporary of timescales, beholden to no ideas of lasting sustainability. Indeed, the mine itself is barely represented in the novel, deemed unworthy of description: "In all the sights to be seen about the world there is no sight in which there is less to be seen than in a gold-mine" (77). Quickly making his fortune, Caldigate returns home, reclaims his inheritance, and soon has a thriving estate, but as he says to one of his tenant farmers, "To tell the truth, my friend, I should not have done very well here unless I had been able to topdress the English acres with a little Australian gold" (170). To top-dress is to improve the soil by sprinkling manure on its surface; Australia's gold thus produces fertility in Britain, while at the site of extraction, on the colonial frontier, it kills the land and what grows there.[1]

Novels like *John Caldigate* suggest how central the far-flung spaces of the wider world are to English literature's conception of resource extraction in the nineteenth and early twentieth centuries. Moving away from the focus on temporal structures that served as the organizing principle in chapter 1, this chapter will explore the spatial imagination of extraction literature and the new orientation toward the horizon of the frontier that such literature exhibits. With the surge of adventure narrative and adventure romance in the second half of the nineteenth century, I argue, extractivism's spatial imaginary was crystallized in generic form. To make this case I focus on five treasure-hunting adventure narratives that variously document the spatio-epistemological rifts of global extractivism: Mary Seacole's *Wonderful Adventures of Mrs Seacole in Many Lands* (1857), Robert Louis Stevenson's *Treasure Island* (1883), H. Rider Haggard's *Montezuma's Daughter* (1893) and *King Solomon's Mines* (1885), and Joseph Conrad's *Heart of Darkness* (1899). All these works depict Latin American and African resource frontiers of the Global South, and each choreographs a complex of spatial movements and material flows—not just anthroturbation (downward, subsurface delving) but also outward movement, across the surface of the earth, into the colonial frontier or back to the imperial center. Geographer Heidi V. Scott has called for more attention to "the dimension of verticality," not just horizontality, in studies of imperial expansion, since verticality is "manifest in practices concerned with the subterranean such as mining" (1853). This chapter describes how horizontal and vertical vectors work together to produce the spatial imaginaries of extractive adventure narrative. Downward delving drives outward movement—a dynamic evident in the treasure-hunting stories that appeared amid the mineral resource scrambles of the industrial era, from the Mexican mining boom early in the century, to the Californian and Australian gold rushes in the middle, to South Africa's Mineral Revolution in the late nineteenth and early twentieth centuries.

Extractive adventure narrative is often premised on a substitution of horizontal for vertical distance in the story of winning underground treasure.[2] Such a substitution is motivated by three factors: first, the unrepresentability of the deep earth from which extractive riches come, as with Trollope's "nothing to see here" account of the Australian gold mine; second, the geological circumstance that the pursuit of high-octane mineral wealth often meant a journey across the earth as well as into the earth because of Britain's relative lack of precious metals and stones; and third, the tendency, inherent in mining under industrial capitalism, for sites to be mined out and become unprofitable, requiring continual movement. These three factors—invisibility, distance, and exhaustion—can be chalked up to geological and planetary conditions but are also produced by the set of social practices that Macarena Gómez-Barris calls "the extractive global economy" (xvi).[3] The geological and the social are thus closely connected in extractive adventure narratives, as they are, indeed, connected in extractivism's legacies for global modernity, such as native dispossession or the resource imbalance that beleaguers the Global South.[4]

Jason Moore has discussed at length the necessity of the frontier to extraction capitalism—the need for not only resource frontiers, but also waste frontiers and labor frontiers.[5] Endless growth means endless appropriation; or, as Moore puts it, "the tendency of surplus capital to rise, and of ecological surplus to fall, constitute an irreconcilable contradiction between the project of capital and the work of the natures that make that project possible. Irreconcilable, that is, within the geographically given boundaries of capitalization and appropriation. The frontier always beckons" (115). This suggests the close relation between exhaustion and imperialism, two dynamics at the heart of extraction adventure narrative. Because of the frontier, Moore conceives of exhaustion as a "place-specific" event: in the history of capitalism, he says, humans have been "remarkably adept at finding ways to overcome such exhaustion" through a "frontier movement" (115). The question remains, however, whether there are frontiers enough—resource frontiers, waste frontiers—to continue on this path of accelerating appropriation, and if so for how long. This question, so live for us today, haunted industrial-era literature too, perhaps even to the extent that it normalized extractive frameworks for thinking about time and space that still shape our thinking today. Exhaustion was a medium-term horizon, one that could be envisaged generationally, and the exhaustion horizon led many writers to dwell on the spatial paradox at the heart of the extraction boom: that extraction requires constant movement, downward and outward, on a planet whose frontiers are beginning to close in.

THE VERTICAL FRONTIER:
CENTRIFUGAL EXTRACTIVISM

Dipesh Chakrabarty has recently described the emergence of "the planet" as a category for humanist thought, one that requires us to position ourselves in a "heterotemporal past" made up of "the history of the planet, the history of life on the planet, and the history of the globe made by the logics of empires, capital, and technology" ("Planet" 1). While his argument's more obvious relevance to literary criticism concerns the work of timescales, I want to suggest, building from his ideas, that extraction narratives place us in a heterospatial world made up of both the critical zone of earth's surface, where humans live, and the deep earth beneath us, where mineral riches abide. If the earth's surface is the site of empire and capital, the deep earth involves planetary processes "vastly larger than what is involved in human calculation" (24). Adventure narrative attends most obviously to the surface world of empire and capital, but the significance it assigns to the planet's interior, and to the deep timescales and non-anthropocentric perspective that such representation demands, conveys, as Chakrabarty puts it, how "planetary processes operating on different scales and involving the actions of both the living and the nonliving are ... interlocked in complicated, complex, and precarious ways" ("Planet" 13). Frequently set on the colonial frontier, adventure narrative is also poised at the precipice of the planet's interior and of deep planetary space. It is thus worth reexamining today in an era of climate change when we are forced to grasp that we live "at the cusp of the global and the planetary" ("Planet" 23).

The epistemological challenge of grasping the planetary manifests spatially in adventure literature's tendency to express the downward delving of industrial mining as an outward movement, across the surface of the earth, reflecting, too, how British extractive imperialism was stretching toward ever-new resource frontiers in the industrial era. The provincial realist extraction novels discussed in chapter 1, when depicting sites of active extraction, tend to avoid going underground and focus instead on the mine shaft or the entrance to the mine as the seat of locational interest, the transactional place where bodies and material move from aboveground to underground and vice versa. In *Jane Rutherford*, the culminating mining accident happens as workers are being lowered into the pit and is described from the perspective of those looking down from the top; the depth of the shaft is such that the fallen miners are transformed into so many sacks of material to be sent back up. In *Hard Times*, Stephen Blackpool dies after falling into an abandoned mining shaft from which he, too, must be hauled to the surface. In *Nostromo*, we learn next to nothing about the experience of Indigenous workers inside the San Tomé silver mine, though we see them entering and exiting the mine in shifts: "the mountain would swallow one-half of the silent crowd, while the other half would move off in long files" (76). The narrative perspective seems incapable of entering the mountain after them.[6]

Why such narrative resistance to entering the mine or pit itself? Clearly the middle-class bias of the novel as genre is at work here, but the material conditions of underground mining could also be said to make it resistant to representation—it is difficult to see underground, and difficult to access.[7] These difficulties seem to have suited a broader desire to overlook what was happening below ground. While inspection of Britain's mines was legally required by the 1860 Mines Inspection Act, as Marx notes in *Capital*, the act was "a complete dead letter, owing to the ridiculously small number of inspectors [and] the meagreness of their powers" (626). In 1865, for example, "there were 3,217 coal mines in Great Britain, and twelve inspectors" (634). Marx cites a miner quoted in the July 1866 *Report from the Select Committee on Mines*: "the inspector has been down just once in the pit, and it has been going seven years" (632). Concealment was, in other words, national policy as well as literary effect, and such a lack of oversight would have been even more evident in overseas British-owned mines.

George Orwell's *The Road to Wigan Pier* (1937) broke strongly from this literary reticence by taking its readers down a mine, and not just down a mine, but across it.[8] Although *Wigan Pier* is not an adventure narrative in any obvious sense, it is worth alluding to in this context because it exemplifies so perfectly how the vertical and horizontal work together in accounting for extractive space. Orwell, in pursuit of documentary accuracy, accompanied coal miners on their daily labors, and his description of the miners' "travelling"—their movement across the mine to reach the extractable coal—injects a horizontal element into a journey that one might expect to be a purely vertical affair. Once down the mine shaft, Orwell recounts, miners must travel through low, cramped passages, with hardly a chance to stand upright, to get to the coal face. Alex Woloch has said that "the intricacy of this episode in *The Road to Wigan Pier* makes it one of the great literary representations of the 'commute'" (131). The miners' traveling is an effect of extractive exhaustion, and it takes up a surprisingly central role in Orwell's account. The coal face is an infinitely receding target: "In the beginning, of course, a mine shaft is sunk somewhere near a seam of coal. But as that seam is worked out and fresh seams are followed up, the workings get further and further from the pit bottom. If it is a mile from the pit bottom to the coal face, that is probably an average distance; three miles is a fairly normal one" (22). Orwell's biblical phrasing, "in the beginning," puts us in the rhetorical register of creation—a mythic moment of resource plenty—but his narrative is entropic: we are digging ourselves into a bigger hole with each passing year, necessitating ever-longer travels to the coal face.[9]

Exhaustion makes the downward delving of mining into a horizontal hunt for new frontiers of extraction, a resource migration with no end in sight. As the Orwell example suggests, this spatial dynamic is evident in the individual mine as in the wider movements of global extractivism. The typical Cornish mine in the nineteenth century, for example, extended horizontally at multiple tiers, like a subterranean layer cake with a hole in the middle:

> A perpendicular pit, or *shaft*, is sunk, and at a depth of about sixty feet a horizontal gallery, or *level*, is cut in the lode. . . . But while this horizontal work is carrying on, the original, or, as it is termed, the *engine*-shaft, is sunk deeper; and at a second depth of sixty feet, a second horizontal gallery, or level, is driven. . . . The main, or engine-shaft, is then carried deeper still; and at the same distance—sixty feet, or ten fathoms—is driven a third, and then a fourth gallery;—and so on to any depth. ("Cornish Mining" 83)

Mining's horizontal range expanded in the nineteenth century as did its vertical range: "collieries grew in number, size, and sophistication," until "towards the end of the century . . . with shallow coals becoming exhausted, larger more capital-intensive collieries were required to exploit the Lower-Series coals, which were often over 2,000 feet (610 metres) below" (Oglethorpe 561). The expanding extractive frontier stretched downward as well as outward, highlighting a key feature of industrial extractivism: its constant centrifugal force, which expanded outward at the global scale as well as at the scale of the individual mine.

OPEN VEINS OF THE GLOBAL SOUTH

This chapter discusses five adventure narratives set in Latin America and Africa that express the outward expansion at a global scale of what I have called centrifugal extractivism. In focusing on these two parts of the world, I take my cue from the literary archive, for Latin American and African settings are prominent in treasure-hunting literature of the period. Latin America became a key setting for the adventure genre while being subject to countless extraction schemes funded by British commercial interests from around 1810 to World War I, a fateful era when the decline of Spanish and Portuguese colonial rule was met by the rise of British industrial might and the hemispheric reverberations of US expansion.[10] Africa was also a key site for extraction adventure stories, especially after the South African Mineral Revolution spurred by the 1867 discovery of diamonds and the 1869 discovery of gold roused the literary imagination while also drawing Britain into two Boer Wars (1880–81 and 1899–1902).[11]

Latin America is a large and diverse part of the world, inclusive of many cultures, languages, and climates, and most of the region was never formally colonized by Britain, but informal empire, as Robert Aguirre and others have described, had profound historical impacts here.[12] Britain was by far the biggest investor in Latin America in the century preceding World War I, and British investors and firms were responsible for extraction and infrastructure projects that transformed the region. Its hold on some parts of Latin America even came to rival in some respects its hold on crown colonies: Chile, for example, in the early 1890s "was sending three-quarters of its exports to Britain and getting almost half of its imports from that country: its commercial dependence at that time was even greater than India's" (Galeano 141–42).

After the decline of Spanish and Portuguese colonialism left an opening for British capital, London limited stock companies lined up to mine Latin America. This was coincident with the early nineteenth-century expansion of steam power in Britain, and Britain's growth on the back of coal-fired capitalism and its race to expand into Latin American mines should be understood as two sides of the same extractive coin. The era saw the "beginning of British private investments in independent and semi-independent foreign nations, and the new nations of Latin America were important centers of attraction" (Rippy 17). The real attraction, of course, was what lay beneath these new nations: gold, silver, copper, and more.[13] Between 1824 and 1825 alone, twenty-six British mining associations were formed to operate in Latin America (Rippy 24); a bust and the late 1825 London market crash would follow (Heath 264), the result of overconfidence in the British steam engine to achieve new levels of productivity in abandoned Spanish mines. In *A General Guide to the Companies Formed for Working Foreign Mines* (1825), Henry English claims that the conditions of Latin American mines are like "our English mines before the general application of the steam-engine" (54), and that "the greatest advantages may reasonably be expected from the introduction of British Capital and Science to the working of the Mines" (11). But as John Phillips, secretary to the ill-fated Real Del Monte Mining Company, would later reflect, steam can "only be economically applied where fuel is to be easily obtained," which was often not the case (7). The 1825 collapse notwithstanding, by the early 1880s British firms working Latin American mines were experiencing a "substantive boom" (Tenenbaum and McElveen 51), and by 1890 there were "150 mining companies in Latin America in which British capital amounting to some £23 million was invested" (Rippy 40). Investment in mines necessitated investment in railways and infrastructure, to ease the removal of subsurface mineral wealth from one part of the world to another.[14]

The explosion of British extraction projects in Africa happened later than in Latin America, but the two regions both appear in adventure narrative with remarkable frequency, suggesting that they catalyzed the extractive imagination in similar ways. Africa, like Latin America, is a large and diverse region, and not all British mining ventures were in present-day South Africa and its environs, but Britain's engagements here did shape its literary accounts of the whole continent. Joseph Conrad's *Heart of Darkness*, for example, takes place in the Congo Free State yet also responds to Britain's conflict with the Boers over the rich mineral resources of South Africa.[15] Such conflict led to the First Boer War (1880–81) and the longer engagement of the Second Boer War (1899–1902), which began a few months after Conrad's novel completed its run in *Blackwood's Edinburgh Magazine*. That these wars were being fought over diamonds and gold was widely understood at the time, despite the usual frantic appeals to imperial ideology. As George Bernard Shaw coolly observed in *Fabianism and the Empire* (1900), a pamphlet he wrote for the socialist Fabian Society, "What has really happened . . . is that a troublesome and poor territory, which the Empire cast off into the hands of a little community of farmer emigrants, has unexpectedly turned out to be a gold-reef; and the Empire, accordingly, takes it back again from the farmers" (22).

The story of the South African Mineral Revolution is fictionalized in J. Byers Maxwell's *A Passion for Gold: The Story of a South African Mine* (1902), a little-remembered novel that takes place in the 1870s when wild speculation led to such extremes as the "salting" of mines (planting gold from other sites as false proof of a mine's productivity). The novel depicts a massive influx of British people and capital in pursuit of South African mineral resources:

> When the Worling Reef Mine was discovered in the Transvaal, the region was not easily accessible, and to most people in England Africa was a land of lions, fever and dangerous jungles. But gold and precious stones, the magnets which, from the dawn of time, have drawn men from the furthermost parts of the world, had already begun to dispel the weird cloud in which South Africa was enveloped. The district in which the Worling Reef was situated had been prospected by adventurers whose reported discoveries created a greater gold mania in England than in South Africa. Worthless and unworked claims were dangled . . . and mad speculation and gambling in shares followed. (111)

Such overenthusiastic speculation leads to losses, and the novel even depicts a panic at the London Stock Exchange where everyone sells their "Worlings" after the mine's manager circulates a false report that the mine is near exhaustion—a stock fraud intended to reduce the price of shares, one that reveals the market's sensitivity to projections of resource exhaustion.

Prior to gold's rise in South Africa, diamonds were the major mineral draw, and according to *Reunert's Diamond Mines of South Africa*, a book produced for the South African and International Exhibition of 1892, "no more important event has happened in South Africa than the discovery of the first diamond" in March 1867, for it "converted what were the most despised possessions of Britain into sources of revenue to the mother country, and fields of ever widening enterprise for her sons" (1). These "fields of ever widening enterprise" really meant new frontiers of deep extraction, for the stones were so valuable that they incited new depths of underground mining, expensive and technologically challenging: "the difficulties of mining increased with increasing depth, requiring constantly larger capital and more elaborate appliances for winning the diamonds" (*Reunert's* 32).[16] The diamonds launched a period of South African history marked by warfare, labor migration, and riotous speculation; after diamonds, gold was discovered in 1869, then coal in 1887. The infrastructures of mining, finance, and transport meant that one extractible commodity led to another, until the land was overrun with investors and engineers in quest of underground mineral wealth.

Using racial capitalism as our analytic lens, it is evident that European profit came from the exploitation of human labor as well as from environmental resources in the extraction of raw material in the Global South, even in the era after the abolition of slavery.[17] This was not only through the regular form of surplus value afforded by waged labor, but through the augmented surplus value created under conditions of racialized hiring practices. As Frederick H. Hatch and J. A. Chalmers recount in *The Gold Mines of the Rand* (1895), white workers and native African workers were differentially paid in South Africa and tasked with different labors in the mine.[18] Most white workers were assigned to skilled labor: "the amount of manual labour falling to the lot of the white hands [is] reduced to a minimum, owing to the employment of natives" (253). The authors note that while the rate of pay for native laborers had recently increased, it is likely "that wages paid to natives will eventually be reduced, through the unremitting efforts of the Chamber of Mines which, since its inception, has directed much energy to the question of the reduction of native wages and the increase of the supply of native labour" (256). Commercial institutions like the Chamber of Mines worked "to secure the concerted action of various mining companies in reducing the wages, and a scale of monthly payments to natives was agreed to by 66 companies, with the result that

FIGURE 2.1. Ford Madox Brown, *The Last of England* (1855). Birmingham Museum and Art Gallery.

in the course of three months the average wage was reduced to 41s. 6d. a month" (256). This illustrates all too starkly how the "cheap nature" of colonial extraction, to use Moore's term, extended from the cheapened labor of Indigenous workers to the cheapened resources those workers dug up. Extractive capitalism was premised on the exploitation of such workers as on the exploitation of natural environments; as Kathryn Yusoff puts it, drawing on Saidiya Hartman's terms, "the fungibility of Blackness and geological resources (as land, minerals, and ores) is coeval, predicated on the ability of the colonizer to both describe and operationalize world-space as a global entity" (Yusoff 32).

The archive of extractive adventure literature, with its heavy representation of Latin American and African settings, raises the question of why, despite the massive impact of the Australian gold rushes on British emigration and British wealth, we do not see a corresponding literary impact from that direction.[19] Trollope's realist novel *John Caldigate* notwithstanding, the settler-colonial conception of the Antipodes does not seem to have excited the same outpouring of extractive adventure stories that British imperial engagements in Latin America and Africa provoked.[20] There is no canonical adventure story of the Australian gold rushes, though they did inspire Ford Madox Brown's haunting painting *The Last of England* (1855) (see figure 2.1). That Brown's melancholy image features white settler emigrants is significant, for it suggests how Australia occupied a different symbolic relation to Britain than did Latin America or Africa, one that was less generative for extractive fantasy (even though the name of the ship in *The Last of England* is "Eldorado," faintly visible on the back of the boat). Africa and South America remain subject to what has been termed "neoextractivism" today, and the inequitable global depletions of extraction capitalism, including its racial inequities, are also evident in the literary archive of the past.[21] These were regions to be drained, not stewarded—and thus were particularly apt settings for treasure-hunt stories.

"GROPING IN THE DARK": KNOWLEDGE AND SPACE IN EXTRACTION LITERATURE

The second epigraph to this chapter refers to South African diamond mining as "a perpetual groping in the dark." Like Trollope's unrepresentable gold mine in *John Caldigate*, the metaphor captures the special form of blind appropriation that we see in accounts of mineral resource extraction in the frontiers of the Global South. Mining, as Jan Zalasiewicz and his coauthors note, is an industrial practice that is generally "out of sight, out of mind" (3), often developing in out-of-the-way places on the provincial or colonial frontier, with much of its labor taking place underground. Yet mining also constitutes a base structure of industrial life: as Louis Simonin wrote of coal in 1868, "the nineteenth century has inaugurated the era of industry, which finds in coal . . . its daily bread" (5). This paradox of extraction, that it is at once so necessary and so invisible, speaks to the particular spatial estrangement of extraction on the colonial frontier, an estrangement that is mediated through the adventure narratives the era gave rise to.

If extractivism, as Stephanie LeMenager has suggested, "means, among other things, abstraction" ("Sediment" 173), it is a goal of this book, against such abstraction, to foreground environmental materiality in my account of extractive literature. Elaine Freedgood argues in *The Ideas in Things* that the mahogany furniture in *Jane Eyre* is "more than a weak metonym for wealth and taste; it figures, first of all, itself" (3); as fictional furniture leads Freedgood to deforestation and other "social and environmental vicissitudes of mahogany," so extraction literature presents us with "as yet unseen connections between historical knowledge and fictional form . . . recorded in the letter and letters of novelistic things . . . that are at once material and figural" (29). In literature of the past the material and environmental effects of industrial extraction are documented, chronicled, represented, explained—or explained away. And just as literature records material changes in the environment, so too is it responsible for such changes: as Amitav Ghosh writes in *The Great Derangement*, "When we see a green lawn that has been watered with desalinated water in Abu Dhabi or Southern California . . . we are looking at an expression of a yearning that may have been midwifed by the novels of Jane Austen. The artifacts and commodities that are conjured up by these desires are, in a sense, at once expressions and concealments of the cultural matrix that brought them into being" (10). Treasure-hunt literature is part of a cultural matrix that encourages overseas extraction, and yet it also conveys the difficult epistemological conditions that mining and extraction pose as material practices. In the five texts on which this chapter focuses, extraction presents problems of knowledge and space that can be summed up in the following terms: concealment, geognosy, and unintended consequence.

Concealment alludes to the perceptual problem of not being able to see underground, which makes mining and extraction difficult to represent and makes underground resources difficult to find. Reflections on this fundamental condition of extraction are a staple of mining literature; for example, John Holland's *The History and Description of Fossil Fuel* (1841) isolates a "chief difficulty which arises in exploring a country in search of coals," which is "the great thickness of alluvial cover, which completely hides the crop or outburst of the strata . . . from our view." The alluvial cover is the soil and other material that overlays subsurface coal beds. Holland fantasizes about peeling away this blanket of obscuration, like dead flesh: "The alluvial cover has been compared . . . to the flesh upon the bones in animals: if the flesh is removed, the whole structure of the bones, their situation and connection, are at once discovered" (175–76).[22] The hiddenness of underground mineral resources is central to the epistemology of extraction, and in this way, extraction literature presents a special case in the work of environmental "concealment" that Ghosh attributes to modern literature.

Problems of knowledge, visibility, and concealment extend to the interpretive techniques that were crucial to early geology at a time when geological knowledge often came from profit-minded searches for new sites of extraction. As Bruce Braun has noted, in the nineteenth century, "the physical spaces daily transformed by mining were intricately intertwined with the epistemological spaces opened up by the discourse of geology" (14). Geology moved at this time from a focus on individual rock and mineral specimens to "the concept of *formations* whereby certain minerals became associated with groups of broadly similar rocks" (Braun 20), and geologists used the German word *geognosy* to describe what Charles Lyell called "the natural position of minerals in particular rocks" (82–83)—in other words, the identification of minerals for extraction by locating associated material. In terms of living species, eighteenth- and nineteenth-century taxonomic systems are now often said to be insufficiently associative and overly individualistic in their categories, but when it came to the nonliving natural world, geologists were highly alert to the relationality of rocks and minerals, for mining speculation demanded as much.[23] As Alexander Humboldt wrote in his essay on Mexico's mines, "Mineralogical facts alone may be presented singly; positive geognosy is a science occupied with the relation and connexion of facts" (*Selections* 145).

Geognosy refers, then, to the techniques by which observers of the natural world, when exploring unknown frontier spaces, identify valuable sites for mining by looking for associated geological features. "Armed with this new ability to 'read' rocks," Braun writes, geologists could travel the world and "mobilize and interrogate nature in new ways" (20), "seeing geologically" (22). Gómez-Barris, in a refinement of Mary Louise Pratt's "imperial eye," uses the phrase "extractive view" to describe this "vertical model of seeing" (6), and such a mode of vision is evident in extractive adventure literature's penchant for landscape description, for landscape features were now understood to be clues as to what lay hidden below. Adventure narratives also remind us, however, that the geognostic gaze was fallible and liable to frequent error. As John Taylor wrote in 1824, "The whole business of mining is experimental; hardly a shaft is sunk or a level driven, but . . . for the purpose of trial" (xviii). This was especially the case on the extractive frontier, as the *Mining Journal* suggests in a 1

January 1898 report on Rhodesia (present-day Zimbabwe): "A great deal has been written upon Rhodesia . . . nothing of which can be regarded as . . . having settled the momentous question of its mineral resources. It is impossible for any one to say, in the present unsatisfactory state of development, what is beneath the surface" (14).

Geognostic fallibility relates to a third problem of knowledge that the adventure genre gestures toward: unintended consequences. For every major overhaul of earthly materials—digging, draining, extracting—was haunted by the possibility of unforeseen danger, often figured in extraction literature through the pervasive trope of the cursed treasure. Climate change, the result of fossil fuel combustion, is the most obvious example of the unintended results that have attended extractivism, but mining disasters of all sorts were constant reminders throughout the period of the unknown dangers of digging deeply into the earth to extract hidden commodities. Even Kurtz's madness in *Heart of Darkness* could be said to allegorize obscure dangers that attend on those who go too deep. Again and again in the texts that follow we see that the cognitive project of adventure narrative cannot be completed until after the return from the journey, if at all, and the journey itself is taken under conditions of cognitive estrangement with unpredictable consequences. Elio Di Piazza notes that "landscape description in literary works" is typically "connected to the observer's cognitive standpoint," but this is complicated in adventure texts because "descriptions follow coordinates that lead to the remodelling of the known world" and "estrangement from the original context" (88).

In the previous chapter, we saw how the hazards of acting in conditions of partial knowledge are expressed in *Nostromo* through dramatic irony: not only does the reader know more than the characters, but the reader can see where the characters are forced to make decisions not knowing what the reader knows. Similar knowledge effects are at work in the adventure texts discussed in this chapter, all of which feature first-person narrators who stress their lack of understanding of the extractive projects in which they are engaged. Mary Seacole, setting up a Panama hotel to cater to California Gold Rush miners, finds herself surrounded by "disgusting excess of license" (26) and wonders "that I had not foreseen what I found," that "my rage for change and novelty had closed my ears against the warning voices of those who knew somewhat of the high-road to California" (28). Allan Quatermain, narrator of *King Solomon's Mines*, insists throughout that he is not a thinker and must leave the work of understanding his adventures to others: "All I can do is to describe it as it is, and the reader must form his own conclusion" (211). Adventure literature foregrounds curtailed knowledge and perspective especially in its conception of space: the difficulty of finding one's way in an unknown country, of reading the signs that point to treasure. The interpretation of quasi-legible maps in *Treasure Island* and *King Solomon's Mines* is one obvious place where we see this space-knowledge interplay, but the narration and narrative structure in adventure texts work toward similar ends. The next sections of the chapter focus on five works of adventure literature, three set in Latin America and two in Africa, that illustrate the interconnected spatial dynamics of exhaustion and frontier and the compromised epistemological conditions in which those dynamics play out in overseas extraction projects.

"A Great Neighbourhood for Gold-Mines": Wonderful Adventures of Mrs Seacole in Many Lands

Mary Seacole played a supporting role in one of the great extractive dramas of the nineteenth century: she ran a hotel in Panama that catered to miners heading to and from the California Gold Rush. She also tried her own hand at gold mining in several failed schemes in the former Republic of New Granada (present-day Panama). Her self-diagnosed "gold fever" is not what she is remembered for today, of course; her reputation rests on her heroic labors nursing soldiers on the field in the Crimean War. A Black woman from Jamaica, Seacole never received the public beatification enjoyed by her white contemporary, Florence Nightingale, yet her wartime labors made her famous enough that her memoir, *Wonderful Adventures of Mrs Seacole in Many Lands*, published in London in 1857, quickly went through two editions (figure 2.2 shows the cover of the first edition).[24] *Wonderful Adventures* is not "adventure literature" in the same way as books like *Treasure Island* and *King Solomon's Mines*, for it is autobiographical rather than fictional, reality rather than romance. But Seacole's use of the word "adventures" in her title and the treasure-hunting milieu depicted in the first half of her book reveal the connections between the adventure romance genre that emerged in the 1880s and the New World travel narratives that predate it.[25] Moreover, the signal spatio-epistemological features of the adventure narrative are present in *Wonderful Adventures* as Seacole negotiates resistant terrain in pursuit of treasure, acting in conditions of incomplete knowledge, in an unfamiliar context, with limited access to the larger forces shaping the narrative events.

These epistemological conditions that attend the treasure-hunting story are evident in the unusual structural features and strange silences

FIGURE 2.2. Cover of *Wonderful Adventures of Mrs Seacole* (London: James Blackwood, 1857).

of Seacole's narrative. There are many secrets in this book, one of which is the identity of the editor, referred to on the title page only as W.J.S.—an identity that remains unknown.[26] In the first edition of Seacole's work, no byline is listed on the cover or title page, and "edited by W.J.S." is all we get. But while Seacole quotes her editor at the beginning of chapter 2, his role is invisible and unnoted for the vast majority of the text. He emerges on the title page to cast confusion over the book's authority and voice, and then retreats into silence, not appearing again (after chapter 2) until a short note in chapter 5 and a brief mention in chapter 12. In terms of establishing conditions of partial knowledge, the importance of first-person narration in all five narratives under discussion here (*Wonderful Adventures, Treasure Island, Montezuma's Daughter, King Solomon's Mines*, and *Heart of Darkness*) cannot be overstated. But it is significant that in all these books, the authority of the first-person narrator is compromised even further by the occasional interjections of an anonymous editor, author figure, or frame narrator. Seacole's gender and race make her unique among these first-person narrators, of course, in that they impose structural, social conditions for this compromised authority, conditions that she must negotiate throughout the narrative.

If her editor is silent, we must also remark Seacole's silence on a number of topics where we might expect her to be more voluble. The description of her marriage, for example, is confined to one paragraph at the end of the first chapter—or, more precisely, a short section of a paragraph. In less than 100 words, she meets and marries Mr. Seacole, they establish a store, he sickens and dies, and the narrative moves on without him. The passage is emblematic of the book's narrative structure, one that is timed to the rhythms of extraction, constantly moving from frontier to frontier in search of the next big thing. This is not a memoir structured by Seacole's personal, emotional, or interior development, where a marriage would presumably be a key event; this is a narrative of accumulation and loss, boom and bust, directed always toward the next frontier.[27] Although *Wonderful Adventures* is not a novel, its structure is typical of the adventure novel as it would later emerge: Elizabeth Chang writes that adventure novels "adhere to a nomadism" whereas "settler novels" are "concerned with a narrowly defined sense of belonging, ownership, association, and claim rooted in a particular connection to a very particular piece of land" (*Novel* 123). This helps explain the different orientations toward extraction that we find in the provincial realist novels discussed in the previous chapter and the adventure narratives discussed here. Nomadism is not unique to adventure literature and is central to related genres like the picaresque, but given that adventure narrative in the industrial era is so frequently focused on the quest for treasure, its roving, on-the-make rhythm fits squarely within the frontier dynamic of extractivism.

Across her narrative Seacole engages in any number of speculations, from selling seashells in Jamaica to peddling pickles in London, but the two ventures that dominate the narrative are, first, her time in Panama, where she serves as support team to the California Gold Rush, and, second, her time in the Crimea running a store during the war. In both cases, Seacole occupies a marginalized perspective on a major world-historical event. We are prepared for this frontier-seeking trajectory in the opening

paragraphs of the book, where Seacole admits, "I have never wanted incli-
nation to rove, nor will powerful enough to find a way to carry out my
wishes" (11).[28] If the narrative progresses by way of "roving," its rhythm
is also extractive: "my fortunes underwent the variations which befall
all. Sometimes I was rich one day, and poor the next" (15). Here Seacole
universalizes the historically peculiar conditions that surround her: the
boom-and-bust culture of risk and speculation that accompanied the
rise of global capitalism more generally and the mineral resource rushes
of the nineteenth century more specifically. The time and place of Seacole's birth situated her historically and geographically between two such
rushes: the British rush on Latin America following the demise of Spanish
and Portuguese rule, and the California Gold Rush in the middle of the
century, which is of direct concern in the text.[29]

Seacole herself was born in Jamaica, which was a British colony until
1962. Although her birth predated the emancipation of slavery in the
empire, William L. Andrews, in the Schomburg Library edition of Seacole's narrative, says that she was born free, "evidently well protected
from the tentacles of slavery" (xxvii).[30] With a Black mother and a white
father, Seacole identifies as "Creole" and refers to her skin color as "yellow" throughout the book, a complex form of racial identification that,
curiously enough, aligns with her narrative's description of gold as "yellow" too. This doubling suggests how Seacole herself is a resource within
the imperial extraction contest; the descendant of slaves of the British
Empire, she has acquired medical skills that continue to serve Britain's
imperial ambitions. Seacole's task in the narrative is to find a way to align
her own independent ambitions with those of the empire—a brief that
she often succeeds in fulfilling. And yet, one of the earliest episodes she
recounts is her first journey to Britain, a trip from colony to metropole
where she encounters "the efforts of the London street-boys to poke fun at
my and my companion's complexion" (13). The episode decenters England
as the origin point of adventure, and instead it becomes the hostile site of
the first voyage out.

Across Seacole's narrative, British, American, Ottoman, and Russian
imperialisms all jostle against the backdrop of rushes for resources and
resource routes. As Seacole spends the narrative in Jamaica, Britain,
Panama, and the Crimea, one notes that all these narrow patches of land,
surrounded mostly or entirely by water, are politically significant largely
because of their accessibility by sea, and strategically located for commercial operations that are global in scope. Seacole's narrative roves, in other
words, using the routes provided by global empire and global capitalism.

That the isthmus of Panama, so far from California, became an important stop on "the great high-road to and from golden California" (17) exemplifies the kind of spatial reorientation that Seacole's narrative effects, where we are moved to view the world as interconnected in surprising ways. In discussing the cartographic literary imaginary, David Harvey says that "the effect of reading such literature is to see ourselves in a different positionality, within a different map of the world" ("Cartographic" 221); so Seacole's narrative remaps the globe by way of mineral resources and resource routes.

Because it was faster for would-be miners to travel by sea to California from the east coast of the United States and from Europe, the route through Panama, requiring a thirty-mile overland trip from one side of the isthmus to the other, became a regular highway to the Golden State, as Seacole describes: "All my readers must know—a glance at the map will show it to those who do not—that between North America and the envied shores of California stretches a little neck of land, insignificant-looking enough on the map, dividing the Atlantic from the Pacific. By crossing this, the travellers from America avoided a long, weary, and dangerous sea voyage round Cape Horn, or an almost impossible journey by land" (17). Here Seacole establishes Panama as a stopping point between Atlantic and Pacific that is populated with, in her words, "the refuse of every nation which meet together upon its soil" (18). Robert Aguirre has described "pre-canal Panama as a site of traversal, a location between," and argues that nineteenth-century representations of Panama "laid the discursive ground for its incorporation into a world transport system whose centers of command lay elsewhere" (*Mobility* 5). We see such a move in Alexander Dunlop's *Notes on the Isthmus of Panama with Remarks on Its Physical Geography and Its Prospects in Connection with the Gold Regions* (1852), where the author, directing his remarks to an audience of British speculators, describes Panama as "a line of communication connecting two great worlds barred from each other," the Atlantic and the Pacific, "each a geographical and a commercial world" (9). Fifty years later, this line of communication is imaginatively opened in William Henry Hudson's 1904 novel *Green Mansions*, when the protagonist, tutoring a wild girl of the rainforest in the geography of South America, decides to "[sink] the Isthmus of Panama beneath the sea" (158).[31]

Despite the long-standing dream of sinking or opening the isthmus, in Seacole's day Panama remained a difficult passage. The horizontal journey overland, necessary for those who sought Californian gold, was, in Seacole's words, "a most fatiguing one," in parts "almost impassable;

and for more than half the distance, three miles an hour was considered splendid progress" (40). Seacole's time in Panama predates the completion of the Panama Railroad in 1855, but her narrative illustrates the industrialization of nineteenth-century extraction as it was happening. She depicts the isthmus with about one-third of the railway completed, as the process of transforming the route to and from the site of extraction (California) into an extraction-based form of transportation (steam railway) was well underway. The process was made difficult, Seacole says, by Panama's intractable natural environment: "It seemed as if nature had determined to throw every conceivable obstacle in the way of those who should seek to join the two great oceans" (17). The complex spatial vectors of global extraction here prove ever more vexed, requiring not only horizontal movements across the earth to find new frontiers for digging, but also vertical extraction into the earth to ease the horizontal travel across it, as in Seacole's account of the carving out of the Panama Railroad. The rail project, unfolding but unfinished during Seacole's stay in Panama, contributes to the narrative's epistemological features of partial knowledge and incompleteness. At times Seacole's description of the railway, which was running by the time her narrative came out, reads as a paeon to industrialism: "iron and steam, twin giants, subdued to man's will, have put a girdle over rocks and rivers, so that travellers can glide as smoothly . . . over the once terrible Isthmus of Darien, as they can from London to Brighton" (18). But she also notes how this smooth horizontal passage, resourced by imperial capital to ease the journey to Californian gold, was built on exploitation: "every mile of that fatal railway cost the world thousands of lives" (19).[32]

Workers carved the railway through the red clay soil that stains Seacole's clothing upon her first arrival in Panama. After the fledgling railroad transports Seacole to Gatun, as far as it could then travel, "the cars landed us at the bottom of a somewhat steep cutting through a reddish clay" (19). Seacole begins her "ascent of the clayey bank," attired "in a delicate light blue dress [and] a white bonnet prettily trimmed." By the time she has climbed the piles of overturned earth, "my pretty dress, from its contact with the Gatun clay, looked as red as if, in the pursuit of science, I had passed it through a strong solution of muriatic acid" (20). The scene anticipates a later adventure novel about extraction on the Latin American frontier: *Soldiers of Fortune* (1897) by American novelist and journalist Richard Harding Davis. Here, in the fictional South American country of Olancho on the Caribbean coast, Hope Langham, daughter of a wealthy New York capitalist, visits a mine capitalized by her father and similarly stains her white dress with the red dirt. In this state, she attracts the gaze of the novel's protagonist, the "master of the mines" who is named, aptly enough, Clay: "his eyes fell on a slight figure sitting erect and graceful on her pony's back, her white habit soiled and stained red with the ore of the mines" (113). Romance ensues. The adventure narrative is generally understood to be a masculine genre, with books like *Treasure Island* and *King Solomon's Mines* possessing few female characters, but in both *Wonderful Adventures* and *Soldiers of Fortune*, respectable women—Mary Seacole and Hope Langham—prove their ability to cut it on the extractive frontier when their light-colored dresses are sullied by red dirt. In the case of Seacole, her race makes her particularly dependent on dress to establish her respectability, and her "due regard to personal appearance" she calls a "duty as well as a pleasure" (20). In Panama, however, Seacole soon learns that she need not fret about dress: she has landed in "a lawless out-of-the-way spot" (20), and, at the crossroads of imperial extraction, there is no way to stay clean.

Panama's value as place in this text has nothing to do with its intrinsic worth and everything to do with its location as crossroads between the extraction site of California and the centers of wealth and power in New York and London. This lack of inherent value is emphasized in the text through various failed extraction schemes in which Seacole engages, none of which yield her the fortune she hopes to find. In chapter 7, Seacole is "attacked with the Gold Fever" when "an opportunity offered itself to do something at one of the stations of the New Granada Gold-mining Company" (62). To get to the station, she must make her way through a resistant landscape, "thick, dense forests, that would have resisted the attempts of an army to cut its way through them." As with the railway cutting, extractivism is again figured as much a horizontal cutting across the earth as a digging into it. Her time in the outpost of Escribanos is, she says, "devoted to gold-seeking" (62), and at one point Seacole thinks she has hit the jackpot: "I once did come upon some heavy yellow material, that brought my heart into my mouth with that strange thrilling delight which all who have hunted for the precious metal understand so well" (64). When she takes some to a gold buyer, however, the metal is declared "valueless." Seacole stays in the Escribanos area for several months, "busy with schemes for seeking for—or, as the gold-diggers call it, prospecting for—other mines" (64). Despite the crocodiles, sharks, and venomous snakes that inhabit the region, the area is thought to be "a great neighbourhood for gold-mines; and about that time companies and private individuals were trying hard to turn them to good account" (63). Small quantities of silver and gold were mined, but no big lodes were found despite legends of treasure: "in times gone by [Panama] had yielded up golden treasures to the Old World. But at this time the yield of gold did not repay the labour and capital necessary to extract it from the quartz" (63). Its mineral resources are exhausted—no longer worth the capital required to secure them—and Panama's primary value is now as crossroads to another extraction site.

Wonderful Adventures thus narrates extractivism's reliance on abstract space, space that is primarily valued for conducting flows of resources elsewhere. The central role of abstract space is also conveyed by the hotel setting, for Seacole joins her brother in running a hotel in Panama for the crowds of rotating guests heading to and from the California mines. By Seacole's account, the busy route is populated by the human tailings, waste, and remainders of various extraction schemes: "In the exciting race for gold," she says, "we need not be surprised at the strange groups which line the race-course" (28). Seacole was on that racecourse herself, of course, and Jessica Damian has argued that Seacole's own gold-mining schemes evince her "commitment to Britain's expansionist projects in Latin America," for "she travels to the region in 1854 with one specific goal in mind: to engineer a transatlantic production and distribution network for gold ore" (28). Damian suggests that Seacole's narrative was intended to be useful to speculators abroad who were eyeing Panama's geography for its extractive potential, and certainly, Seacole's rendering of Panama can be read in the context of a broader print-based conversation at this time encouraging Latin American extraction by providing description of the land and its features that would be of interest to mining specula- tors.[33] But we must also note that Seacole withholds many details about the gold-mining expeditions in which she is engaged, and that her expe- dition is ultimately unsuccessful. She leaves Panama for England because "I had claims on a Mining Company which are still unsatisfied," and "had to look after my share in the Palmilla Mine speculation" (67). Once in Lon- don, however, the project seems a fruitless endeavor: "that visionary gold- mining speculation on the river Palmilla, which seemed so feasible to us in New Grenada," was considered in London "so wild and unprofitable a speculation." Without missing a beat, Seacole moves on to what she calls a "very novel speculation," running a store at the front of the Crimean War, and thus she "threw over the gold speculation altogether" (69).

The broader backdrop of the industrial extraction boom provides, I want to suggest, the environmental-economic context for the episte- mological conditions that govern Seacole's narrative and other treasure- hunting stories of the era. These narratives capture the historical mood and mentality that underlay a rush of mining projects, many of which proved to be ill founded, and they illustrate the epistemological conditions that underlay the rise of imperial extractivism. Adventure narratives like Seacole's are formally and discursively suited, in other words, not only to encourage extraction in the far-flung frontiers of the world, but also to convey the experience of being an unknowing participant in en masse agencies that were transforming the planet at the dawn of the Anthropocene.

"A Mine of Suggestion": Treasure Island

Robert Louis Stevenson's *Treasure Island* is a decidedly fictive adventure romance, while *Wonderful Adventures of Mrs Seacole* is a nonfiction adventure memoir, yet the terrains of these two treasure-hunt narratives are surprisingly similar. Both feature Caribbean settings, and both conceive of this region from the standpoint of the California Gold Rush and the reorientation of space that it effected, a reorientation that both authors experienced firsthand.[34] Mary Seacole witnessed the thick of the Gold Rush while running a hotel in Panama; Stevenson saw its tailing off. He came to California in 1879, following wife-to-be Fanny Osbourne, and wrote *Treasure Island* two years later. As many critics have noted and Stevenson himself acknowledged, the flora and fauna of Treasure Island actually resemble those of northern California, not the Caribbean where the novel is ostensibly set, suggesting the impact of California on his writerly imagination. Live oaks, rattlesnakes, redwood trees, and sea lions are some of the Californian species that somehow populate Stevenson's Caribbean island. *Treasure Island* is perhaps the ur-text—certainly the nineteenth-century ur-text—for the treasure-hunting adventure story, and Stevenson's California experience had prepared him to think about resource extraction in terms of booms and exhaustion.[35] In *Silverado Squatters*, his memoir of life in the Napa Valley, he wrote that these "new lands" were "already weary of producing gold" and were "begin[ning] to green with vineyards" instead (205).

If *Treasure Island* and *Wonderful Adventures* share a similar spatial orientation—the Caribbean through the prism of Gold Rush California—they also share narrative features that highlight the limited perspective from which their settings are apprehended. Jim Hawkins, Stevenson's first-person narrator, is recounting events that happened when he was a child, and his control over the text is strangely limited, mirroring Seacole's authorial relation to her invisible editor. Hawkins is writing his account at the behest of others: "Squire Trelawney, Dr Livesey, and the rest of these

gentlemen . . . asked me to write down the whole particulars about Treasure Island" (9).[36] Livesey serves as a kind of editor, inserting his own
version of events in chapters 16–18 when Hawkins abandons the stock-
ade. (The cross-cutting of narrators in this section demonstrates the limited knowledge available to both Hawkins and Livesey, neither of whom
can report the whole story.) Unlike with Seacole's narrative, we do know
the identity of the editor figure here, but we do not know the extent of
Livesey's incursions into Jim's narration. In describing the voyage to the
island, Jim reminds us of his obscure narrative constraints: "We had run
up the trades to get the wind of the island we were after—I am not allowed
to be more plain" (59). Later, in the midst of a skirmish with the pirates,
the voice of an unnamed editor (or author) intrudes via footnote to provide
additional information not known to Jim in the moment he is describing:
"The mutineers were soon only eight in number, for the man shot by Mr
Trelawney on board the schooner died that same evening of his wound.
But this was, of course, not known till after by the faithful party" (113).
Since Jim is writing his narrative well after the events it depicts, he would
surely have known of the pirate's death by the time he is writing, thus the
footnote's only ostensible purpose is to signal, again, Jim's limited perspective on the adventures he is describing. In a crucial way, the reader is even
less informed than Jim, for Jim tells us at the beginning of the novel that he
will be holding back the most important information of all, "the bearings of
the island," because "there is still treasure not yet lifted" (9).

The treasure map is the key to the novel, the document on which the
accumulation of the hidden wealth depends, and at times possession of the
map is positively equated with possession of the treasure: "You'd be as rich
as kings if you could find it" (31).[37] Yet Jim provides readers with an expur-
gated version of the map where the most important piece of information—
location—is left out. This is concealment in its most overt form, expressed
in the novel's opening sentence and governing its spatio-epistemological
conditions. The treasure map was not printed with the initial serialized
version of *Treasure Island*, which ran in the children's journal *Young Folks*
from October 1881 to January 1882, but it did appear in the 1883 book
edition, and Stevenson later attested to its importance in his conception
of the narrative: "I had written [*Treasure Island*] up to the map. The map
was the chief part of my plot" ("My First Book" 295). His language suggests
the central role of maps and charts in galvanizing the extractive imagination: he refers to the treasure map as "a mine of suggestion" (297), "an
inexhaustible fund of interest for any man with eyes to see" (289). If the

overriding fantasy at the heart of *Treasure Island* is inexhaustible wealth, the fantasy of finding that wealth on the frontiers of the world is closely connected, and the map brings these two desires—the extractive and the imperial—together. In Conrad's *Heart of Darkness*, Marlow famously describes the imaginative affordances of maps in encouraging imperial speculation: "when I was a little chap I had a passion for maps. I would look for hours at South America, or Africa, or Australia, and lose myself in all the glories of exploration" (8). Jim Hawkins is similarly entranced by the treasure map and expresses his enchantment in terms that anticipate Marlow's: "I brooded by the hour together over the map. . . . I approached the island in my fancy, from every possible direction; I explored every acre of its surface" (41). Here the fantasy of treasure is inseparable from the lure of the frontier, and frontier space and the resources it promises are collapsed altogether, as in Squire Trelawney's reference to "the port we sailed for—treasure, I mean" (42).

Britain's own lack of precious metals, apart from its more utilitarian mineral resources, created the geological conditions for the particular form of extractive desire that we see here, and nineteenth-century writers on geology often saw in the earth's unequal distribution of precious resources evidence of a divine plan for British imperialism. In an article titled "Gold" from the 20 October 1853 issue of *True Briton*, precious metals are imagined to have been hidden in the New World, like Easter eggs, to provoke European "industry": "the discovery of America in 1492, first opened to mankind the secret of the treasures, which the earth had in store to excite his industry, or to reward his intelligence." "Industry" and "intelligence" are John Taylor's watchwords, too, in his 1824 introduction to Humboldt's *Political Essay on the Kingdom of New Spain*, where Taylor marvels at "the extraordinary spectacle of a vast country [Mexico], teeming with the precious metals, applying to the inhabitants of a northern island, comparatively barren in native wealth, for assistance in the extraction of its treasures. It is, indeed, a striking proof that the richest gifts of nature are useless without the industry and intelligence requisite to bring them into action" (viii). Far from being "barren" of native wealth, Britain was extremely rich in coal, coal that was conveniently located near shipping ports—a geological boon that drove its early industrialization and development of extractive methods that it would then import into Mexican mines.[38]

The tendency to read preexisting beliefs into the geological distribution of extractible materials was a defining feature of natural theology earlier in the century. Thomas Gisborne's *The Testimony of Natural Theology to*

Christianity (1818), for example, found in the difficult extraction of under-ground commodities proof of the Christian doctrine of original sin:

> Consider, that the beds of coal and the metallic veins are deeply sta-tioned below the surface of the earth; that they are buried under strata of powerful resistance. . . . Is it conceivable that men, innocent, happy in the full enjoyment of the paternal favour of God . . . should be doomed by their Heavenly Father to seek the mineral production . . . ? Assuredly we may without hesitation conclude, that, if to innocent and favoured man minerals were of importance, they would be provided for him by Divine goodness in stations easy of detection and of access. . . . It may be said that minerals were formed and deposited in the earth at the Creation . . . before there had been the opportunity of transgres-sion on the part of man. . . . The Deity, when placing mankind in a state of innocence upon the globe, devised and carried into execution, in its very structure and composition, provisions and prospective arrangements unadapted to the then existing state of men, but suited to the situation of men in the event of their falling from holiness. . . . His Omnipotence foresaw such a fall, and made preparations for it. (95–101)

Gisborne's circuitous reasoning attempts to reconcile the intense vertical challenges of mineral extraction with the idea of divine creation. Much like Ruskin in the first epigraph to this chapter, he ponders why such resources must be buried so deeply and inaccessibly in the planet. Simi-lar arguments were hazarded to explain the horizontal distance between Britain and precious metals on the colonial frontier, and even late in the century, after the decline of natural theology, these kinds of formulations continued to pervade the discourse on extraction. Simonin, for example, posited in 1868 that "gold and silver were, by the design of the Creator, placed along a sort of metalliferous equator, as if to attract irresistibly civilized man to the colonization of those countries, which otherwise he would not attempt" (362). Precious metals have thus "helped in the colo-nization of different countries" (289). The vertical and horizontal impedi-ments to extraction functioned equivalently in such discourse as barri-ers to British accumulation—barriers destined to be annihilated through industrial means.[39]

Overcoming the horizontal as well as the vertical resistances to the accumulation of mineral wealth made the virtues of imperialist capital-ism, these sources suggest, self-evident; thus the subterranean frontier was companion to the horizontal frontiers of the New World: two earthly barriers to the winning of wealth.[40] "The frontier always beckons," as

Moore puts it, and its call can be heard in every line of *Treasure Island*: the urge for accumulation is palpable across the text, and not just in the pirate characters, who are defined by a compulsion for treasure that overrides loyalties and affiliations of all kinds. Everyone in this narrative, not just the pirates, is motivated by wealth. The pirates care for nothing else, as the Squire explains early on: "What were these villains after but money? What do they care for but money? For what would they risk their rascal carcasses but money?" (36). Five lines later, ironically enough, the squire is putting his own carcass on the line for the chance at treasure. Doctor Livesey asks, "Supposing that I have here in my pocket some clue to where Flint buried his treasure, will that treasure amount to much?" The Squire responds, "It will amount to this; if we have the clue you talk about, I fit out a ship in Bristol dock, and take you and Hawkins here along, and I'll have that treasure if I search a year" (37).

Once on the island, after some reversals, the pirates lay hold of the map and set off to find the treasure, "carrying picks and shovels—for that had been the very first necessary they brought ashore" (164). Nearing the spot where "seven hundred thousand pounds in gold lay somewhere buried," every man of them is burning with "the thought of the money . . . their whole soul was bound up in that fortune" (172). The treasure hunters are guided by a map, of course, but when they come to the long sought-after spot marked, on the map, "bulk of treasure here," what they find instead is an open pit, empty and exhausted: "Before us was a great excavation, not very recent, for the sides had fallen in and grass had sprouted on the bottom. In this were the shaft of a pick broken in two and the boards of several packing-cases. . . . All was clear to probation. The *cache* had been found and rifled: the seven hundred thousand pounds were gone!" (173). This initial moment of surprise, when the treasure seems to have been lifted, calls forth intense anxieties of exhaustion. The map and the treasure, after all, have been conflated repeatedly throughout the text, a collapse of sign and signifier that previews the collapse of the treasure and its value in pounds ("the seven hundred thousand pounds were gone!"). This latter collapse is indicative of how Stevenson's treasure of gold bars and coins—materials already mined and processed—nevertheless signifies as an extraction story.

The entire conceit of the "buried treasure" story, in fact, is to take premined and preprocessed materials and put them underground again, so that the intensity of labor involved in finding and processing mineral wealth is significantly reduced and the journey becomes primarily a horizontal rather than a vertical quest. Such a story functions economically to deny the value of labor—specifically, the labor of mining and processing metals—but it also functions politically as a parable of imperialism and ecologically as a fantasy of a more openhanded, benevolent nature, a nature that does not withhold its treasures from sinful humans, but offers them up prêt-à-porter. Marx, though ever mindful of miners' labor, exhibited a similar tendency to present extracted mineral wealth as the free gift of nature; it is a tendency that goes back centuries, as we have seen, and is theological in its roots. He describes, in *Capital*, "ore extracted from their veins" as "objects of labour spontaneously provided by nature" (Marx 284). In the "extractive industries," he specifies later, "the material for labour is provided directly by nature" (287). The fantasy at stake here, and in Stevenson's novel, is that of an unlimited store of underground resources, there for the taking, with no accompanying cost.[41] Jason Moore's *Capitalism in the Web of Life* helps us account for the work of nature in producing such resources and for the appropriation of "cheap nature" as part of the historical tendency of capitalism. The restless global reach toward the frontier, in Moore's reading, is bolstered by the long-standing idea that nature's gifts come with no strings attached.

Treasure Island seemingly explodes the fantasy of openhanded nature by initially offering only an empty, exhausted pit to its treasure seekers: "The buccaneers, with oaths and cries, began to leap, one after another, into the pit, and to dig with their fingers. . . . 'Dig away, boys,' said Silver . . . 'you'll find some pig-nuts'" (174). Small edible roots, pig-nuts are all that seems to be buried in the island's subsurface. But we soon learn that Ben Gunn, a marooned pirate, has already "found the treasure . . . dug it up . . . [and] carried it on his back, in many weary journeys . . . to a cave" (176). Entering Gunn's cave, Jim finds "a large, airy place, with a little spring and a pool of clear water, overhung with ferns" (177). Ferns represent the primeval past, the Carboniferous era when coal was formed—a garnish more appropriate for a coal seam than a cache of gold, perhaps, but one that frames the novel's precious metal resources as the extended work of nature over deep time. In the cave Jim beholds "heaps of coin and quadrilaterals built of bars of gold. That was Flint's treasure that we had come so far to seek, and that had cost already the lives of seventeen men. . . . How many it had cost in the amassing . . . perhaps no man alive could tell" (178). Jim sours on the treasure as he grasps the coins' and bars' material history as colonial commodities and products of extraction—no free gifts, these. Countless horizontal and vertical frontiers were assailed in the making of this hoard, which includes coins "English, French, Spanish, Portuguese" and "doubloons and double guineas and moidores and sequins, the pictures of all the kings of Europe for the last hundred years, strange Oriental pieces. . . . Nearly every variety of money in the world must, I think, have found a place in that collection" (179).[42]

Jim is assigned the job of "packing the minted money into bread-bags" (179), recalling Squire Trelawney's excited declaration earlier in the novel that they'll have "money to eat" if they find the treasure (39). The common comparison between bread and extracted mineral wealth at work here calls to mind the claims of Thomas Malthus, earlier in the century, which prompted fears of exhaustion of a much more basic resource: food. By the second half of the century, the agricultural revolution and advances in fertilizer had assuaged such worries about food shortages in Britain, but fears of extractive exhaustion, of a diminishing supply of coal, gold, and other underground resources, in many ways echoed a Malthusian structure of feeling around scarcity.[43] When Jim packs the gold into bread bags in *Treasure Island*, it suggests how the social base has shifted from agriculture to mining, and how fears of scarcity have shifted from food to metals. Gold is the new bread of life.

Treasure Island remains overridden, however, by a Malthusian sense of overcrowding, which compels the journey abroad to new frontiers. For Jim is replaced within his family almost immediately after joining the crew of the *Hispaniola*, and the realization that he is so much surplus is a blow to his ego. After his father's death, Jim had seen himself as an indispensable protector for his mother, but when the opportunity comes for him to go as cabin boy on the treasure-hunting expedition, the Squire "found [his mother] a boy as an apprentice . . . so that she should not want help while I was gone." The sight of the replacement boy pains him: "It was on seeing that boy that I understood, for the first time, my situation. . . . At the sight of this clumsy stranger, who was to stay here in my place beside my mother, I had my first attack of tears" (44). *Treasure Island*, like *Wonderful Adventures*, is a boom-and-bust narrative aligned with the rhythms of extraction, but it is also, as Jim's tears remind us, a children's book— a story of childhood precarity that taps into the child reader's fears and desires. Even further, it is a book about children's role as resources within their families and communities, a book concerned with social reproduction as a factor in extraction capitalism. The novel's representation of Jim's central role as child laborer—first in his family's inn, then as cabin boy on the *Hispaniola*—highlights capitalism's ever-hungry need for new frontiers of appropriation.[44] But while set in the late eighteenth century, the novel also tells the story of how the cheap resource of child labor has, by the 1880s when the novel is published, dried up, and of how this shift connects to its overseas extraction plot.

While we have long recognized a close tie between the nineteenth-century movement against child labor and the development of children's literature as a genre, we have not tracked the specific ways such texts narrate the closing of the child labor frontier as a resource for capitalism in the Western world.[45] Jason Moore describes "the historically transient character of cheap labor frontiers" (229) and argues that each wave of historical capitalism has depended on the opening of a new frontier of cheap nature, the "process of getting extra-human natures—and humans too—to work for very low outlays of money and energy" (304). These waves also build on one another historically: David McDermott Hughes argues, for example, that the "quantitative, scientific approach to slavery" that emerged on eighteenth-century Caribbean plantations was a precursor for the fossil fuel regime, a "project of imagining slaves as energetic objects" (32, 38).[46] Children were an important labor frontier in early industrial capitalism, and as Marx describes, the rise of the machine also entailed the rise of child labor: "In so far as machinery dispenses with muscular power, it becomes a means for employing workers of slight muscular strength, or whose bodily development is incomplete, but whose limbs are all the more supple" (517).[47] Malm highlights the particular dependence of the textile industry on indentured child labor from 1760 to 1830: parish apprentices were shipped out from London to remote mills, where they worked without remuneration—a vivid example of the appropriation of the child worker as cheap nature.

In the 1830s and 1840s, public outcry and state regulation began to limit child labor, including in the mines. The 1842 Mines and Collieries Act prohibited boys under ten from working in the mines (this was later amended to age thirteen), a proscription intended to rectify a situation where, according to the Children's Employment Commission, "instances occur in which children are taken into the mines to work, as early as four years of age," and where "eight to nine is the ordinary age at which employment in these mines commences" (*Condition* 19). Legislation such as the 1842 Mines and Collieries Act and the 1833 Factory Act prompted enterprising owners to seek a new frontier of cheap labor, and Malm argues that they found it in coal and the steam engine: "The shift to steam was a major cause of the decline of the apprenticeship system, but causation also went . . . the other way" (136). The cheap labor of children was replaced by the cheap energy of ancient plant life. Read in this light, the rise of fossil-fueled capitalism can be connected to the social movement to limit child labor, and capitalism's redirected search for cheap nature.

Treasure Island is, on the face of it, an adventure story of imperial extraction, a journey to a Caribbean island to dig up buried wealth. But the novel—a landmark work in the history of children's literature—also narrates the closing of the child labor frontier, which, along with lust for gold, the emancipation of slavery in the empire, and fears of resource exhaustion, ignited a relentless hunt for new avenues of exploitation. *Treasure Island* represents the ever-shifting frontiers of nature on which capitalist acceleration depends, and it foregrounds the imperial politics of this shifting frontier at a historical moment when cheap nature was often colonial nature. Its quest-for-treasure plot captures how Britain moved beyond its own borders to enfold new natures, and the ecological violence of this enfolding is visible on the surface of Treasure Island itself, where marks of deforestation and soil erosion are left behind by the English pirates who exploit it: "we could see by the stumps what a fine and lofty grove had been destroyed. Most of the soil had been washed away or buried in drift after the removal of the trees" (101).[48] The island is disappearing, its soil washing into the sea—an unwitting anticipation of the unintended consequence of sea-level rise under extractivism-induced climate change.

"The Secret Stores of the Empire": Montezuma's Daughter

Like *Treasure Island*, *Montezuma's Daughter* by H. Rider Haggard is a historical adventure novel, but it is set in Mexico during the early years of the Spanish colonization of the Americas and uses this sixteenth-century setting for a pointedly presentist purpose: to rationalize British ascendancy in the region after the decline of Spanish rule.[49] At the time Haggard serialized *Montezuma's Daughter* in the *Graphic* in 1893, Britain was heavily invested in Mexican mines: "The par value of English capital invested in Mexican mining organizations was in excess of £8.5 million at the end of 1890 and almost £11.7 million at the end of 1911" (Rippy 97). Individual mining schemes were more precarious than these totals suggest, however, and the majority were not profitable.[50] Investment continued nonetheless on the strength of well-publicized bonanzas and the legend of Mexican mineral wealth, to which this novel contributes.[51] *Montezuma's Daughter* is thus an adventure novel about one of the most famous treasure hunts in history—the Spanish quest for gold in the Americas—but it tells this story at a moment of British extractive eminence in the region. In this way it situates Britain's involvement with Mexican mines in triangulated relation to the legacy of Spanish Empire, calling for what Sukanya Banerjee describes as a "transimperial" critical method, one that "posits a relation of comparison, connection, and contiguity between different imperial constituencies" ("Transimperial" 927).

Such a comparison is evident from the outset, as *Montezuma's Daughter* begins with its narrator, Thomas Wingfield, hearing the news of the defeat of the Spanish Armada: "the strength of Spain is shattered . . . and England breathes again. They came to conquer, to bring us to the torture and the stake. . . . God has answered them with his winds, Drake has answered them with his guns. They are gone, and with them the glory of Spain" (1). The victory here is not only Sir Francis Drake's, but one for Haggard's readers: English imperial ascendancy over Spain. The moral righteousness of this ascendancy is clear from the parade of Spanish cruelties through which the novel drags its readers, from the torture chambers of the Inquisition to a nun condemned to be entombed alive in a convent, with her newborn child, for breaking her vows. One scene takes place on a Spanish slave ship, loaded with African men, women, and children soon to be forced into labor in the mines and plantations of Latin America. Here Thomas Wingfield is himself briefly enslaved: "he is a finely built young man and would last some years in the mines," says one officer, regarding him (85). The scene is a pointed reminder that slavery powered Spain's preindustrial mines in Latin America; by contrast, modern British mining operations appear benign. As with child labor in *Treasure Island*, *Montezuma's Daughter* implicitly presents steam power as the sequel to slave labor, despite the fact that not all British-owned mines operating in Latin America were innocent of using slave labor in the steam era, and despite the legacy of exploitative, racialized labor practices that followed emancipation.[52]

British mining pursuits in Mexico depended heavily, for the recruitment of investors, on widespread faith in the steam engine and in Britain's technical superiority over Spain. Such "superiority," one observer notes, was visible in British "machinery for draining . . . substituting steam-engines for the more simple contrivances" (*Observations* 12). These steam engines, "in fact, all the *materiel* of a mining establishment," had to be conveyed from Britain (G. H. 18). In an 1826 scene in the industrialization of Mexican mining, Philip Taylor depicts the steam engine in the role of a foreign guest of great power, newly arrived to this land: "The first steam-engine erected in the Real del Monte, was put in action on the 12th of last August, at the Mine of Moran. So novel a sight to the natives of Mexico, naturally attracted vast numbers of all ranks; and having heretofore seen no other means of raising water from their mines than such as were adapted to the comparatively feeble power of men or mules, they were of course astonished at the gigantic and untired efforts of one of these great servants to the arts" (241).[53]

Despite such set pieces of techno-imperialism, transporting steam engines to the site of the mines proved difficult given the lack of roads and railways, and fuel was often wanting.[54] But when British mining pursuits in Latin America initially did not live up to expectations, many blamed Spain, attributing the failures to Britain's moral superiority as an imperial power. Slavery and forced labor made for one vector of comparison, as in this 1827 account: Spain "had never paid the labour of extraction" and instead accumulated precious metals "at first by open plunder, and long afterwards by dooming the Indians to a life of forced labour" ("Cornish Mining" 91). Spain's poor stewardship of Latin American resources was a second vector of comparison, and British writers frequently blamed Spain for having spoiled the rich mines of the New World—an ironic line of argument, given their own motives. There had been Spanish mines in Latin America as early as 1526, but it was not until after 1700 that Spain "began to work them strong" (Dahlgren 20), and British writers often dwelled on the state of these flooded and derelict eighteenth-century mines. The pamphlet *Observations on Foreign Mining in Mexico* even suggests that Spain purposely drained Mexico's best veins when it sensed the coming fall of its empire: "as the security of the persons and property of the Spaniards and Royalists became more precarious they . . . thoroughly exhausted the bonanzas in sight" (10–11). Here a discussion of the end of Spanish Empire shadows forth fears of the end of British Empire, fears that were bound up with projections of extractive exhaustion. Such fears are also at stake in Haggard's novel.

Wingfield, the narrator of *Montezuma's Daughter*, is himself the son of a Spanish mother, but his father warns him from his childhood that "there is a country called Spain where your mother was born, and there these devils abide who torture men and women, aye, and burn them living in the name of Christ. . . . Hate all Spaniards except your mother, and be watchful lest her blood should master mine within you" (11). Of all Spain's faults, its greed for gold is, ironically enough, at the top of the novel's list. When Wingfield meets Montezuma, the Mexican emperor destined to be defeated by Hernán Cortés, he warns him "of the Spaniards . . . and their greed of gold" (133). They have come, Wingfield says, "to take the land, or at the least to rob it of all its treasures" (133). Later, after Montezuma's demise, Wingfield reflects on the Spanish gold lust, "Never did I see such madness as possessed them, for these poor fools believed that henceforth they should eat their very bread off plates of gold. It was for gold that they had followed Cortes" (224).[55]

Mexico, and specifically Mexico City, appears in Haggard's novel as a veritable El Dorado of extractible wealth, akin to the lost city of gold that was an early impetus for British imperial treasure hunting and that Sir Walter Ralegh never found.[56] When Wingfield first meets Guatemoc, for example, Montezuma's son with whom he forms a bond and an alliance, he seems to personify the treasures of the Mexican subsurface: "His body was encased in a cuirass of gold. . . . On his head he wore a helmet of gold surmounted by the royal crest, an eagle, standing on a snake fashioned in gold and gems. On his arms, and beneath his knees, he wore circlets of gold and gems, and in his hand was a copper-bladed spear" (107; see figure 2.3 for the illustration of this passage).[57] But Mexico also appears in the novel in the guise of a bleeding woman, as in a long description of the landscape of Mexico City and the two volcanic mountains that stand over it (114). This was a vista that had been held up for aesthetic and geognostic appreciation in Humboldt's *Political Essay on the Kingdom of New Spain*: "Though the object of the work which I now publish is more to describe the territorial riches than the geological constitution of New Spain, I have thought proper . . . to give a more lively idea of the beauty of the situation of the city of Mexico" (*Political* cxx). In *Montezuma's Daughters*, however, Wingfield sees a vision of one of these mountains bleeding, literalizing the "open veins of Latin America" metaphor used in Eduardo Galeano's classic study of extractive imperialism: "the light still lingered on the snowy crests of the volcanoes Popo and Ixtac, staining them an awful red. . . . Either it was so or my fancy gave it the very shape and colour of a woman's corpse steeped in blood and laid out for burial. . . . in that red and fearful light the red figure of the sleeping woman arose, or appeared to rise, from the bier of stone. . . . There it stood a giant and awakened corpse, its white wrappings stained with blood" (135–36).

The passage suggests that the mountains of Mexico, those rich mounds of hidden treasure, are dying, or have died, under European colonialism, but it also emphasizes Wingfield's epistemological limitations as narrator. Note his use of "or" in the passage to qualify his vision: "Either it was so or my fancy"; the sleeping woman "arose, or appeared to rise." Wingfield cannot trust his senses, and thus neither can we. As with Seacole and Hawkins, he is an ostensibly truthful narrator, but limited in his capacities as witness: "Now I, Thomas Wingfield, saw these portents with my

I saluted him in the Indian fashion

FIGURE 2.3. Maurice Greiffenhagen, "I saluted him in the Indian fashion." Illustration from *Montezuma's Daughter* (London: Longmans, 1898), 108.

own eyes, but I cannot say whether they were indeed warnings sent from heaven or illusions springing from the accidents of nature" (139). Writing his narrative at the request of Queen Elizabeth, Wingfield is, like Jim Hawkins, under a degree of external compulsion with respect to the narrative he tells, and as is typical of the adventure genre, he frequently calls attention to those parts of the story he will not relate: "I shall dwell but briefly on all the adventures which befell me during the year or so that I remained in Spain, for were I to set out everything at length, this history would have no end" (45); "all the details of this war I do not purpose to write, for were I to do so, there would be no end to this book" (214); aboard the slave ship, "the sights and sounds around me were so awful that I will not try to write of them" (87); of his torture at the hands of the Spanish, "I will not renew my own agonies . . . by describing what befell me" (232). During one of the battles in which he fights against the Spanish on the side of the Anahuac people, Wingfield asks, "How can I tell all that came to pass that night? I cannot, for I saw but little of it. All I know is that for two hours I was fighting like a madman" (193). With his limited perspective, Wingfield struggles throughout to balance adventure narrative's promise to place the reader in a deluge of incident and its geognostic compulsion to provide an extractivist overview of place. Given his ignorance, the task of providing readers with an account of the Mexican setting often falls to the author, who intrudes frequently by way of footnotes signed "—AUTHOR," which gloss, correct, and explain Wingfield's descriptions of Mexico.

Wingfield marvels that someone so lowly as himself could be caught up in the movements of imperial history: "Little did I, plain Thomas Wingfield, gentleman, know, when I rose that morning, that before sunset I should be a god, and after Montezuma the Emperor, the most honoured man, or rather god, in the city of Mexico" (122). The fantasy of omnipotence and omniscience in which the narrator indulges with this elevation to divine status is, however, undercut by his failure to understand his real position, for Wingfield soon learns from Otomie, his wife to be, that ascension to the godhead is traditionally followed by murder at the hands of the state: "every year," she explains, "a young captive [is] chosen to be the earthly image of the god Tezcat" and then "offered as a sacrifice to the god whose spirit you hold" (128). Wingfield's apparent rise to divinity is a net in which he has become haplessly enmeshed: "Now it must be understood that though my place as a god gave me opportunities of knowing all that passed, yet I, Thomas Wingfield, was but a bubble on that great wave of events. . . . I was a bubble on the crest of the wave indeed, but at that time I had no more power than the foam has over the wave" (140).

This "great wave of events"—Spain's invasion of the New World—would lead, of course, to genocide. As Deckard notes, "sixty years after the arrival of the Spanish, two-thirds of the Aztec and Inca population had been decimated, dying of pox, slaughter, or over-work in the mines" (28). This is the larger story of Haggard's novel, which drips with blood as well as gold, and which represents the Spanish quest for American mineral wealth as a death mission on an epic scale. And yet, considering that the novel appeared in 1893, when Britain held the vast majority of mining investments in the region, the novel's aim of favorable comparison to an earlier imperialism remains troubled. Imagining conspicuous extraction as a display of power, Montezuma, in Haggard's account, "sent embassies to Cortes, bearing with them vast treasures of gold and gems as presents, and at the same time praying him to withdraw" (139), but Guatemoc, wiser than his father, curses the wealth of the land, which he calls "the bait that sets these sea sharks tearing at our throats": "would that the soil of Ana-huac bore naught but corn for bread and flint and copper for the points of spears and arrows, then had her sons been free for ever" (200). For Gua-temoc, underground riches undermine Indigenous autonomy, no matter whether Spain or Britain extracts that wealth.

In Haggard's depiction, when Cortés and his army take Mexico City, Montezuma loses "all the hoarded gold and treasure of the empire" (148)—but not all the treasure, as it turns out. In an ironic reversal of the treasure-hunt adventure story, the novel's key chapter "The Burying of Montezuma's Treasure" narrates the replacement of extracted resources back underground—sealed off, forever, from European greed. The events of this chapter are based on a legend associated with La Noche Triste, 30 June 1520, the night of a key battle between Cortés and Montezuma.[58] In *Montezuma's Daughter*, Wingfield and Guatemoc play out that legend and hide away the treasure, "the secret stores of the empire" (197). Ana-huac power, much like the power of industrial Britain, is tied to its "secret stores"—here gold rather than coal—and though their empire is slipping away, the Anahuac resist handing it over to the Spanish invaders: "This was the last triumph that Guatemoc could win, to keep his gold from the grasp of the greedy Spaniard" (231). In this episode of reverse extraction, "great jars and sacks of gold and jewels" are returned to the earth, disap-pearing down a shaft rather than being pulled from its opening. Guatemoc and Wingfield restore the treasure in terms that approximate but reverse the movements of miners at work: "Guatemoc took torches in his hand, and was lowered into the shaft by a rope. Next came my turn, and down I went, hanging to the cord like a spider to its thread, and the hole was very deep" (198). Underground passages stretch out from the shaft, as in a mine, and jars and sacks of treasure are lowered down and stored in the subterranean galleries.

This scene of Indigenous resistance ironically previews the dynamics of the Indigenous anti-extractivist movement today, where activists are fighting in many parts of the world to "keep it in the ground." When the treasure is buried, Guatemoc declares, "Curses on it, I say; may it never glitter more in the sunshine, may it be lost for ever!" (200)—a resource curse of another sort. Like the treasure that remains buried on Treasure Island, whose bearings Jim will not tell, or the diamonds left underground in *King Solomon's Mines*, this fictional cache of extractible resources will be tampered with no more. Wingfield withholds the information on where Montezuma's treasure is buried, though he will later suffer torture at the hands of the Spanish who seek its location: "I swore in my heart that [the Spanish] should never finger the gold by my help. It is for this reason that even now I do not write of the exact bearings of the place where it lies buried . . . though I know [them] well enough" (202). Echoing closely Jim's language in *Treasure Island*—to the extent that one sees Stevenson's influence on Haggard here—the passage suggests that some treasure is best kept underground.

Haggard himself, of course, was fully enmeshed in the imperial-extractive relations that this scene could be said to resist, and ironically enough, he had even sought for Montezuma's legendary treasure on a journey to Mexico to research this novel.[59] *Montezuma's Daughter* is dedicated to Haggard's friend John Gladwyn Jebb, with whom he conducted the search for the lost treasure; Jebb owned a silver mine in Chiapas, which the two men visited together, and under Jebb's influence Haggard had also invested in Mexican copper mines (Pearson 30). This reminds us that whatever Indigenous resistance Haggard's novel documents, imagines, or even celebrates, *Montezuma's Daughter* remains part of a print genre that works to generate extractive desire. As Luz Elena Ramirez argues, the novel allows Haggard "to imagine a place for the English in Mexico's history, if only through fiction," which supports "nineteenth-century advertisements about mining Latin America's riches" (48). Indeed, one of the effects of the reverse extraction scene in "The Burying of Montezuma's Treasure" chapter is to suggest that Mexico's extractible wealth is *still there*, not yet exhausted—much in the same way that Jim Hawkins tells us at the opening of *Treasure Island* that "there is still treasure not yet lifted" on Treasure Island (9). In his dedication to Jebb, who died before the publication of *Montezuma's Daughter* but who read it in proofs before he died,

Haggard writes, "You know even where lies the treasure . . . the countless treasure that an evil fortune held us back from seeking. Now the Indians have taken back their secret, and though many may search, none will lift the graven stone that seals it" (vii). From this opening dedication, extractive desire offers a motivation for reading the novel, and extractivism is coded into the adventure romance genre that Haggard helped develop.

In the dedication as in the novel itself, however, two conflicting desires seem to be at stake: the desire to find the treasure and the desire for no one to find the treasure, for it to stay locked underground until the end of time.[60] Such ambivalence toward a treasure deemed "cursed" is characteristic of the adventure genre more broadly, but in Haggard's case it can also be traced to a more particular circumstance, for it was during Haggard's trip to Mexico to research this novel that his nine-year-old son, Jock, died. Jock had been left home in England, and his sudden death from measles was the "evil fortune" referred to in Haggard's dedication, which kept Jebb and Haggard from their plan to hunt down Montezuma's treasure (Pocock 86–87).[61] This curious intermingling of a child's untimely demise and the extraction of underground resources—an entwinement discussed in chapter 1—held sway over the novel Haggard eventually produced, for Wingfield, too, endures in the course of the novel the death of his children with Otomie. Richard Pearson has suggested that the buried treasure at the heart of *Montezuma's Daughter* represents Haggard's own buried grief, "a potent symbol of the text's concealment and repression of mental trauma" (31). The connection the novel draws between imperial extraction and untimely death, however, has precedence within extraction literature and also references a broader historical-environmental context.

Thomas Wingfield's long-awaited marriage to Lily Bozard at the end of the novel—a marriage deferred for almost the whole of the narrative—resembles the marriages described in chapter 1: it is a late marriage, childless, and full of disappointed hopes. The two are drawn together in their youth and make an engagement that their families do not support, but the accomplishment of that engagement is serially delayed by a seemingly endless relay of plot complications that stretch out for close to twenty years: Wingfield's vow to seek revenge on his mother's murderer, his journey to Spain, his capture by a slave ship and near drowning, his adoption into an Aztec tribe in Mexico, his torture at the hands of the Spanish, his marriage to Montezuma's daughter Otomie, his three children with her, his central role in the Aztec stand against Cortés, and finally, the death of Otomie and their children and his return to England alone at midlife. Wingfield and Lily marry shortly after his return, but the marriage is not what Lily, at least, had hoped for: "But one child was born to us, and this child died in infancy, nor for all [Lily's] prayers did it please God to give her another" (5). This is "the secret sorrow that ate [Lily's] heart away" (5), that she had "wor[n] [her] youth away," and "grow[n] old" waiting "twenty years and more" (6). This sad outcome is relayed at the very beginning of the novel.

There are many ancestral lines that come to an end in *Montezuma's Daughter*—English, Mexican, and Spanish lines. The continual sounding of this note of exhausted generation signifies beyond Haggard's son's death to a larger story of exhausted extraction, declining empires, and frontiers that are closing in rapidly. *Montezuma's Daughter* tells several empires' stories at once: it looks back to the Spanish conquest of Mexico and the height of Spanish imperialism, looks out on the nineteenth-century British dominance of Latin America, but also shadows forth the inevitable end of all empires through its narration of the fall of Montezuma and the Anahuac civilization. Just as the beginning of Haggard's novel narrates the end of the story, looking back to the end of Mexico's empire is also a means of looking forward to the end of the British Empire and to the prospect of resource exhaustion that many feared would bring that end about.

"Trading, Hunting, Fighting, or Mining": King Solomon's Mines

Montezuma's Daughter is dedicated to Haggard's mine-owning, treasure-hunting friend Jebb, but *King Solomon's Mines*, the earlier novel that made Haggard famous, opens with a much less personal dedication: "This faithful but unpretending record of a remarkable adventure is hereby respectfully dedicated by the narrator, Allan Quatermain, to all the big and little boys who read it" (37).[62] Signed not by Haggard, but by the novel's protagonist, the dedication seems to call for a child reader, and likewise the binding of the book's first edition emulated that of *Treasure Island*.[63] Haggard admired Stevenson's text and its runaway success on the book market and set out to consciously imitate it, though *King Solomon's Mines* (which beat even *Treasure Island* in its popular success) was widely read by both adults and children.[64] The overlap between children's literature and adventure literature that we see with both novels is significant, however, for if we imagine treasure-hunting adventure stories as generic expressions of extractivist frameworks for understanding space, empire, and resources, we can assume those frameworks would be particularly foundational for the child reader, whose elemental ideas of the wider world might be shaped by early encounters with such patterns of genre and representation. On a more formal level, too, the overlap between adventure and children's literature is important because of adventure literature's reliance on the epistemologically constrained narrator; child narrators can easily inhabit such a role, as with Jim Hawkins, but even adult narrators of adventure literature are touched by the genre's association with naivete.

We see this with Haggard's narrator in *King Solomon's Mines*: Allan Quatermain is an adult, "fifty-five last birthday" (41), but shares Jim Hawkins's ingenuous quality of apparently artless narration. This is a feature of adventure narrative; the genre depends on plain-speaking narrators who report extraordinary events. These "faithful" and "unpretending" narrators are often subject to ambiguous external constraints, and their narrative authority and perspective is structurally limited within the text. Such is the case with Quatermain, an honest witness to events he seems incapable of understanding. In the first paragraph of the novel, he reflects, "I wonder why I am going to write this book: it is not in my line. I am not a literary man" (41). Nathan Hensley has described Haggard's prose as "a machine for converting thought into deed, internal deliberation into physical action" (198). Quatermain, like Haggard's prose, is better at doing than thinking, but he is a doer with curiously little faith in his own capacity for directed agency: "I am a fatalist, and believe that my time is appointed to come quite independently of my own movements" (63).

With narrators like Quatermain, we can see how adventure narrative is perfectly suited to represent human activities en masse—ecological activities, imperial activities, extractive activities—where individuals together form a collective, even planetary force of which they may not even realize they are part. In a discussion of the adventure romance genre, Cara Murray notes that it "emerged in the 1880s precisely when the pace of modernization on a global scale was being stepped up by the New Imperialism" (150), and she describes how it works to create "the spatiotemporal configurations necessary for modern economic developments" (151). My goal in this chapter has been to foreground the extractive significance of the treasure hunt, which is typically at the root of the adventure story, and thus to show how vertical resistances and horizontal distances are bound together in adventure narrative as in extractive capitalism. This spatial dynamic is different from that of the realist novel, another genre that, as Ghosh argues, "shape[d] the narrative imagination in precisely that period when the accumulation of carbon in the atmosphere was rewriting the destiny of the earth" (7). Realism created a universe based in probability and predictability, while in adventure literature the improbable is almost guaranteed to happen. Adventure narrative has profoundly shaped our environmental imaginations, however, through its representation of underground treasure and the resource frontier, exemplifying how literary worldbuilding—realist or not—fosters the spatial conceits through which we encounter the earth.

We return, then, to the question of knowledge, space, and setting, and to adventure narrative's peculiar affordances for expressing the meagerness of individual human perspective in the face of planetary amplitudes, distances, and durations. In *King Solomon's Mines*, Quatermain actually begins the novel by expressing regret about its modest scope: "Now that this book is printed, and about to be given to the world, the sense of its shortcomings . . . weighs very heavily upon me." The account appears to him hopelessly partial: "I can only say that it does not pretend to be a full account of everything we did and saw" (39). He has decided that "the best plan would be to tell the story in a plain, straightforward manner" (40), but like the other narrators discussed in this chapter, his narration is subject to editing and correction that diminish his authority. A few pages into the novel, for example, an anonymous editor intrudes via footnote to cast doubt on Quatermain's frame of reference: "Mr. Quatermain's ideas about ancient Danes seem to be rather confused . . . probably he was thinking of Saxons" (44). With the narrator's interpretive standpoint almost immediately called into question—for a misunderstanding of race, significantly enough—the epistemological foundations of the story are shaky from the outset. Such a narrative structure is, in part, a codification of the imperial worldview into genre: an ignorant narrator depicting a place and a people he does not understand and producing effects in that setting that he cannot foresee. But it is also a codification of an ecological worldview, for while the extraction boom of the industrial era was global in scope and had impacts that extend into the present day, it was carried out through a large number of individual mining projects, many if not most of which were unsuccessful. It was a frontier-seeking process that proceeded on the environmental logic that when one mining site fails or expires, another can be sought elsewhere. Planetary amplitude offset individual human error, but on the other hand, individual human actions across vast imperial space added up to a planetary force.

Haggard's *King Solomon's Mines* offers just such a vision of colonial extraction on a rich frontier, and this particular frontier of South Africa, as discussed earlier, was in the midst of a mining boom at the time the novel appeared. Haggard had lived in the region while holding a position as attaché to the special commissioner for the Transvaal from 1875 to 1881, during which time he "seems to have raised the Union Jack in the 1877 annexation ceremony of the Transvaal" (Hensley 216). In a series of articles published in *Longman's Magazine* in 1899 he depicted the "wastes" of the region as a rich storehouse for an exhausted Europe: "underneath their surface lie all minerals in abundance, and when the coal of England and of Europe is exhausted, there is sufficient stored up here to stock the world" ("Farmer's" 434). That the mines belong, properly, to the world, or at least to England and Europe, is taken for granted, but in his novel *King Solomon's Mines*, Haggard would struggle to reconcile the winning of colonial plunder with the deep and arduous labors of extraction. The challenging spatial conditions of underground extraction underpin Haggard's novel in two key ways: through its acutely gendered ambivalence about wealth achieved without the vertical toil of extractive labor, and through its effort to convey the geological processes through which precious resources formed in such faraway frontiers to begin with.

The story of Britain's "discovery" of South African diamonds is a story of luck and coincidence, not dogged determination, and provides some context for Haggard's eagerness to provide a pseudo-extractive, horizontal (rather than vertical) journey into the African interior. As the story is recounted in *Reunert's Diamond Mines of South Africa*, John O'Reilly's finding of the first South African diamond in March 1867 was domestic happenstance: "he was returning from a hunting-trip across the Vaal River; and resting for the night at Schalk van Niekerk's farm . . . he noticed a beautiful lot of river pebbles on the table, out of which he picked the 'first diamond.' . . . Van Niekerk's farm was soon 'rushed'" (17). Later, again, "pure accident led to the discovery of the Dutoitspan and Bultfontein Mines in 1870, and of De Beers and Kimberley in the following year" (17) (see figure 2.4 to see what it looked like a year later). Diamonds would be followed by gold, coal, and more, and by 1910 the total value of metal and mineral exports from South Africa was £44,455,519, or 81.3 percent of its exports (*Industrial Prospects* 48).[65] These metal and mineral exports included a diverse portfolio from asbestos to zinc, but the trio of gold, diamonds, and coal accounted for over 98 percent of the commodities in this group (50–51).

The quest on which Quatermain embarks in *King Solomon's Mines* is quite patently a quest for diamonds, and the pretense of searching for Sir Henry's brother makes for a rather halfhearted narrative frame, as other critics have noted.[66] Heidi Kaufman sees the lost brother story as Haggard's attempt to sidestep the "moral problem" of greed and feelings of

THE ROADWAYS, KIMBERLEY MINE, 1872.

FIGURE 2.4. Photograph, "The Roadways, Kimberley Mine, 1872,"
from *Reunert's Diamond Mines of South Africa* (London: Sampson Low, Marston;
Cape Town: J. C. Juta, 1892), 20.

"imperial guilt" around pursuit of the diamonds, noting that we must
contextualize the novel within "the frontier mentality and get-rich-quick
stories that flourished throughout the 1870s and 1880s with regard to the
discovery of diamonds, as well as the subsequent mining frenzy" (530).
But if the novel's evident focus is the diamonds, and its evident context is
the South African Mineral Revolution, it is crucial to note that Haggard's
diamonds have already been mined in an ancient past. Quatermain, Sir
Henry, and Good have come for the diamonds, but they have not come to
extract the diamonds themselves, nor to compel native laborers to extract
them. The novel avoids direct representation of extractive labor and hori-
zontalizes the work of extraction by featuring a quest for treasure that is
already mined and processed—much like *Treasure Island* does.

The novel reaches for tropes of extractive labor to justify the men's
sudden acquisition of wealth, however, in line with nineteenth-century
gendered narratives of extraction and labor. By the time *King Solomon's
Mines* was published, the labor movement and other social forces in England had cast mining as a distinctly masculine calling, especially since
women and children under ten had been banned from underground mining labor since 1842. Quatermain confesses early in the narrative to the

appeal of labor-less extraction and treasure that has already been dug up: the "treasure which those old Jewish or Phoenician adventurers used to extract from a country long since lapsed into the darkest barbarism took a great hold upon my imagination" (51).[67] Still, he takes care to define himself from the novel's opening paragraphs as a worker and a miner— certainly "not a literary man": "At an age when other boys are at school, I was earning my living as a trader in the old Colony. I have been trading, hunting, fighting, or mining ever since" (41). The inclusion of "mining" in this list of manly pursuits legitimizes his claim on the diamonds by establishing his credentials in extractive labor.

Many critics have discussed the gender politics of *King Solomon's Mines*, arguing that the novel, as Merrick Burrow writes, "constitutes a fantasy within which the protagonists' masculinity is redeemed and elevated through immersion into 'barbarian' experience . . . in the African interior" (74). That African interior, as represented by the novel's famous map, closely resembles a female body—down to the two mountains called "Sheba's breasts"—a textual detail that has led Anne McClintock and other critics to see "the journey to King Solomon's mines [as] a genesis of racial and sexual order" (McClintock 241) where "Quatermain contains the eruptive power of the black female" (243). While McClintock, in her landmark analysis, does compare the diamond mine to a womb (257), the novel's extraction plot has for the most part not figured into critical accounts of its gendered landscape, which is perhaps symptomatic of literary critics' general neglect of the mineral substrata that undergird social relations. Rebecca Stott has noted that the adventurers' "actions involve cutting into a passive landscape," which she reads as a "sexual allusion" through "the metaphor of virgin forest" (152), but the "cutting" actions that she describes also evoke extractive labor, and I would suggest that the extractive fantasy the quest enacts is a fantasy version of the paradigmatically masculine labor of mining. We can tie this reading back to *Montezuma's Daughter*, which also features a geognostic landscape presided over by two breast-like mountains.[68] What is the milk exuded by nature's breasts? Diamonds, gold, and silver, at least in these two novels.

Even beyond Haggard's feminized landscapes, diamond mining was peculiarly gendered since women were the primary users of diamonds. Randolph Churchill, visiting South Africa at the height of the Mineral Revolution, thus detected a whiff of unmanly endeavor surrounding the whole project of diamond mining: "In all other mining . . . purposes are carried out beneficial generally to mankind. This remark would apply to gold mines, to coal mines, to tin, copper, and lead mines; but at the De

Beers mine all the wonderful arrangements . . . are put in force in order to extract from the depths of the ground, solely for the wealthy classes, a tiny crystal to be used for the gratification of female vanity in imitation of a lust for personal adornment essentially barbaric if not altogether savage" (48). At a key moment in *Heart of Darkness*, Conrad depicts an African woman fully decked out in "ornaments," "necklaces," "yellow metal," and "the value of several elephants' tusks" (75–76)—similarly suggesting that female vanity is not only a motivator for mineral extraction projects but an "essentially barbaric" motivator at that.

Perhaps to mitigate such suspect associations, *King Solomon's Mines* presents Quatermain's journey into the interior of southern Africa as a cutting into the natural world and into resistant colonial terrain—one that is modeled on the hard labor of mining. Here the vertical turns horizontal through a spatial trick by which entry into the interior of the country mirrors extractive entry into the interior of the earth. Proceeding from the coast inward, the adventurers must force their way into a thick and resistant land: "the karoo bushes caught our shins and retarded us"; "the atmosphere was thick and heavy" (87). In the heat of the desert, they dig a hole to elude the sun, burrowing into the ground like miners: "We set to work, and with the trowel we had brought with us and our hands succeeded . . . in delving out a patch of ground about ten foot long by twelve wide" (89). Like miners, they find reward by penetrating the resistant land, and the deeper into the interior they go, the more they find: "as we went the country seemed to grow richer and richer" (124). When at last they make it to the mines, Quatermain is struck by "an intense curiosity" to know "who was it that had dug the pit" (204). He wonders at how the "long-dead men" (199), from whose labors he will profit, had carved the ancient mine "by hand-labour, out of the mountain" (208) centuries before dynamite and steam emerged to industrialize extractive industry and multiply its environmental effects. He ponders a rude carving, a bit of ancient graffiti, "doubtless the handiwork of some old-world labourer in the mine" (207). The tools of these laborers, such as trowels "of a similar shape and make to those used by workmen to this day" (214), still lay about the long-abandoned mine—the ghostly presence of extractive labor past.[69] Ultimately, however, the proxy labor quest that the novel has unspooled allows Quatermain to conclude that the diamonds "were ours." He insists they "had been found for *us* thousands of years ago by the patient delvers in the great hole" (217). The overland journey to the mines justifies this sense of ownership, as the labor of the quest substitutes for the labor of the "patient delvers."

DOWN AND OUT [129]

Even after the adventurers find the diamonds, however, they are not finished with their pantomime of mining labor, for thanks to Gagool, the ancient finder of witches, they are soon trapped in the cavern. The circumstances of their entombment are sensational, yet underground entrapment was an everyday risk that industrial-era miners faced in their labors. High-profile cases of entombed miners such as Thomas Shaw, who was sealed underground in the Homer Hill Colliery of the West Midlands in 1877, primed Haggard's readers to associate mining with the risk of subterranean entrapment. The Shaw case happened eight years before *King Solomon's Mines* was published, and the story spread far beyond the local context, even reaching the House of Commons.[70] In Haggard's novel, the men trapped underground ironically fear "death from exhaustion" (223), but ultimately they escape by tracing their way through "the ancient workings of a mine, of which the various shafts travelled hither and tither as the ore led" (227). As this escape through mining shafts suggests, Quatermain's entire quest mirrors extraction's forms: the difficult work of pushing through a resistant land to reach the underground treasure, the remnants of past mining labor found along the way, the experience of being trapped underground and making one's way through horizontal underground shafts, the threat of depletion toward exhaustion. The form of this quest is thoroughly extractive, a proxy for the mining labor the adventurers did not do to acquire the colonial treasure.

The novel thus sets up an equation by which the adventurers' labor aligns with ancient miners' labor to justify the adventurers' acquisition of the diamonds, but it also places that labor within extended timescales that stretch into the past as well as the future. For the diamonds' value comes not only from the labor expended to extract them, but also from the work of nature, the long geological processes by which the diamonds formed over centuries, and *King Solomon's Mines* strives to account for these planetary processes—so vast that they dwarf the temporal span between Quatermain and the ancient diamond miners. The geological configurations that made for such precious stones and metals did not occur evenly across the globe, and as we have seen, this geographical disparity is central to the colonial treasure narrative. It is because of England's desire for such precious commodities as diamonds and gold, after all, that Sir Henry's impoverished brother, in the backstory to the novel, "started off for South Africa in the wild hope of making a fortune" (49).[71] Nineteenth-century natural theologians, as discussed earlier, sought to understand geology as an order of Providence, a divine encouragement for colonialism, and their formulations would persist in later, more secular accounts of extractive imperialism.

By the 1880s, when Haggard was writing, natural theology was in the descendant, yet his novel still strives to read the protracted formations of geological time in such a way as to justify imperial appropriation. This is evident in the various appeals in *King Solomon's Mines* for a European geognostic gaze to be trained on the African landscape it describes. Early in their journey to the mines, Quatermain observes a distinct break between fertile and desert land and says, "it would be difficult to say to what natural causes such an abrupt change in the character of the soil was due" (80). The remark invites future geological diagnosis of the landscape and the riddles it poses. Similarly, when Quatermain enjoys a sublime vision, "inexpressibly solemn and overpowering," of "huge volcanoes—for doubtless they are extinct volcanoes," the mountainous formations veil themselves before his eyes: "strange mists and clouds gathered and increased around them, till presently we could only trace their pure and gigantic outline swelling ghostlike through the . . . gauzy mist" (94). The extinct volcanoes seem to summon an extractive gaze to unveil them, especially since diamonds are often located in the vicinity of volcanoes, thrust up from deep recesses by the tumult of eruption. Quatermain's own geological knowledge is limited, yet his eye for the formations that point to extractive wealth is keen, as with his first glimpse of Solomon's diamond mine: "'look there,' I said, pointing to the stiff blue clay which was yet to be seen among the grass and bushes which clothed the sides of the pit, 'the formation is the same [as the diamond mines at Kimberley]. I'll be bound that if we went down there we should find 'pipes' of soapy brecciated rock'" (203). The novel thus envisions the geological timescales that minerals like diamonds inhabit while also revealing how geology, the very science that establishes such timescales, is bound up with extractivism.

At times the novel reframes human endeavor in the vast scale of geological time and deep planetary processes, as when the adventurers enter the mine and find a cathedral of stalactites, some "not less than twenty feet in diameter at the base," others "in process of formation" (206). Such a process is, Quatermain realizes, "incalculably slow," somewhere around "a foot to a thousand years, or an inch and a fraction to a century" (207). Earlier in the novel, Quatermain had taken comfort in the idea of humans' lingering presence on the earth by way of our monuments: "man dies not whilst the world, at once his mother and his monument, remains" (165). This Anthropocenic feat of transforming the earth into a monument to man is finally exemplified inside the ancient cavern where Quatermain finds not just a mine but a tomb with human remains transformed into human-shaped geological formations: "This was the way in which the Kukuana people had from time immemorial preserved their royal dead. They petrified them. What the exact system was, if there was any, beyond placing them for a long period of years under the drip, I never discovered, but there they sat, iced over and preserved for ever by the silicious fluid" (210). Here we see a vivid example of the human etched into the geological record, and in this case it is not just "man" but specifically the Black African man whose body memorializes humanity—but only in a form where race becomes illegible. Following Yusoff's arguments about race and geology, indeed, we might say that the preserved Kukuana royals do the work of representation only in being recounted by the white narrator: "As land is made into tabula rasa for European inscription of its militant maps, so too do Indigenes and Africans become rendered as a writ or ledger of flesh scribed in colonial grammars" (Yusoff 33).

Finally it is the diamond mines themselves, not just the petrified bodies within them, that represent the project of turning Earth into a monument of the human. For while Quatermain and his companions do eventually reap an "enormous" fortune from the few diamonds they retrieve from the mines, the vast majority of the treasure is sealed underground, abandoned until "the end of all things": "Perhaps, in some remote unborn century," Quatermain reflects, "a more fortunate explorer may hit upon the 'Open Sesame,' and flood the world with gems. But, myself, I doubt it. Somehow, I seem to feel that the millions of pounds' worth of gems . . . will never shine round the neck of an earthly beauty" (232). In an ending that anticipates the lost treasure hidden underground in *Montezuma's Daughter*, *King Solomon's Mines* draws on the diamonds' capacity to outlive the earth itself—"the end of all things"—to envision a posthuman world marked indelibly by human delving into the planetary subsurface.

"To Tear Treasure Out of the Bowels of the Land": Heart of Darkness

King Solomon's Mines was so popular that it inspired a parody, *King Solomon's Wives; or, The Phantom Mines*, by Hyder Ragged, pseudonym for Chartres Biron. Its opening lines parody the intrusive editor of adventure fiction, who interrupts the narrator's second sentence, prompting an annoyed retort by the narrator: "Am I writing this book, or you?" (13). The passage's mocking attention to the devices of adventure writing, and to the question of who is in control of the story, narrator or editor, captures the compromised, limited narrative perspective that we find in extractive adventure narratives. Such generic tendencies form the basis for Joseph Conrad's formal experiments in *Heart of Darkness*, a novel that takes the matter and structure of the treasure-hunting adventure tale but upends them, subjecting the genre to a metafictional appraisal that confronts its deep entwinement in deceptive resource ideologies, all by means of a notoriously uncertain narrator, Charles Marlow.[72]

Marlow's ambiguous perspective proved so amenable to Conrad's purposes—"a most discreet, understanding man," Conrad calls him in his 1917 author's note (112)—that he became a transtextual character, appearing also in "Youth" (1898), *Lord Jim* (1900), and *Chance* (1913). In *Heart of Darkness*, Conrad's second Marlow text, Marlow's boyhood "hankering" after the blank space on the map calls forth the romance of the resource frontier, the untapped horizon of mysterious plenty (9), but his actual arrival in this imagined place proves profoundly unsettling and unmooring: "For a time I would feel I belonged still to a world of straightforward facts; but the feeling would not last long" (16).[73] Conrad's use of impressionist technique—emphasizing the sensory or affective impressions of a place or event on a given perceiver, rather than the place or event itself—and delayed decoding, to use Ian Watt's term (175), render into fiction a thoroughly subjective version of events, one apparently liable to all kinds of erroneous or glancing impressions.[74] The novel's frame narrative and anonymous narrator, who hazily encircles Marlow's story while revealing little about himself, likewise ground *Heart of Darkness* at its most fundamental level in profound epistemological uncertainty.

Celebrated for its elliptical style and complex narrative perspective, *Heart of Darkness* could be said to formally encode the peculiar kind of mystification and disorientation with which one encounters another part of the world on a mission to extract resources, an aspect of the novel that floats to the surface when we read it in the context of the treasure-hunting adventure story. Throughout this chapter I have explored how adventure narratives exude extractivism in their generic features, but Conrad's novel does this with a heightened degree of ironic awareness and stylistic aplomb.[75] Indeed, in his 1917 author's note, Conrad actually equates his novel *to* colonial resources, underscoring the close relationship between imperial adventure narratives and the commodities in which they traffic: "it is well known that curious men go prying into all sorts of places (where they have no business) and come out of them with all kinds of spoil. This story, and one other . . . are all the spoil I brought from the centre of Africa" (112).[76]

Conrad uses the word "spoil" here in its oldest sense of booty or plunder, but the term can also refer to the cast-off skin of a snake, or to a process of rotting or decay—two fugitive meanings, to use Freedgood's expression, that suggest his narrative interest in probing the distinction between
the living and the nonliving in conceptualizing colonial resources and
their textual mediation. Such probing is most obvious in the prominent
narrative role of ivory, for ivory is, after all, the dead teeth of elephants.
Fossil fuels are also the organic remains of once-living creatures, but much
less obviously so than ivory, since the creatures from which they form
inhabited another geological age. Ivory, the lost teeth of recently living
megafauna, asks us to think about the relative animacy or inanimacy of
colonial resources, thereby evoking, too, the long history of African slavery within the global imperialist project—a project that was built on the
peculiar anti-animism of turning a human into an object.[77] Here it might
be helpful, with Yusoff, to use "the inhuman as an analytic with which
to scrutinize the traffic between relations of race and material economy
and to think race as a material economy that itself emerges through the
libidinal economy of geology (as the desire for gold, mineralogy, and metallurgy)" (7).

At least since Chinua Achebe's influential 1977 essay "An Image of
Africa," critics have debated the source of *Heart of Darkness*'s persistent
objectification of Black bodies: Conrad, Marlow, or both? In pursuing
questions of race, material resources, and the inhuman in this text, we
might note, too, that Marlow conceives of the objectification of persons as
of a piece with the extractive logic through which all African resources are
perceived: observing a chain of Black prisoners as he travels the colonial
frontier, Marlow describes them as "raw matter" (18), reminding us that all
colonial commodities, whether human or mineral, must first be extracted
from the web of life before being reanimated in the process of commodifi-
cation.[78] Cheap nature, to use Jason Moore's term, is crucial to the project
of global capitalism, and it encompasses the human and the nonhuman,
the living and the nonliving.

If it is evident that Conrad is rooting his tale in, while also ironizing,
the narrative and structural tendencies of the adventure genre—a genre
that emerged from the accelerated plunder of global mineral resources
that attended the rise of industrial extraction—it may be less obvious how
Heart of Darkness qualifies as a treasure-hunting tale of mineral extraction, for the resource at stake here is ivory, whereas all the other extraction
narratives I have considered focus on subterranean resources. Just as novels like *Treasure Island* and *King Solomon's Mines* elucidate the logic of

extractivism by depicting treasures that are already mined and processed and have since been reburied, so too with *Heart of Darkness*'s ivory—ivory that is buried underground. The cache of ivory that Kurtz amasses is "mostly fossil," as the manager "disparagingly" remarks, prompting Marlow to explain, "it was no more fossil than I am; but they call it fossil when it is dug up" (59). Conrad's depiction of fossil ivory ties the novel quite explicitly to the buried treasure plot at the heart of the adventure genre, and to the dynamics of extractive exhaustion that drive that plot.

As several critics have noted, the primary commodity being extracted from the Congo in the 1890s was not ivory, but rubber, raising the question of why Conrad focuses on ivory, and why, specifically, fossil ivory.[79] In large part, the choice of ivory reflects, I think, the same narrative encoding of exhaustion that I have identified in other extraction literatures of this period. As Jeffrey Mathes McCarthy and others have discussed, by the time Conrad was writing, the ivory trade in Africa was severely diminished owing to the overhunting of elephants. This may explain the absence of elephants in *Heart of Darkness*, an absence that has been emphasized by a number of critics.[80] By turning to ivory as the primary treasure in this ironic treasure-hunt narrative, Conrad "describes an economy whose fructifying power is extinct, and thereby renders an environment tipping toward collapse" (McCarthy 42). Indeed, even in 1890, when Conrad worked in the Congo for eight months, the ivory trade was already on the wane, and "traders turned to digging for fossil ivory" because it was the only game in town (McCarthy 43).[81] Ivory's near exhaustion speaks to the book's engagement with the nature of colonial extraction as a finite process with a constantly moving target. It embodies the fears of resource exhaustion that hovered over discussions of mineral extraction, just as it raises unsettling questions about the imperial muddling of living and nonliving resources. Beyond its symbolic freight of exhaustion and animation, ivory's whiteness also made it an apt representational carrier for the European imperial project, the "white man's burden" fantasy of Enlightenment benevolence and the undergirding realities of racial capitalism and resource extraction that subtend it.

That Marlow's commercial networks focus on fossil ivory, then, enmeshes the novel in the narrative structures of extractivism that we have identified in related texts: first, the sense of digging deeply into the interior of a country, in a pseudo-extractive labor process, to gain the treasures of the frontier; second, a narrative emphasis on the planetary subsurface, rendered in *Heart of Darkness* through constant intertextual references to hell and other mythologies of the underworld; and, finally,

the quest structure of a journey out to the colonial frontier to take something and a return to Europe with the spoil. A twist, however, is that in
this case the spoil seems to be Kurtz himself. Marlow journeys to the interior of the country to extract a living man.[82] Further complicating the
living/nonliving binary in reference to colonial commodities, by the time
Marlow finds him, Kurtz has practically transformed into ivory, after having "collected, bartered, swindled, or stolen more ivory than all the other
agents together" (58). An agent of exhaustion, Kurtz has acquired "heaps
of it, stacks of it. . . . You would think there was not a single tusk left either
above or below the ground in the whole country" (59). The epistemology of impressionism turns, as McCarthy notes, on "the blurred boundary
between subject and object" (63), and in Kurtz's case, the blurred boundary between exhauster and exhausted works to communicate the agencies
of nonliving extractable resources and the way they mediate human relations. Kurtz's head, in Marlow's description, has become "like a ball—an
ivory ball" (59), and when Kurtz dies on the return journey, his "fate it was
to be buried," like fossil ivory, "in the mould of primeval earth" (85). There
are shades of Stephen Blackpool, lying at the bottom of Old Hell Shaft, in
Marlow's imagery of an underground, buried Kurtz: "I looked at him as
you peer down at a man who is lying at the bottom of a precipice where
the sun never shines" (86). The parallel suggests that this particular visual
trope, gazing down at a man at the bottom of a dark pit, is paradigmatic
of the extractive imaginary.

Marlow's extracted treasure (Kurtz) dies, just as Kurtz's extracted treasure (ivory) was once living tissue—an elephant's tooth. "Can Rocks Die?"
asks Elizabeth Povinelli in a chapter title of *Geontologies* (30), a book-
long reflection on the life/nonlife distinction and how it is deployed under
capitalist knowledge structures. With ivory, the question is particularly
vexed, all the more so when one of your prime samples of the stuff is "an
animated image of death carved out of old ivory"—that is, Kurtz (74). Let
us pause briefly over the ambiguity of this famous description: Kurtz is an
"image," but the image is "animated," and thus alive; on the other hand,
he is an "animated image of death," that is, an apparently living image
of nonlife. What is more, the image of nonlife is made of "old ivory"—
the long-dead tissue of a once-living animal. The passage hovers over the
living/nonliving divide, and the animacies of ivory take on a religious or
fetishistic quality in the narrative. Across the company stations through
which Marlow must pass, "the word 'ivory' rang in the air, was whispered,
was sighed. You would think they were praying to it." But if ivory is an idol
that vexes the line, like Christ himself, between living and dead, there is

yet "a taint of imbecile rapacity" that "blew through it all, like a whiff from some corpse" (27). Spoil indeed.

Beyond his representation of ivory, Conrad's debt to the spatial imaginary of extractive adventure narrative, and even to the speculative genre of hollow earth fiction that I discuss in the next chapter, is apparent from the outset of his story: leaving Europe, Marlow says, "I felt as though, instead of going to the centre of a continent, I were about to set off for the centre of the earth" (15).[83] This reference to Jules Verne's *Journey to the Centre of the Earth* (1864) underscores the downward and outward movement of extractive adventure narratives, where movement across the globe is figured as a journey into the treasure-rich planetary subsurface, correlating the imperial and the extractive. Later, Marlow encounters a band of "sordid" and "greedy" gold hunters in the Congo interior who call themselves "the Eldorado Exploring Expedition" and seek "to tear treasure out of the bowels of the land" (37), dramatizing, again, the interactive horizontal and vertical vectors of the resource frontier. Many other references to underground settings and subsurface extraction appear across the novel, such as when Marlow describes the climate as an "earthy atmosphere as of an overheated catacomb" (16) or compares Africa to "the gloomy circle of some Inferno" (19). He observes the "mounds of turned-up earth" littering the African shore, "a waste of excavations" coupled with "objectless blasting" from an effort to build a railway (17–18). Elsewhere he sees "a vast artificial hole somebody had been digging" with no apparent purpose at all: "It wasn't a quarry or a sandpit. . . . It was just a hole" (19). At the front lines of extractive imperialism, anthroturbation becomes cultural practice, a ritual engagement with the planetary subsurface rehearsed even where unproductive.

Heart of Darkness begins far away from these front lines, of course, in a boat on the Thames with an anonymous frame narrator who recalls all the great "hunters for gold" who set out on this river before him, or returned, like Sir Francis Drake in the *Golden Hind*, with "round flanks full of treasure" (5). This opening reveals a well-beaten path between resource frontier and imperial center, even if the destination of the frontier keeps changing.[84] It sets the novel, as Jesse Oak Taylor says, at "the contact zone between the metropolitan economy and the material resource base on which it feeds" ("Wilderness" 21). *Heart of Darkness* initially contrasts the imperialists of Marlow's age with earlier colonists like the Roman conquerors of Britain, who were just "men going at it blind" (7); but the Europeans in Africa prove to be just as blind and even more destructive, thanks to industrialization, which amplifies human engagements with nature to a wider magnitude of disruption. Marlow's own blindness, experienced in the fog on the Congo River, comes from "a white fog, very warm and clammy, and more blinding than the night" (48). The fog makes Marlow and his companions feel claustrophobic, as though "buried miles deep in a heap of cotton-wool . . . choking, warm, stifling" (53).

This feeling of being buried alive suggests that Marlow's journey to the Congo interior is, like the journey in *King Solomon's Mines*, a horizontal voyage that figures a downward delving into the planetary subsurface. Marlow describes his spatial movements in terms of vertical depth rather than horizontal distance: moving up the river, he says, "We penetrated deeper and deeper into the heart of darkness" (43). A profound spatial disorientation takes effect when, blinded by the fog on the river, he loses his capacity to locate himself, and is vertically as well as horizontally disoriented: "Were we to let go our hold of the bottom, we would be absolutely in the air—in space" (52). "Going up that river," Marlow says, "was like travelling back to the earliest beginnings of the world" (41). This last simile is often quoted to exemplify Anne McClintock's idea of anachronistic space, the representation of colonial terrain as prehistorical, premodern, or otherwise outside of time (McClintock 40–42), but since the time line of the earth has been fathomed geologically by way of stratigraphic layers of rock, Marlow's figuration is also extractive—for a journey back in time is, geologically, a journey downward through stratigraphic layers. The stratigraphic imagination in *Heart of Darkness*, in other words, is turned sideways: as the boat gets further down the river, it is digging deeper into the past, horizontally so.[85] The voyage to the interior is conceived as a movement through layers of time that are marked by the resistant atmosphere of the earth's critical zone. Here the air is "thick, heavy, sluggish" (41) and must be fought through. The journey upstream naturally requires more time and more fuel than the journey back to the coast, but the difficult struggle of the journey to the interior also mirrors the extractive pursuit of underground wealth: it is a journey of "excessive toil" for an "accursed inheritance" (43).

The fuel that is used to power the steamboat is wood, and Marlow says that during the journey, the search for wood fuel and the tending of the engine keep him diligently focused on the "surface-truth" of "leaky steam-pipes" (44–45). Minding the steam engine prevents him from confronting the "inner truth" of his experiences: "I had to keep a look-out for the signs of dead wood we could cut up in the night for next day's steaming. When you have to attend to things of that sort, to the mere incidents of the surface, the reality—the reality I tell you—fades" (42). The passage calls attention to the way that fuel, steam engines, and the stuff of industrialization—rivets, to take another key *Heart of Darkness* example—generally fade to the background of literature, clothed as mere "surface-truths," unworthy of interpretation. This tendency keeps us from minding the fuel and the engines, from paying attention to the meanings and histories they carry into the text, but in *Heart of Darkness*, Marlow's naming of the steam engine as "surface-truth" actually emphasizes how key to his story is the search for fuel for the boiler. His journey to the inner station is taken up with the tending of the fire and its fuel, and all his impressions come through this perceptual sieve—that which he manages to grasp through the haze of wood-powered steam.[86] So we all experience the world through the semitransparent veil of industrial infrastructure, which obscures as much as it reveals.

The distraction that the engine offers, which keeps Marlow from a more direct acknowledgment of his plight, touches on Allen MacDuffie's recent reading of *Heart of Darkness* as a narrative that anticipates the everyday denial that often seems psychically necessary in life under climate change. As MacDuffie writes, "to read *Heart of Darkness* as a climate change novel is to notice the ways in which Conrad both critically exposes and implicitly reinforces the logic of domination that is at the root of the crisis" ("Charles" 561). In a similarly doubled interchange, Conrad figures the steam engine itself and the work of its tending as that which hinders Marlow's awareness of his environment—a ruinous feedback loop. To experience extraction-based life is to become acclimated to the habits that facilitate environmental catastrophe by learning to pass over—to deny as "surface-truths"—the everyday fossil fuel infrastructure that has overheated the atmosphere. Marlow, grappling with the difficulty in communicating his story to his fictional auditors on the boat, suggests that their life conditions are infelicitous to the message he is trying to convey: "Here you all are . . . a butcher round one corner, a policeman round another, excellent appetites, *and temperature normal—you hear—normal from year's end to year's end*. And you say, Absurd!" (58–59, my emphasis). Marlow is talking about body temperature, not ambient temperature, but the passage speaks to our present predicament and our temperatures that are decidedly not normal. Perhaps we are the audience he has been waiting for.

CONCLUSION

Andrew Lang, who was an admirer and promoter of the adventure romance and especially the work of Stevenson and Haggard, wrote that the value of this kind of literature came from its well-traveled authors:

these "men of imagination and literary skill have been the new conquerors." They "have gone out of the streets of the over-populated lands into the open air" and "escaped from the fog and smoke of towns" (200). I have suggested in this chapter that the fog and smoke of industrial Britain's coal-polluted skies are more closely related to the rise of the adventure genre than we have realized. It is not simply that overseas adventure is an escape from the smoke, but rather, the adventure is part of the same forces that produce the smoke. Treasure-hunting adventure narratives are a literary genre of extraction-based life, and they foreground in their very structures and forms the spatial dynamics of extractivism and its crucial reliance on the frontier.

Adventure narrative's downward and outward movements are mutually determining, and just as the drive for more underground mineral wealth pushed outward movement across the imperial world, so the imperial project exerted ever greater demands for extracted resources. The insolubility of these circumstances registers in the unsteady epistemological grounds of the genre that depicts them. In the narrative forms of adventure literature—its constrained and limited narrators, subject to correction and contending with concealments and unknowns—we see how overseas resource extraction emerged as a global project of collective force, carried out by individual practitioners unaware of the planetary implications of their collective agency. As with the provincial realist novels discussed in chapter 1, the looming fear of exhaustion penetrates the narrative forms of the adventure genre, which dwell on the unwinnable chase of underground resource extraction and the spatial paradox on which it rests.

CHAPTER THREE

Worldbuilding Meets
Terraforming

ENERGY, EXTRACTION, AND
SPECULATIVE FICTION

*"I am indeed, every day of my yet spared life, more and more grateful that
my mind is capable of imaginative vision, and liable to the noble dangers
of delusion which separate the speculative intellect of humanity from the
dreamless instinct of brutes."*

JOHN RUSKIN, "THE STORM-CLOUD OF THE NINETEENTH
CENTURY" (1884), 7

"The work to come involves dreams."

MICHAEL MARDER, *ENERGY DREAMS* (2017), 164

SPECULATIVE FICTION OFTEN relies on the narrative device of the
dream, and this chapter reads speculative fiction as an archive of the
energy dreams of the past. If "the work to come involves dreams," as
Michael Marder suggests, we are right to take the environmental-political work of speculation seriously, and many nineteenth-century thinkers
would have agreed. John Ruskin's "The Storm-Cloud of the Nineteenth
Century" (1884) famously came to the brink of theorizing anthropogenic
climate change from fossil fuel emissions by identifying a "modern plague-
cloud" (30) through speculative observation rather than instruments. In
the above epigraph, Ruskin attributes his lecture's prognostications not
merely to his own observations of the weather, but to the powers of "speculative intellect," "imaginative vision," and a capacity for dreaming that

sets him apart from "dreamless" brutes. Setting aside the question of animals' actual faculty to dream, I want to stress that Ruskin's dream is an energy dream, or perhaps an energy nightmare, for, as he says, "there are at least two hundred furnace chimneys in a square of two miles on every side of me" (33).[1] Speculative literature often took flight from such ruminations on the coal economy—ruminations sparked by the exponential growth in subsurface extraction and the growing social dependence on it in the industrial era.

Jules Verne's *Journey to the Centre of the Earth* (1864), an early example of the subgenre of science fiction known as hollow earth fiction, exemplifies this tendency when the narrator, Axel, pauses on his subsurface journey to describe the "huge layers of coal" he encounters underground, which, he says, "will be used up by over-consumption in less than three centuries, if the industrial nations are not careful" (102).[2] Energy extraction and the imminence of coal exhaustion haunt nineteenth- and early twentieth-century imaginings of worlds otherwise and worlds to come. Diminishing coal presented a vision of material constraint and looming shortage, even in the midst of plenty, that was worrisome for the future and invited speculation on what a new energy base might look like. E. A. Wrigley, historian of the Industrial Revolution, has articulated this paradox of the coal economy that we find at the heart of such narratives: "indefinite dependence on fossils fuels is impossible because the size of the *stock* available is reduced each time a ton of coal is dug . . . but the use of fossil fuel can provide an interlude during which exponential growth is possible" (*Path* 18).[3] The boomtimes of industrial growth were haunted always by the specter of exhaustion.

All the texts I discuss in this chapter are set in extractive spaces or imagine a post-extractive society, and all are imbued with this overarching sense of a disappearing underground stock, whether it be coal or another subterranean resource. The texts range generically from hollow earth narrative to utopian and fantasy fiction: Edward Bulwer Lytton's *The Coming Race* (1871), Rokeya Sakhawat Hossain's "Sultana's Dream" (1905), William Morris's *News from Nowhere* (1890), H. G. Wells's *The Time Machine* (1895), and J.R.R. Tolkien's *The Hobbit* (1937). Five very different texts in the speculative mode, all dream about energy beyond extractivism, and all convey through their secondary worlds the role of energy in shaping culture, environment, and society. Speculative fiction proves to be an apt venue for theorizing energy because it specializes in depicting systems.[4] Drawing on the work of Claude Bienvenu, Jean-Claude Debeir, Jean-Paul Deléage, and Daniel Hémery define energy as that which "must be supplied or removed from a material system to transform or displace it" (2). In building secondary worlds, speculative authors experiment with adding or taking away features of the primary world, thereby dramatizing the broader transformations or displacements that follow from such changes. Energy mimesis is thus a feature of speculative fiction's form, empowering the genre to tell stories of energy and energy change.

The question of energy's relation to literature mirrors that of energy's relation to dreams, for in both cases, materiality is the rub. If we are bound, in our current interregnum between dithering and derangement, to energy's inconvenient material forms—its infrastructures, path dependencies, and fixed capital—dreams and fiction seem mercifully free of such tethers.[5] But the concept of energy is itself all too prone to mystifying such material constraints. Heidi C. M. Scott observes that the term "energy" denotes a "material entity known as fuel" but is also "used to obfuscate between fuel and metaphorical concepts such as the 'energy' of intellect, ingenuity, collaboration" (*Fuel* 7). In its broadest sense, "energy" is a spark that makes things happen and thus bears on discussions of narrative order or narrative cause and effect; fuel, on the other hand, is "brute matter," a "physical substance used in specific eras of history to provide energy" (*Fuel* 7). The conceptual gap between these different senses of "energy" can be traced to mid-nineteenth-century scientific discourse, where, as Bruce Clarke notes, "an earlier science of physics borrowed the word *energy* from a traditional register of literary and philosophical significances." Faced with the need to name "new conceptions of natural objects and processes," early physicists imported the term "from its theological and humanistic provenance" to the new field of thermodynamics, and thus the "already overdetermined term *energy* became even more charged with powerful semantic currents" (2).

In the past as today, the overdetermined conceptual category of "energy" has helped obscure fuel's materiality. Allen MacDuffie writes that the "highly mystifying fantasies about energy" that arose with industrialism "persist today and have helped make [the energy crisis] as intractable as it is" (*Victorian* 12). This suggests a need for greater critical attention to energy as "brute matter," to fuel in its gross material embodiments— its dirty and dangerous extraction, production, and combustion. And yet literature can never be merely a record of such gross embodiments, for it is also an archive of enabling fantasy, of hopes and plans for renewal and change. This chapter will situate speculative energy fiction at the crossroads of what is and what we dream could be instead—the crossroads of mystification and possibility, of fooling ourselves and saving ourselves. For if literature's dream capacities seem to offer potential relief from energy determinisms, this is not to overlook Imre Szeman's bracing claim that literature has been complicit with fossil capitalism in key ways, that "instead of challenging the fiction of surplus" it "participates in it just as surely as every other social narrative in the contemporary era" ("Literature" 324). The fact that Szeman here speaks specifically of *contemporary* fiction is significant, however, and the case I want to make in this chapter and in the book as a whole is that there is something distinct to be gained by looking back to literature from the first century of industrial extraction, when energy path dependencies were still hardening into the global system of fossil fuel exploitation that we inhabit today. Szeman says we are now possessed of an "apparent epistemic inability or unwillingness to name our energy ontologies" (324), but such ontologies did not arrive whole-cloth with the birth of steam: as these literatures of the past demonstrate, they emerged through narrative and discourse, through a complex negotiation of material reality, abstraction, and fantasy.

Speculative fiction was a field for such negotiation, but it also offers glimpses of other possible ontologies that fell by the wayside or never came to be. Speaking specifically of utopian fiction, Fredric Jameson has said that its "deepest vocation is to bring home . . . our constitutional inability to imagine Utopia itself . . . not owing to any individual failure of imagination but as the result of the systemic, cultural and ideological closure of which we are all . . . prisoners" (289). This touches on a long-standing point of contention in utopian studies: whether "depictions of utopia can form a kind of counter-ideological force or are themselves always and inherently ideological" (Kreisel, "Teaching" 164). Whether speculative fiction offers a vision of a world outside extractivism or, as Jameson would have it, merely a revelation of the prison of extractivism itself, my suggestion is that historical speculative fiction presents a somewhat different case, for its temporal distance provides a further vector of estrangement that is politically clarifying for long-duration systems like fossil capitalism.

The genres of science fiction and fantasy are often said to have originated in the nineteenth century while utopian fiction is said to have had its heyday at the turn of the century.[6] Critics have variously traced this historical surge in speculative fiction to economic, political, and cultural shifts.[7] To begin to write an origin story rooted in energy shifts and the boom in industrial extraction, I would note that Imre Szeman pinpoints 1870 (just before *The Coming Race* was published) as "the moment when more energy begins to be extracted from fossil fuels than from photosynthesis" and 1890 (when *News from Nowhere* was first serialized) as "the year in which more than half of global energy comes from fossil fuels" ("Conjectures" 284). A premise of this chapter is that the post-1870 acceleration of the energy-extraction complex must be central to our understanding of the concomitant rise in science fiction, utopia, and fantasy, as the frequency of underground settings and energy novums in these genres would suggest. No mere genres of excess, the cornucopia of surplus energy and material in these texts is ever shadowed by speculation about its diminution and exhaustion.

THE WORK OF ENERGY AND
THE WORK OF NATURE

This chapter's focus on energy builds on the previous two chapters' concerns with time and space, for two of the major gains that drove the transition to fossil energy are its capacities for saving time and for speeding transport across space. Such capacities have been central to speculative fiction, too, from H. G. Wells's time machine to *Star Trek*'s warp drive. Darko Suvin identifies "voyage time," a neat encapsulation of the time-space relation, as a key structuring principle for early science fiction (*Metamorphoses* 149). Beyond saving time and conquering distance, substituting for human labor was another factor in the transition to fossil energy and another key concern of the speculative imagination. More time, more space, less work: such are the purported benefits of fossil fuels, and such are the common fantasies of speculative fiction. Clearly the rise of extraction-based life and the proliferation of speculative genres go together.

Energy is produced, of course, through many means besides fossil fuels: sunlight, running water, animal power, nuclear power. In this way energy could be said to express the work of nature, the capacity for work that both the living and the nonliving possess.[8] As Patricia Yaeger puts it in her landmark column on energy and literary periodization, "thinking about literature through the lens of energy, especially the fuel basis of economies, means getting serious about modes of production as a force field for culture" (308). By offering this "modes of production" framework for the energy humanities, Yaeger helped shape a now-common approach that encompasses human work, the work of nature, and the infrastructure and fuel systems in which all this labor takes place—the entirety of energy's productive milieu.[9] A recent book by Cara New Daggett, however, argues that "the thermodynamic rendering of energy—as the measurement of productive, valuable work—has arguably become so dominant . . . as to crowd out other possible ways of imagining energy" (20), with the effect of consigning labor to the domain of "energy governance" (16) and aligning fossil fuels with an idealized work ethic (196). She advocates instead "a partnership between post-carbon and post-work politics" (199)—a partnership already imagined, as we shall see, in energy utopias like *The Coming Race*, "Sultana's Dream," and *News from Nowhere*.

The Greek origins of the word "energy" reveal its close connection to the idea of work: "composed of the prefix *en-* and the noun *ergon*, *energeia* can be literally translated as 'enworkment,' putting-to-work" (Marder 3).[10] *Capital*, Marx's landmark analysis of labor under capitalism, communicated the crucial relation between the two concepts by defining "labor" in terms of "natural force," or energy. Although his understanding of "natural force" is anthropocentric and geared toward human mastery of nature, Marx described a dialectical relation between human labor and the natural world that illuminates how our ways of being in the world are produced in part by the energy materialities that surround and flow through us:

> Labour is, first of all, a process between man and nature, a process by which man, through his own actions, mediates, regulates and controls the metabolism between himself and nature. He confronts the materials of nature as a force of nature. He sets in motion the natural forces which belong to his own body . . . in order to appropriate the materials of nature in a form adapted to his own needs. Through this movement he acts upon external nature and changes it, and in this way he simultaneously changes his own nature. He develops the potentialities slumbering within nature, and subjects the play of its forces to his own sovereign power. (283)[11]

In many ways Marx anticipates here the conclusions of the speculative fictions I discuss in this chapter: that humanity "simultaneously changes [its] own nature" as it "acts upon external nature and changes it." But this analysis, as Jason Moore has also stressed, lacks an account of the work of nature, instead maintaining that "all those things which labour merely separates from immediate connection with their environment are objects of labour spontaneously provided by nature" (Marx 284), including "ore extracted from their veins" (284). Authors of speculative fiction commonly attribute the distinctive features of the human and humanoid societies they depict to their energy means, mirroring Marx's dialectical understanding of humans, nature, and the metabolic energy that passes between them; but such fictions often attend to the work of nature in producing energy, too—the metabolic energies that pass within and between nonhuman and even nonliving parts of the natural world.[12] Industrial-era speculative fictions are alert, in other words, to the agencies of energy systems outside of or beyond human appropriation and metabolization.

The question of how far energy's agencies extend has led several recent critics to suggest that modern subjectivity itself is an effect of fossil fuels: Dipesh Chakrabarty, for example, writes that the "mansion of modern freedoms stands on an ever-expanding base of fossil-fuel use" ("Climate" 208), and Bob Johnson locates "the origins of modern life" in the "role that fossil fuels have played . . . in fueling our affective attachments to modernity" (1–2). Such energy-determinative reasoning represents a useful recognition of our constrained position within energy systems of historical depth and global reach, systems whose physical properties have had unforeseen impacts on the earth system. And yet, such reasoning can also exacerbate the sense of being trapped within a destructive regime. Frozen in carbonite like so many Han Solos, how can we transition to a postcarbon society? Speculative fiction offers a possible space to challenge conceptual path dependencies and to exercise our dream faculty, which may be significant if, as Marder writes, "the new enworkment of energy cannot get off the ground without dreamwork . . . an exercise in imagining what another energy, or other energies, independent of the default extractive-destructive paradigm, would look like" (164).

Industrial-era speculative fiction often imagines "the new enworkment of energy," but it is also attuned to energy exhaustion, to the sense of being bound to an energy and resource system that is inherently depletive. A goal in the first two chapters of this book was to show how the exhaustion trajectory became fixed in industrial-era thought and discourse to the extent that it reshaped literary genre, but beyond this argument, I have also sought to pose the question of how an accommodation of exhaustion-based life has shaped our response to the changing climate we inhabit today, one where each metric ton of carbon emitted now promises to hang in the atmosphere for many years to come, someone else's problem.[13] In the nineteenth and early twentieth centuries, forecasts of resource exhaustion demanded a confrontation with generationally proximate calamity, and speculative fiction, more than provincial realism and adventure fiction, excelled at connecting the extractivist regime of the day with future calamity.

Consider, for example, M. P. Shiel's *The Purple Cloud* (1901), a novel devoted to imagining near-term human extinction. After Adam Jeffson survives an apocalyptic cloud of poisonous gas, he decides to take on the "long, multitudinous, and perplexing task of visiting [Britain's] mineregions" in search of survivors (108). For six months he descends the shafts of England, Scotland, Wales, and the Isle of Man, guided by Ordnance Survey maps and topographical books (100). He visits graphite, cobalt, manganese, lead, copper, iron ore, and coal mines, as well as "underground stone-quarries" and "underground slate-quarries" (111). He lives many months "in the deeps of the earth, searching for the treasure of a life" (107). Though his search fails, the novel's detailed tour of Britain's subterranean republic evinces the prominence of extractive settings in the end-time imagination, as do Jeffson's speculations about the nature of the planet's interior: "I have some knowledge of the earth for only ten miles down: but she has eight thousand miles: and whether through all that depth she is flame or fluid, hard or soft, I do not know, I do not know" (157). "Nothing," he concludes "could be more appallingly insecure than living on a planet" (166).

As Jeffson's musings convey, industrial-era speculative genres are poised at the precipice of the planetary while still operating in the realm of the global, to use Chakrabarty's terminology discussed in chapter 2. They engage with deep time and planetary processes—the work of nature—indifferent to human lives and human perspectives yet turn on the timescales and systems most relevant to humans.[14] The resemblance of industrial-era prognostications of resource exhaustion to the structure of feeling that we inhabit today in relation to climate change suggests that such prognostications are a significant ontological, epistemological, and discursive inheritance. It is the task of the energy humanities to work through this inheritance, to build a cultural history of climate change, and to find resources in the past for reimagining and restructuring environmental relations today.

SPECULATIVE ENERGY AND SPECULATIVE FINANCE

Before I turn to the speculative narratives at the heart of this chapter, I want to briefly discuss the historical relation between speculative finance and speculative fiction and how it bears on an energy-focused analysis of speculative genres, for the very word "speculative" calls to mind the importance of investment to the capitalization of large-scale industrial mining projects, and this economic note will sound frequently in the pages to follow.[15] The particular financial exigencies of the mining industry help shape the stories we tell about it and affect the capacities of these stories to imagine beyond the current system. Industrial extraction is a capital-intensive engagement with the natural world, one that requires large outlays of investment that must be recouped over years of profit, with the prospect of local exhaustion always lurking around the corner. Investment in the mining industry and its bosom companion, the railway industry, were understood in the industrial era to be particularly volatile financial ventures: in H. G. Wells's novel *Tono-Bungay* (1909), for example, the protagonist's uncle is ruined by a "financial accident" related to the "intellectual recreation that he called stock-market meteorology" where he would "trace the rise and fall of certain mines and railways" (78). Bankruptcy ensues.

The era's mining press is packed with supposition on how long individual mines would last and how much they had to give. This information, intended to guide investors, reads in some instances like meteorology (still a young science of prediction at this time) and some like outright puffery. In our current moment, we see similar dynamics at work in the discussion of fossil fuel investments, although the calculations must now account, too, for the prospect of climate change and the time line for the transition to low-carbon energy. In ongoing debates about the proposed Carmichael Mine in Queensland, Australia, for example, speculation about the coal mine's future life and yield have factored into decisions about the federal subsidy of this large-scale project, but the mine's life span and yield are being calculated in some cases for up to sixty years, even as coal mines are being decommissioned elsewhere and climate scientists are urging that we transition off coal much sooner than that.[16] Today's fixed capital, tomorrow's stranded asset.

The industrial-era mining press was similarly occupied with predicting mines' future life and future yield, and a closer look at the language around mining shares and stockholding in this context reveals shared conceptual tendencies across speculative finance and speculative fiction. Just as setting and worldbuilding serve as the grounds of estrangement in speculative fiction, so mine speculation thrived in estranged settings, and distance sometimes seemed to make the mine grow richer. Because the further away the mine was, the more its prospects seemed to grow, W. J. Wybergh encouraged readers in his 1922 writings on South Africa to take all estimates as to the length of time until exhaustion in a given South African mine "as highly speculative" (91). Mines in Britain were not immune from these dynamics, especially in provincial mining areas like Cornwall or Wales where distance from financial centers remained an estranging factor, even as their advocates would proclaim the "safety" of these mines over foreign investments. In Henry English's *A Compendium of Useful Information Relating to the Companies Formed for Working British Mines* (1826), for example, the Welsh Iron and Coal Mining Company is pushed precisely on the basis of its supposedly nonspeculative qualities: "This is no speculative undertaking; it is no problematic or visionary scheme" (73). But Cornwall is promoted, on the other hand, for its "mineral fecundity," "vast riches of nature," and mines that are "beginning to develope [*sic*] themselves" (27).

As this reference to Cornwall's mines "develop[ing] themselves" suggests, the mining press often mystified labor, energy, and material resources, suggesting, again, the discursive links between speculative fiction and speculative finance. It is not just that speculation has "two distinct semantic-conceptual registers: cognitive and economic" (Uncertain Commons 9), such sources suggest, but that each has shaped the other domain. The *Mining Journal*, for example, a publication that catered to both speculators and engineers, reported on financial and material aspects of extractive industry, but at times its rhetoric verged into a speculative mode not unlike science fiction. An advertisement in the 7 March 1868 issue, for example, promises a new energy source: "LOCOMOTIVE AND ALL OTHER STEAM ENGINES SUPERSEDED by a NEW MOTIVE POWER, possessing extraordinary economy in weight, space, and fuel." The advertisement avows that "a splendid fortune may be quickly realised by any spirited capitalist who will assist in the development of these very important inventions" (181). Though the nature of the new motive power is left unstated, the advertisement voices a desire for new energy forms that is also evident in speculative fiction like *The Coming Race* and "Sultana's Dream."

Another glimmer of speculative storytelling appears in an 1868 article from the *Mining Journal* that imagines new species emerging from the extractive spaces of the underworld in a manner not unlike *The Time Machine* or *The Hobbit*. Underground frontiers were being rapidly uncovered during the industrial extraction boom, and in "Subterranean Caverns, and Their Inhabitants," Dr. E. H. describes the skeletons and bones of animals, humans, and extinct quadrupeds that have been found in the "dark recesses" of underground caverns, "hidden from the light of the sun, perhaps, for hundreds of years." Such Gollum-like species invite imaginative speculation: "in our anxious desire for the answer our minds are often carried beyond the pale of mental jurisdiction, and reason ceases to be our guide." The piece looks forward to *The Time Machine* and *The Hobbit* while looking back to Verne's *Journey to the Centre of the Earth*, revealing a discursive thread that connects speculative fiction with the evolutionary speculations prompted by industrial mining and its unveiling of subterranean frontiers. This suggests how the underground imaginary, extractive imaginary, and energy imaginary were coproduced by multiple discourses, including finance and science, and how literature provided formal and temporal patterns that enable us to see these intersecting and multiple imaginaries.

ALLEGORIES OF EXTRACTIVISM

Reading for energy and extraction in speculative fiction requires that we go beyond the "energy systems approach to literature" that has been usefully described by Justin Neuman, one that asks us to read "out from the traces of energy use encoded within a text to the externalities and converter chains that exceed those traces" and to read "in to analyze the discourses and images that condition subjective experiences of energy use and also contribute to its invisibility" (152). While these reading strategies continue to be useful, in speculative fiction we are also operating in a domain of energy allegory, a stratigraphic layering of meanings that requires reading not just out and in, but also above and below. Such allegorical layers, Bruce Clarke has suggested, create hypothetical models that place similar demands on readers as scientific reasoning: identifying allegorical structures in narrative, he says, "parallels the modality activity of scientific theorizing" (18).[17] But if allegory is akin to scientific modeling, it is also spatially akin to the idea of the underworld: a hidden world below that parallels and mirrors yet remains separate from the world above. Allegory's "spatial expression," Clarke argues, is "through a complexly polarized world of upper and lower realms" (28). Such "spatial stratification" means that allegory shares discursive and conceptual grounding with extractive industry, as I will suggest with respect to hollow earth and subterranean fiction especially.

We are now at a moment of widespread critical attention to speculative genres and their Anthropocene dimensions, but the turn to speculative fiction in the era of climate change is not just the critical imposition of another layer of allegory, for there is, as I have suggested, a close connection between the upwelling of these genres and the surge in mineral resource extraction, along with its familiar shadow of resource exhaustion, in the industrial era. Throughout this book I have sought to show how literary genre and form correlate to emergent aspects of extractivism; in the context of widespread debate about coal exhaustion in the latter decades of the nineteenth century, speculative fiction came to depict energy transition as source of social transformation, as we will see in the first three texts discussed in the chapter: *The Coming Race*, "Sultana's Dream," and *News from Nowhere*. Written in an era of exponential growth, these three speculative works anticipate in various ways the drawbacks of large-scale appropriation. They imagine new forms of energy that do not require underground extraction, and they attempt to imagine beyond the contours of extraction-based life. The last two texts I discuss in the chapter, *The Time Machine* and *The Hobbit,* instead hearken backward to various engagements with the premodern pastoral as they grapple with the future-oriented energy anxiety common to speculative fiction of this era. The time machine and the ring of power also anticipate the new energy agencies and new energy horizons looming at the dawn of the atomic age. The secondary worlds in all five of these texts offer an allegorical, comparative account of extraction-based life, which is evident in the prominent role they all assign to the underground/surface dynamic.

I will begin with a novel that has been viewed as a landmark in the development of science fiction, *The Coming Race*, published in 1871 when the speculative fiction boom was beginning and when extractivist path dependency was solidifying as a global, imperial force. Jason Moore describes the post-1870 moment as an "era of peak appropriation" enabled by coal and the colonies, just on the other side of a historical "tipping point" when "capitalism as a planetary system became possible through the production of a globe-encircling railroad and steamship network," and planetary work and energy were appropriated on an unprecedented scale (136–37). This moment of peak appropriation was also, as I have described, a peak era for utopian and speculative fiction, and critics such as Matthew Beaumont have read the era's speculative impulse as a structure of feeling symptomatic of an emergent consciousness. One such emergent consciousness that we find in this speculative outpouring of extraction and energy narratives, I argue, is a nascent anti-extractivism, a latent sense of disquiet that the circuits of capital have become so beholden to stocks of finite underground material. It is a disquiet powerful enough to fuel the imaginary building of new worlds.

"Natural Energetic Agencies": The Coming Race

Edward Bulwer Lytton's *The Coming Race* (1871) is part of a subgenre of speculative fiction known as hollow earth fiction—narratives that take place in imaginary worlds under the surface of the earth and borrow from scientific speculation about the composition of the earth's interior.[18] This tradition of fiction, thought to have begun with Ludwig Holberg's *A Journey to the World under Ground* (1741), flowered in the nineteenth century, with Bulwer's novel recognized as its "crowning achievement" (Sinnema 9). A goal of this chapter is to connect the emergence of hollow earth fiction, exemplified especially by *The Coming Race* but also to an extent by *The Time Machine* and *The Hobbit*, with the extraction ecologies of the industrial era that the rest of this project traces through the archive of literature. Questions about what lay underground and how much remained to be exploited drove political and geological debates about exhaustion at this time, fueled by a fear that when the coal ran out so too would modern life. A hollow earth novel like *The Coming Race* speaks to this particular nature-culture conjunction not only by peeling back the surface layer of the earth to imaginatively reveal the unknown regions of the vertical frontier, but also by inventing a new energy supply capable of superseding fossil fuels altogether.

Certainly Bulwer's novel hit a nerve at the time it was published, running through five editions in its first year and proving highly influential.[19] Suvin argues that it showcases "techniques of realistic credibility imported, probably for the first time, into the utopia-derived societal anatomy—a genuine novelty, to which both Bellamy and Wells are indebted (and through them all utopian fiction and all SF since)" ("Extraordinary" 242). *The Coming Race* achieves this "realistic credibility" through different means than the adventure narratives discussed in the previous chapter. Unlike such novels as *Treasure Island* or *King Solomon's Mines*, there is no guiding map to the underworld here, nor a fictional editor, nor an attempt to situate the fantastical narrative in a nest of false documents that create an alternative print reality. As we saw in chapter 2, there is a textual-material basis to the adventure genre's assertions of realistic credibility, but here the connection between the textual world and the reader's world is made speculatively, not materially. Still, in the case of Bulwer's novel, speculation would become material in a strikingly literal way after publication, for *The Coming Race* inspired the brand name of Bovril, a British beef extract meant to supply strength and stamina, and one that contributed to the misperception that beef, an energy-intensive food to produce, is a necessary source for energy and nutrition in the human diet. That Bovril is the most widely recognized legacy of Bulwer's novel suggests how fiction can support capitalist narratives of surplus, as Szeman has suggested ("Literature"), but it also suggests more generally that literature—even wildly speculative literature—mediates the natural world in ways that have material, energetic impacts.

Energy has a key role in Bulwer's novel, for *The Coming Race* depicts the underworld civilization of the Vril-ya, who are named for their mysterious energy source, vril. Vril has no "exact synonym" and proves difficult to explain in overworld terms; as the narrator says, "I should call it electricity, except that it comprehends in its manifold branches other forces of nature, to which, in our scientific nomenclature, different names are assigned, such as magnetism, galvanism, &c." Variously connected to electricity, mesmerism, ether, and magnetism, vril is a fluid "capable of being raised and disciplined into the mightiest agency over all forms of matter" (59), one that partakes of "natural energetic agencies" (54).[20] It far exceeds the powers of any known fuel in Bulwer's day, allowing the Vril-ya to influence the weather (and in this sense it is not so different from fossil fuels after all) as well as other people's minds and "bodies animal and vegetable" (54).[21] Vril serves in the narrative as what Suvin has called the "validation," a feature of science fiction that provides a "principle of believability," in this case, a new energy source that can ostensibly account for the different features of this underground society ("Victorian" 153). Bulwer's narrator, who is unnamed but becomes known to the Vril-ya as "Tish," meaning "small barbarian" (115), explains that the political and social development of the Vril-ya was shaped by their discovery of vril. The decline of democracy, which they "looked back to as one of the crude and ignorant experiments which belong to the infancy of political science" (59), was brought about "by the gradual discovery of the latent powers stored in the all-permeating fluid which they denominate Vril" (59). The effects of this discovery were "remarkable in their influence upon social polity" (59–60), and the influence was so significant that it collapsed the word for energy and the word for civilization: "the word A-Vril was synonymous with civilisation; and Vril-ya, signifying 'The Civilised Nations,' was the common name by which the communities employing the uses of vril distinguished themselves" (60).[22]

Despite the narrative emphasis on this near-magical imaginary fuel, Bulwer's interest in extractive energy is apparent from the novel's outset. Vril is not a fossil fuel, but the novel's underground setting and its mining prologue locate its intervention in the domain of extraction-based life. Elizabeth Chang has argued that the hollow earth narrative exemplifies "fictional genre as environmental form," and that it attunes its readers to the ways that "plot and character both develop from the foundational demands of the environment of their imagined world" ("Hollow Earth" 388). These narratives also emphasize how wider social and political forms develop from foundational environmental demands, and in the case of Bulwer's novel, the underground setting prompts reflection on the social forms that emerge within extractivism. Consider, for example, the narrative's temporal lag: the narrator purports to be "commit[ing] to paper those recollections of the life of the Vril-ya" many long years after visiting their world (81), mimicking the strange temporalities of fossil fuels, which form underground long before being, as it were, expressed. Extractivism's global reach is evident, too, in the novel's exposition. Bulwer's narrator is from the United States but is sent to Liverpool for "literary education" and "commercial training," from whence he visits a mine in an unknown location: "I was invited by a professional engineer, with whom I had made acquaintance, to visit the recesses of the—mine, upon which he was employed" (33).[23] This coyness regarding the location of the underground chasm that leads to the land of the Vril-ya recalls the constrained epistemological conditions of adventure fiction, discussed in chapter 2, as does the narrator's motive "for concealing all clue to the district of which I write" and "refraining from any description that may tend to its discovery" (33).

While visiting the unknown mining region, the narrator becomes rapt with the hidden world below: "I accompanied the engineer into the interior of the mine, and became so strangely fascinated by its gloomy wonders, and so interested in my friend's explorations, that I prolonged my stay in the neighbourhood, and descended daily, for some weeks, into the vaults and galleries hollowed by nature and art beneath the surface of the earth" (33). So far, such an opening stages what Neuman calls an "energy recognition scene," where writers "imagine their way across the vivid and tangible materiality of . . . the energy systems on which they depend" (152). Such a narrative unveils the work of "nature and art" that together produce underground resources. But Bulwer quickly moves from revelation to speculation about what lies beyond the mines. The novel exhibits the same sort of frontier thinking that we saw in the adventure stories of the previous chapter, but here the frontier is downward and downward, not downward and outward. The downward search for new frontiers of extraction ultimately brings our narrator to the land of the Vril-ya. His engineering friend "was persuaded that far richer deposits of mineral wealth than had yet been detected, would be found in a new shaft that had been commenced under his operations" (33). While "piercing this shaft," the men come across the fateful chasm—a chasm through which the engineer descends first, and sees, remarkably, a road and gas lamps: two key markers of civilized life, deep underground. The two men decide to descend again together and are lowered down by cage with the help of six miners (35). Descending still further with a rope and grappling hooks, the engineer falls and dies—a reminder of the constant threat of accident in underground mines. The narrator sees "my companion an inanimate mass beside me, life utterly extinct" (37) but is spared such an end. Trapped in "the bowels of the earth" (37), he is rescued by the Vril-ya.

The underground people of the Vril-ya are fed by agriculture and powered by vril, yet their world, in color and description, is built from underground mineral wealth rather than the resources of surface life. That for
which we need to dig is for them readily plentiful: "Precious metals and
gems are so profuse among them, that they are lavished on things devoted
to purposes the most commonplace" (110). The Vril-ya wear jewels on their
heads (39), use thin metallic sheets in lieu of paper (49), have floors tiled
with "blocks of precious metals" (40), and tessellate their walls with "metals, and uncut jewels" (41). Despite all this mineral wealth, they possess no
notion of extraction. The narrator, describing to the Vril-ya how he came
to be in their lands, must explain the very idea to them: "I answered, that
under the surface of the earth there were mines containing minerals, or
metals, essential to our wants and our progress in all arts and industries,"
and that he had inadvertently happened upon their world "while exploring one of these mines" (52). With metal so plentiful, the Vril-ya employ,
in lieu of coinage, a literal system of fossil capital: "their money currency
does not consist of the precious metals, which are too common among
them for that purpose. The smaller coins in ordinary use are manufactured from a peculiar fossil shell, the comparatively scarce remnant of
some very early deluge, or other convulsion of nature, by which a species
has become extinct" (118). The detail suggests how geological and evolutionary knowledges inform the hollow earth genre. The vegetation that the
Vril-ya cultivate is not green, but "of a dull leaden hue or of a golden red,"
and the comparisons to lead and gold figure even agricultural production
in terms of underground mineral resources (37). There are lakes and rivers
of water underground, but also of "naptha," liquid petroleum. The "nether
landscapes" offer views of "vast ranges of precipitous rock" (47), and the
trees resemble "gigantic ferns," the major botanical source for coal (38).

Like miners underground, the Vril-ya are untethered from the solar
and lunar rhythms that dictate time on the surface of the earth. They keep
their underground realm lit at all hours because of a "great horror of perfect darkness," and they experience "no change of seasons" (101). Even their
vegetation grows according to no cyclical rhythm: "at the same moment
you see the younger plants in blade or bud, the older in ear or fruit" (101).
Although the Vril-ya are not an extraction-based society, their social
rhythms mirror extractive time in that they are disconnected from surface
cycles of time, seasons, and regeneration. Clarke reads the novel's sunless
world in the context of mid-nineteenth-century theories of the death of
the sun: "classical thermodynamics pulled the sun off its throne and put
the old orb out to work" (25), thus the novel wants to convey "that human

life could be possible without sunlight and coalfire" (48). For industrial-era readers, however, the idea of a sunless world had a more obvious parallel: the underground world of miners. Walter Morel in *Sons and Lovers*, discussed in chapter 1, must resort to a watch to know the time while working in the coal mine. The Vril-ya, "being excluded from all sight of the heavenly bodies, and having no other difference between night and day than that which they deem it convenient to make for themselves," utilize a different system of time entirely. The narrator, like a traveler unadapted to a new time zone, continues to resort to his own watch and "compute their time with great nicety" (103).[24]

The Coming Race's subterranean setting and its approximation of aspects of miners' lives allows the narrative to explore, in speculative fashion, what it means to have a society based in underground extraction. Mining is thus at the heart of the novel's setting and its envisioning of a social order built around its energy source. The conceit of an entire civilization—down to its language and political structures—emerging quite evidently as superstructure (Vril-ya) from an energy base (vril) suggests how Bulwer's speculative novel theorizes the determinative qualities of energy regimes. Vril is not a fossil fuel but is at least as indispensable and powerful a force for the Vril-ya as coal was for industrial Britain, and it mirrors coal's exhaustibility, at least indirectly: "this fluid, sparingly used, is a great sustainer of life; but used in excess . . . rather tends to reaction and exhausted vitality" (86). That overuse of energy leads to exhaustion is a calculation that holds true for vril as for coal, even though the stocks of vril are apparently unconstrained. Despite these allegorical correspondences, however, *The Coming Race* takes great care to place industrial energy regimes in the past and to stress vril's superiority over fossil fuels. The narrator learns, for example, that the Vril-ya are "familiar with most of our mechanical inventions, including the application of steam as well as gas" (59), but that such "vehicles and vessels worked by steam" are now housed in a museum. "Such were the feeble triflings with nature of our savage forefathers," one of the Vril-ya sniffs (92).

Many critical discussions of *The Coming Race* interpret vril as a metaphorical expression for psychological, linguistic, or political dimensions of the Vril-ya rather than as a fuel; such readings, while persuasive in their way, disregard energy as a systemic factor of wide import, suggesting the distance we have come in recent years toward recognizing the social significance of energy materialities.[25] Bulwer theorizes an even deeper relation between the Vril-ya and the energy source for which they are named, however: a Lamarkian inheritance through which energy use has adapted their very bodies. The Vril-ya carry a staff—something like a magic wand—that allows them to use vril at will, and they are biologically adapted to wield the staff effectively through a "visible nerve" that has evolved in their wrists and hands. As Zee, one of the narrator's primary informants in the nether world, explains, the nerve "has been slowly developed in the course of generations," and who knows but that "in the course of one or two thousand years" surface humans may be able to develop such a feature as well. This raises the question of whether the Vril-ya's adaptation to vril in any way allegorizes humans' adaptation to fossil fuels. Evolution has not yet molded us to our energy choices, at least in an obvious way, and extraction-based life remains a mere blip in the evolutionary timescale; the timescale Zee describes for the Vril-ya adaptation to vril reminds us, in fact, that the age of fossil fuels has been short enough to preclude overt biological adaptation. But in speculating that a species *could* evolve to better align with an energy regime, *The Coming Race* suggests two things: that humans have the potential to become ever more bound to subsurface fuel sources, and that our relation to fossil fuels is presently less fixed and more modifiable than we may realize.

Several of the texts I discuss in this chapter are similarly concerned with the question of evolutionary adaptation to an energy regime, but one distinct aspect of *The Coming Race* is that the Vril-ya do not actually resemble miners. Short stature was a benefit in extractive labor, as an illustration from *The Condition and Treatment of the Children Employed in the Mines* (1842) makes clear (see figure 3.1), and as George Orwell says in *The Road to Wigan Pier* (1937), "when the roof [of the mine] falls to four feet or less it is a tough job for anybody except a dwarf or a child" (23). Unlike the dwarves and hobbits of *The Hobbit* or the Morlocks of *The Time Machine*—subsurface creatures all—the Vril-ya are as "tall as the tallest men below the height of giants" (39). Subterranean life has not stunted their growth, and they even possess artificial wings that allow them to soar to underground heights in the chasms below. Their dissimilarity to miners, despite living underground, suggests the Vril-ya's evolution beyond extraction-based life, and the fact that the female Vril-ya are larger than the male conveys, too, their evolution beyond the gendered divisions of extractive labor.

The Coming Race is a labor utopia as well as an energy utopia, and it depicts, like the subtitle of *News from Nowhere*, "an epoch of rest," a post-carbon society that has transformed beyond the restless pace and alienating labors of extraction-based life.[26] Unlike Morris, however, Bulwer imagines liberation as freedom *from* work rather than pleasure *in* work. Anticipating Oscar Wilde's "The Soul of Man under Socialism" (1891),

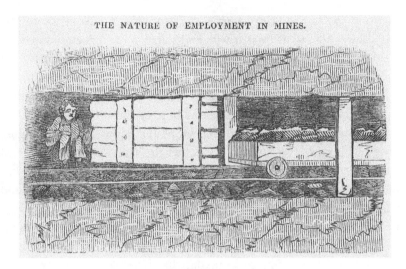

THE NATURE OF EMPLOYMENT IN MINES.

FIGURE 3.1. Illustration from *The Condition and Treatment of the Children Employed in the Mines* (London: William Strange, 1842), 41.

The Coming Race imagines utopian work as performed by machines and automata. The Vril-ya are "an indolent set of beings after the active age of childhood"; "they rank repose among the chief blessings of life" and "are not fond of rapid movements requiring violent exercise" (85). Nor do they need violent exercise: their energy system allows all that fossil fuels allow and more, without the work of extraction. Vril's capacities for terraforming, for example, allow the Vril-ya to "rend their way through the most solid substances, and open valleys for culture through the rocks of their subterranean wilderness" (59). Vril has also changed the climate, though the change is viewed positively: "many changes in temperature and climate had been effected by the skill of the Vril-ya," and "the agency of vril had been successfully employed in such changes" to the effect that "a temperature congenial to the highest forms of life could be secured" (70).[27]

As this invariable temperature suggests, *The Coming Race* suffers from the typical utopian complaint—lack of action—and Christopher Lane has called it a novel that makes us "heartily grateful for social acrimony" (617), but the torpor of the Vril-ya does not quite equate to peacefulness. Vril has the destructive power to decimate "a capital twice as vast as London" from a great distance, and in this sense, Bulwer's energy utopia is also an energy dystopia, one that anticipates nuclear power and weapons of mass destruction capable of instantaneous annihilation of entire population centers. Zee recounts a chilling legend of the Vril-ya, an origin story for their civilization that also prognosticates their future: "we were driven from a region that seems to denote the world you come from," Zee tells the narrator, "in order to perfect our condition," and "we are destined to return to the upper world, and supplant all the inferior races now existing therein" (88). The Vril-ya are not just "the coming race," then; they are the race that is coming for us. Describing a hypothetical conflict between a society of thirty million that lacks vril and fifty thousand of the Vril-ya, Aph-Lin, one of the Vril-ya host, says "it only waits for these savages to declare war, in order to commission some half-a-dozen small children to sweep away their whole population" (109). Thirty million approximates, not coincidentally, the number of people who lived in Britain (including Ireland) at the time this novel appeared.

Despite, or because of, its eschatological vision, Bulwer's novel struggles with the question of how to end, of how to cultivate narrative closure in a world of repose that yet awaits its future. Fittingly enough given its extractive setting, the novel resorts finally to a crisis of reproduction, one that emerges suddenly and dominates the last phase of the book. Zee falls in love with the narrator, and he fears annihilation if he is even suspected of returning her affections since he is an evolutionary threat to Vril-ya society. As Aph-Lin, Zee's father, says, "the children of such a marriage would adulterate the race" (130). The primary dysgenic risk the narrator poses is that he has not adapted the nerve to operate the vril staff, an "evolutionary gap" that separates him from Zee (Komsta 162). The novel's crisis thus turns on the unsuitability of the narrator's own energy ontology for social and biological reproduction. The drama of human extinction, hovering in the horizon of the future, is present in the novel's vista in more ways than one, and at the end of the narrative, writing years after his return from the nether world, the narrator reflects back on the "people calmly developing, in regions excluded from our sight . . . powers surpassing our most disciplined modes of force." He prays "that ages may yet elapse before there emerge into sunlight our inevitable destroyers" (168). If this drama of delayed violence and future human annihilation is inevitable, it remains for the future to contend with and thus structurally resembles narratives of environmental exhaustion.

"We Do Not Fight for a Piece of Diamond": "Sultana's Dream"

Jameson writes that "Utopias seem to be by-products of Western modernity, not even emerging in every stage of the latter" (11), but here I take up a utopian work that emerges from Bengali modernity—a feminist energy utopia that engages questions of gender and extractivism while exploring the imperialist dimensions of energy regimes.[28] "Sultana's Dream" was published in 1905, during an efflorescence of utopianism in the Anglo-American literary sphere, but appeared in a publication at the margins of that sphere: the *Indian Ladies' Magazine*, an English-language periodical from Madras, India.[29] Both Anglo and Indian women wrote for the journal, but a regular section called "Our Special Indian Lady Contributors' Column" always featured work by Indian women, and it was here that "Sultana's Dream" appeared (see figure 3.2).[30] The author, Rokeya Sakhawat Hossain, was a writer, activist, and champion of women's education in Muslim Bengal who grew up in an elite Muslim family where women lived in purdah and had little access to education.[31] She learned to read and write through secret lessons from her brother, Ibrahim Saber, and with his help married a man, Syed Sakhawat Hossain, who supported her intellectual pursuits (Ray 432). Hossain wrote "Sultana's Dream" with her husband's encouragement and went on, after his death, to found the Bengali Muslim Women's Association and the Sakhawat Memorial Girls' School.

The utopian world depicted in "Sultana's Dream" reverses the gendered social reality that structured Hossain's life and illustrates more completely than *The Coming Race* the feminist potential of post-extractive energy. In Ladyland, as it is called, men live in purdah while women govern and maintain public life, a change so dramatic from Hossain's reality that it has been taken as the fundamental novum of the story. "Sultana's Dream" has attracted a fair amount of attention from critics interested in Hossain's feminism and her early depiction of a sex-segregated utopia, a full decade before Charlotte Perkins Gilman's *Herland* (1915), but less discussed in the criticism, except as a relatively inconsequential novum, is the solar-powered energy system on which Ladyland runs.[32] Hossain's narrator, who visits Ladyland in a dream, first encounters and describes this energy transition when visiting a kitchen where "I found no smoke, nor any chimney either. . . . It was clean and bright. . . . There was no sign of coal or fire." When the narrator asks her guide (whom she calls Sister Sara) how the people of Ladyland cook, Sister Sara describes the solar power that has supplanted combustion and shows the narrator "the pipe, through which passed the concentrated sunlight and heat" (83).[33]

As the significance of sunlight in Hossain's worldbuilding suggests, this is a utopia of the air rather than the earth, in contrast to Bulwer's subterranean, sunless world of the Vril-ya. Not only does Hossain's narrator fly through the air in a Ladyland "air-car"; she even falls out of it when she wakes from her dream: "we got again into the air-car, but as soon as it

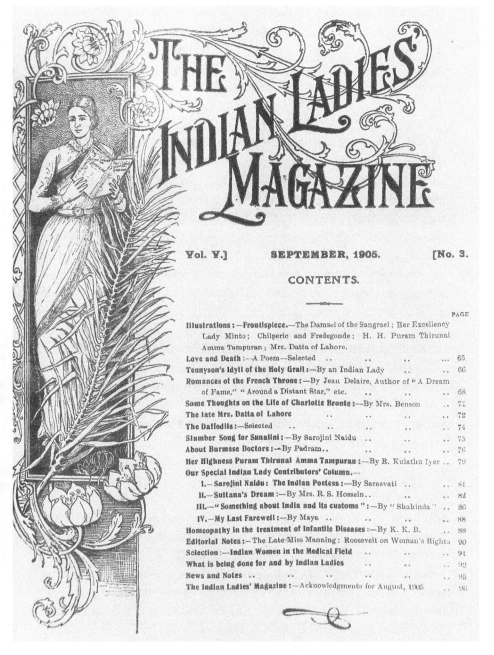

THE INDIAN LADIES' MAGAZINE

Vol. V.]   SEPTEMBER, 1905.   [No. 3.

CONTENTS.

FIGURE 3.2. Table of contents from the *Indian Ladies' Magazine*, September 1905.
©British Library Board, P.P.3778.bc.

began moving I somehow slipped down and the fall startled me out of my dream. And on opening my eyes, I found myself in my own bed room still lounging in the easy chair!" (86). The feeling of falling while dreaming, typically followed by a hypnic jerk, is a common one, and Hossain draws on the sensation to situate her utopia in the realms above rather than the realms below. The narrative is structurally bilevel as well: like many utopias, "Sultana's Dream" is premised on a dialogue between narrator and guide, which the narrator recounts in the course of exploring Ladyland; such a structure models dialogical thinking and the sharing of social perspectives across experiential divides. The narrator is amazed, for example, that the women of Ladyland have "invented an instrument by which they could collect as much sun heat as they wanted" and keep "the heat stored up to be distributed among others as required" (84). (Ladyland has not just solar power, one notes, but solar power with battery storage.) They use solar energy for cooking and heating and, apparently, for the electricity that powers their agriculture and "aerial conveyances."[34] Remarkably, these aerial conveyances, or flying cars, run not just on electricity but on "hydrogen balls . . . to overcome the force of gravity" (86), anticipating George W. Bush's ill-fated 2005 Hydrogen Fuel Initiative by exactly a century.[35]

In Hossain's story dreams are the immaterial channel through which new energy systems come into view, but the story also imagines energy systems as the subtending material basis for dramatic realignments of social, natural, and ideational relations. Sister Sara explains that Ladyland's energy arrangements are the direct result of the Queen's decree years ago "that all the women in her country should be educated" (83–84), and one of Ladyland's two women's universities developed the solar energy scheme, indicating the close connection between Ladyland's feminism and its energy revolution. At the time, Sister Sara explains, "the men of this country were busy increasing their military power" and scorned the women's solar power research, which they considered "a sentimental nightmare" (84). Solar power turns out to be less pacific than the men expect, however, for when the country was attacked by a neighboring kingdom, Sister Sara recounts, the women of Ladyland successfully deployed the sun as a weapon, which proved to be a turning point in gender relations. The women "directed all the rays of the concentrated sun light and heat toward the enemy," for whom the "heat and light were too much . . . to bear," and they "were burned down by means of the same sun heat" (85). Afterward, the Queen ordered the men of Ladyland to remain in seclusion, where they would henceforth stay in the *mardana*—a male version on the female *zenana* (82).[36]

"Sultana's Dream" offers a fictional escape from an extraction-based modernity: from its energy basis in coal, its masculinist labor practices, and its ecologically devastating impacts on the natural world. The escape is figured in the story's opening, when the narrator moves from a mine-like condition of cramped darkness, in purdah, to a sunlit, open world. The story starts at night, but the narrator awakens to morning in utopia, where she moves from the confined quarters of purdah to the bustling, sunlit streets of Ladyland. Disoriented, the narrator's initial experience mirrors miners' accounts of temporal confusion while working below ground, detached from the solar rhythms that organize life on the earth's surface. The outdoor world in which the narrator finds herself she can only compare to a garden, but Sister Sara says, "Your Calcutta could become a nicer garden than this, if only your countrymen wanted to make it so."[37]

Solar power is an apt foundation for a garden utopia, and while such an energy regime may have seemed like magic to some readers in 1905, it did have a theoretical basis in scientific discourse at the time and had been proven with the use of the first solar engines in the 1870s and the first solar panels in the 1880s.[38] In *Solar Heat: Its Practical Applications* (1903), published two years prior to "Sultana's Dream," the US writer Charles Henry Pope expressed hope that "some stir [would] come to inventors, to ingenious men, handy men, to anybody who can find out new ways of 'catching the sunbeams' and extracting gold from them" (22), since solar power, unlike fossil fuels, is "a pure gain to humanity. It subtracts nothing; the world will not be in the least impoverished to-morrow by the fullest use of [it] to-day" (153). Many who dabbled in early solar power technology were motivated, as this suggests, by fears of fossil fuel exhaustion and angst about a depletion-based energy regime. Since only a fraction of the sunlight that hits the earth is used in photosynthetic (energy-producing) processes, a desire to more actively exploit solar energy is long-standing (Wrigley, *Path* 7). A.S.E. Ackermann, in a long 1915 article entitled "The Utilisation of Solar Energy," attributes the need for solar research to "the obvious fact that there is a limit to our supplies of coal and oil" (540). Fifty years earlier, the author of an 1866 article titled "The Continuity of Nature" speculated about "future means of supplying heat," given "the prospective exhaustion of our coal-fields," and envisioned the emergence of something like solar imperialism: "As the sun's force, spent in times long past, is now returned to us from the coal which was formed by that light and heat, so the sun's rays, which are daily wasted, as far as we are concerned, on the sandy deserts of Africa, may hereafter, by chemical or mechanical means, be made to light and warm the habitations of the denizens of colder regions" (132). In an 1895 article titled "Harnessing the Sun," Peter Townsend Austen decries coal as "a wasteful source of energy" and "inconvenient" since it "has to be mined, freighted, and stored" compared to solar power (160). Extraction and exhaustion appear in this discourse as the primary negative values of fossil fuel.

Early writers on solar power often date it to Archimedes's use of mirrors to set fire to the Roman fleet when it attacked Syracuse in 212 BC, a legend that would seem to have inspired Hossain in depicting Ladyland's harnessing of the sun for military defense.[39] More recent histories of solar power trace its origins even earlier to passive solar constructions that required little in the way of technology.[40] During the industrial era, early devices intended to more actively exploit solar energy typically employed mirrors, water, or both, and the first patent for such a device was issued in London in 1854 (Pope 32). The first working solar-powered engines were developed a few years later by a Frenchman, August Mouchot, who demonstrated his reflector and pumping engine in France in 1875 (Pope 35), after taking out his first patent for a solar device on 4 March 1860 (Ackermann 543) (see figure 3.3). Mouchot's research was supported by the French government, which hoped to use his solar-powered engines in the deserts of Algeria, and in 1877 France sent him to Algeria to conduct solar power experiments on the ground (Pope 34–37).

Colonialism was, as this suggests, a key factor in the demand for solar power, given the difficulties of transporting British coal to the furthest reaches of the empire. Where coal was lacking, demand for wood as fuel for steam engines, and to build the railways that moved extracted material to the ports, led to the decimation of many a forest at the empire's frontier; in the early days of South Africa's Mineral Revolution, for example, "the only fuel available was native timber, and it is estimated that over a million large trees have been cut down to supply the Kimberley wood market. The whole country within a radius of a hundred miles has been denuded of timber, with the most injurious effects on the climate" (*Reunert's* 8). The demand for fuel was such that exorbitant prices were paid to import Welsh coal to South Africa until a local source was found near Johannesburg; today, this region's coal-fired power stations are responsible for the highest concentration of nitrogen dioxide pollution anywhere in the world.[41]

Industrialism was, then, only as portable as its fuels, and fuel provision and transport proved to be limiting factors in extractive imperialism. Frank C. Perkins wrote in *Scientific American* in 1908 that the development of an effective solar engine would mean that "one of the most perplexing

GRAND GÉNÉRATEUR SOLAIRE INDUSTRIEL
PAR PROF. A. MOUCHOT (1878)

FIGURE 3.3. Mouchot's solar engine, from Charles Henry Pope, *Solar Heat: Its Practical Applications* (Boston: Charles H. Pope, 1903), 41.

problems of tropical countries will have been solved," and a 1914 article in the *Journal of the Royal Society of Arts* makes the same point in the British context: "In the wide extent of the British Empire there are many places within or near the tropics, where there is a combination of excessive sunheat, scarcity of fuel, and aridity of surface," ideal conditions for the "economical" working of solar power ("Egyptian"). The British Raj was one place that seemed to offer prime conditions for solar power. A resident of Nadiad, India, wrote in *Scientific American* in 1908 that "my residence is in a tropical part of India . . . where fuel is scarce and dear," and that "in this part of the country (about 300 miles north of Bombay) there is a great opening for cheap power in small units" such as solar engines can provide (Bishop). William Adams, deputy registrar of the High Court of Bombay, was one of the first to make headway in this research area, and he published *Solar Heat: A Substitute for Fuel in Tropical Countries* in Bombay in 1878. Adams had engineered two solar-powered devices, a pump and a cooker, and received a medal from the Sassoon Institute of Bombay for his efforts (Ackermann 549).

In the end, steamships, railways, and oil would beat out solar power in the race to fuel the colonies, as coal became easier to move with new globe-girdling transportation infrastructure and as oil sourced outside Europe took on a new prominence in twentieth-century global capitalism, but Hossain's "Sultana's Dream" invites us to reconsider the early historical edge of solar power experimentation as well as its connection to imperial expansion and the infrastructures of fuel transport.[42] Imperialist ends motivated early solar experiments, but Hossain decolonizes solar power, imagining a women-ruled Indian utopia that harnesses the sun for its own sovereign aims. In "Sultana's Dream," we first encounter solar power in the kitchen, which may suggest Adams's Bombay solar cooker as an influence, but it is notable that Hossain positions solar energy in the feminine sphere of the home while also destabilizing the gender of such spheres altogether by substituting the mardana for the zenana. Though it is not clear how much solar experimentation Hossain was aware of and how much she intuited on her own, the plausibility of her energy novum distinguishes it from the more fantastical fuel imaginaries of *The Coming Race* or *News from Nowhere*.[43]

As in Bulwer's fictional world, labor demands in the post-extractive society of Ladyland are markedly reduced, countering one of the most obvious social benefits of fossil fuels: their promise to diminish hard manual labor. The women of Ladyland no longer work inside the home, having turned over domestic labor to the men, and outside the home they manage to run the country on two hours of work a day, leaving plenty of time for leisure pursuits.[44] When the narrator asks how this can be, Sister Sara explains that women, unlike men, don't "[waste] six hours every day in sheer smoking" (83), a joke that reminds us that not only is smoke from coal missing in this utopia, but so is its correlate: male labor outside the home. Because underground extractive labor was almost exclusively male at this time, it follows that a feminist utopia would be differently powered, and that a reversal of gendered power structures would require a post-extractive, post-smoke energy imaginary. "Sultana's Dream" does not directly explain the link between extraction-less energy and diminished male authority, but the contours of its utopian worldbuilding make the point clearly enough: extraction-less solar power reorganizes labor arrangements and hence also gender relations.

The gendering of extractive labor as male can be connected both to specific developments in the nineteenth century and to long-standing "cultural cosmologies [that] establish mining as the proper activity of men," as Jessica Smith Rolston writes. "Within a series of symbolic oppositions" that have a long history in many different cultures, she says, "orebodies (a well-defined mass of ore-bearing rock) and mine spaces are understood as female in contrast with the laboring bodies of masculine miners" (11). Designations like the "mother lode" of the California Gold Rush or the "Veta Madre" of Mexican silver mining convey an idea of the mine as "feminine and womb-like, enclosing the masculinized work of the male miners," or perhaps being violated by it (11).[45] Carolyn Merchant notes that the Reverend John Scott, for example, in his description of the California Gold Rush, "used the imagery of sexual assault to advocate mining the female earth for metals" (112). The masculinity of mining was reinforced, too, by a gendered division of labor that hardened ideologically with the rise of industrial capitalism, even though we can find many examples of nineteenth-century women mine workers (Rolston 10).

Female underground labor in the mines was legally prohibited in Britain following the 1842 Mines and Collieries Act, but the situation was less clear-cut in the British Raj, which was a British colony yet had its own legal structures. In Britain, the 1850 Coal Mines Inspection Act had required regular inspections (though, as noted in the previous chapter, the letter of this law was rarely honored), but in India under British rule, an inspector of mines was not appointed until much later, 1894. In his first annual report, this inspector, James Grundy, emphasized that in India there were no legal principles in place to address "the very important subject of child and female labour in the under-ground workings of mines" (Grundy 1). He said that in some Indian mines, women did work underground, though they were a minority; these women workers generally engaged in support tasks such as carrying coal or water rather than in direct extraction, cutting or hewing coal. With typically racist-imperialist reasoning, Grundy insists that such work "is natural to [the women], they appear to be born to carry loads on their heads, and it is not in any sense like the unnatural work of the female workers under-ground in England before the prohibiting Act was passed" (74).

Compared to other parts of the empire, India was not thought to be a particularly rich colony for mining, despite famed diamonds like the Koh-i-Noor, which was displayed at the Great Exhibition of 1851 shortly after being ceded to Britain as a symbol of India's natural wealth. In a report on India's Mineral Exhibits at the Colonial and Indian Exhibition of 1886, the *Mining Journal* grumbled that "the minerals of the Indian Empire at the Exhibition are both meagre and disappointing, especially when viewed by the light of the many wonderful contributions to the mineral resources of the British Empire made by some of the younger and less powerful dependencies" ("Mineral Exhibits"). Newer targets of British imperial exploitation such as South Africa were renowned for their underground mineral wealth and envisaged as perfect sites for extractive imperialism, but India was primarily valued as a plantation colony for tea, indigo, cotton, and more.[46] The *Mining Journal* went on to say that "the average Indian coal . . . is considerably inferior to English" and that India's "total output for 1883 was 1,315,976 tons" of coal, which we might compare to 142,385,416 tons of coal mined in England and Wales that same year (*Cork Industrial Exhibition* 53).[47] Most of India's coal was from Bengal, and "meagre and disappointing" though it may have been, it was in great demand for railways and industry on the subcontinent; by the mid-1890s, Bengali coal was not only being used domestically but was also being exported to Ceylon, China, Africa, and the West Indies (Grundy 110).

Such factors provide context for the solar power imaginary of "Sultana's Dream," a story that suggests how gendered social hierarchies and imperial ambitions are underlaid and augmented by the labor and energy demands of fossil capitalism. Energy arrangements in Hossain's story are, however, a matter of dreams as well as fuels, and Ladyland's post-extractivism is rooted in technological advancement as well as ideology, desire, and acceptance of the material limits of a given place. As the Queen explains to the narrator, Ladyland lives in peace because "we do not covet other people's land, we do not fight for a piece of diamond though it may be a thousandfold brighter than the Koh-i-Noor" (86). "Sultana's Dream" thus connects extraction not only to energy but to imperial war and Britain's violent race to the extractive frontier (particularly evident in the aftermath of the Second Boer War, when this story was published). Ladyland's relation to the natural world remains instrumental, as the "garden" ideal suggests, and in some respects it does not seem all that different from that of the West since Enlightenment: "We dive deep into the ocean of knowledge," the Queen says, "and try to find out the precious gems [that] Nature has kept in store for us" (86). An extractivist idea of earth's resources persists in the comment, and Fayeza Hasanat, for one, argues that "Sultana's Dream," despite its groundbreaking feminism, cannot extricate itself from a view of the natural world grounded in patriarchal disgust for generative materiality.[48] Hossain's utopia demands closer consideration from the perspective of the energy humanities, however, for its optimistic vision of the social improvements to be gained beyond extraction-based life and for its revelation of the ways that gender hierarchies are underscored by the labor and energy demands of fossil capitalism. In Hossain's feminist solar power utopia, literature's dream capacities offer a space for imagining how new energy formations might intersect with new gender formations.

"A Man from Another Planet": News from Nowhere

In the Kelmscott Press edition of William Morris's *News from Nowhere*, published two years after the novel's original 1890 serialization in the socialist newspaper *Commonweal*, red glosses at the top of each page summarize the action. A red gloss in chapter 9 reads "A man from another planet" (77). By locating this seemingly backward-looking agrarian utopia, printed on an archaic handpress, no less, in the futuristic realm of the "planet romance" subgenre, the gloss gives color and prominence to a running planetary motif in the novel, one that might be otherwise easy to miss.[49] William Guest is a visitor to Nowhere who arrives on the wings of a dream, wakes up in the socialist future, and tells his primary informant in the new world, Old Hammond, to speak to him "as if I were a being from another planet" (102).[50] Hammond goes on to refer to Guest as "a man from another planet" no less than seven times. Feeling himself a true alien in this future postcapitalist world, Guest reflects, "I really was a being from another planet" (177). His outlandishness is evident when he attempts to pay Dick, the first person he meets, with the nineteenth-century coins in his pocket, producing confusion since Nowhere has no concept of money. The coins have aged with the time travel, though Guest has not: "the silver had oxydised, and was like a blackleaded stove in colour" (60). The tarnished silver marks Guest as a visitor from the past age of extraction capitalism, come to a future imbued with "the spirit of the new days," the essence of which, as Hammond explains, is "delight in the life of the world; intense and overweening love of the very skin and surface of the earth on which man dwells, such as a lover has in the fair flesh of the woman he loves" (174). From a Victorian London tunneled through with underground railways, the smoky gloom of which are depicted in the opening pages of the novel, and from an economy premised on the value of extracted underground metals, Guest has arrived in a land devoted to the earth's "surface," "skin," and "flesh": a post-extractive utopia.[51]

Guest is not really from another planet, of course, but another time. *News from Nowhere* offers a vision of the future, imagining, like "Sultana's Dream," a better life through energy transition. In "Sultana's Dream" the earth's surface is exploited through the use of solar power; *News* is far less techno-utopian but is likewise an anti-extractive utopia that exalts in earthly surfaces and the downstream impacts of energy change.[52] If gender is the primary social vector of extractivism for Hossain, for Morris the vector is capitalism. *News* depicts a postindustrial, postcapitalist society that has practically done away with fossil energy altogether, to everyone's benefit. Having evolved beyond coal power, the residents of Nowhere live happily on their own steam with the help of water power, animal power, and a cryptic, clean-burning energy source that fills in the narrative gaps.[53] Like *The Coming Race*, which is more squarely premised on a fantastic fuel, *News from Nowhere* is another energy utopia that depicts "an epoch of rest," as the book's subtitle phrases it: a postcarbon future that has successfully revolted against frenetic modernity and the exploitative conditions of labor that inhere under industrial capitalism. Bulwer envisions a release from tedious manual labor through vril, but Morris imagines liberation as pleasure in work and bodily exercise, a pleasure enabled by social arrangements that promote unalienated labor.

How did this transformation of energy and labor come about? Nowhere has experienced an event described as "the great change in the use of mechanical force" (116), an event that parallels the society's other so-called great change—its socialist revolution. Unlike the socialist "great change," which is described in copious detail in a long chapter at the center of the novel, the "great change in the use of mechanical force" is mostly unarticulated. It seems to have entailed a widespread rejection of fossil fuels, for Hammond describes it as a change of "habits," a word that references consumer behavior but also the social forms that shape consumption. Mining still happens in Nowhere, but on an artisanal rather than industrial scale: "whatever coal or mineral we need is brought to grass and sent whither it is needed with as little possible of dirt, confusion, and the distressing of quiet people's lives" (116). Here Morris furtively abstracts Nowhere's residual extractive labor through the use of passive voice ("is brought to grass"), and abstracts coal and mineral consumption, limited though it is, as well ("is needed"). Such mystification of the materiality of energy and extraction is likewise evident with Nowhere's new energy source, a clean, mysterious power that propels the "force-barges" on the river. As Guest travels around Nowhere, he sees no coal combustion, but he observes the force-barges navigating the Thames "without any means of propulsion visible to me." He reasons that these "'force-vehicles' had taken the place of our old steam-power carrying," and Dick explains that Nowhere uses force-vehicles on land as well as water. Guest decides to inquire no further, however, sure that he "should never be able to understand how they were worked" (203). Here Morris relies on speculative fiction's capacities for obfuscation when he rubs up against a problem for which he has no answer—how to wean society from fossil fuel combustion. Bradon Smith has argued, however, that the significance of speculative fiction's capacity for "the envisioning of alternative energy systems" is not "that these alternatives need be plausible or even possible" but that they offer "an imaginative space in which to think through the experience of dramatic infrastructural, social, and environmental change" (140).

News from Nowhere positions the reader in a postcoal environment, offering sensual detail to convey what such an environmental change would feel like. Rowing up the Thames in the dark near Reading, for example, Guest surmises the transformation that must have happened in the area, for "everything smelt too deliciously in the early night for there to be any of the old careless sordidness of so-called manufacture" (204). In the opening frame of the novel, by contrast, when Guest leaves a London meeting of the Socialist League and takes the Victorian underground railway home, he "stew[s] discontentedly" on "that vapour-bath of hurried and discontented humanity, a carriage of the underground railway." Morris again uses the word "habit" to suggest how individual behavior is shaped by infrastructural forces, referring to the railway as "the means of travelling which civilization has forced upon us like a habit" (53). The train is heading toward Hammersmith, so we know it is the Metropolitan line, which was still using steam engines underground at this time; there were countless complaints about smoke and steam in the tunnels, and the line switched to electric carriages in 1906 (Halliday 22–23). Morris's novel thus begins with an underground setting, filled with coal smoke, a steamy hell of discontent. No wonder that when Guest wakes the next morning, on the surface of the earth in the socialist future, the absence of pollution is one of the first things he notices: "How clear the water is this morning!" (56), he says of the Thames. Dick, whom he meets at the riverside, is likewise "as clean as might be" (57). Looking down the Thames, Guest sees that "the soap-works with their smoke-vomiting chimneys were gone; the engineer's works gone; the lead-works gone" (58). Later, touring Nowhere, he passes a workshop where glass and pottery are made, and observes that there is no smoke coming from the kilns and furnaces. "Smoke?" Dick asks. "Why should you see smoke?" (94).[54]

News from Nowhere presents an environmental alternative to extraction-based life, and as these dispatches from the future suggest, Morris himself was well aware of the widespread impacts of coal pollution. In 1890, the same year that Morris wrote *News from Nowhere*, his friend and fellow socialist Edward Carpenter published an article titled "The Smoke-Plague and Its Remedy," which described coal pollution's damaging effects on the health and mortality of the poor. Carpenter blamed the profit motive for such atmospheric conditions: "After a hundred years of commercialism we have learned to breathe dirt as well as eat it" (204). Because of coal smoke, "the climate of England is not so fair as it used to be," Carpenter says, and "we have made it so" (207)—an amazing intuition of anthropogenic agencies and their uneven impacts. In the cities of England, he continues, "populations go their obscure way through dirty streets under a dirty sky" (206).[55] Like Carpenter, Morris traced the roots of the coal economy to the capitalist drive for productivity, and indeed he was an early adopter of the view that capitalism is fundamentally detrimental to the environment because it is premised on never-ending growth, growth that is impossible for natural systems to sustain.[56] Such a position departed from the fundamental assumptions of classical political economy, for the very idea of a self-regulating environmental balance was, as Fredrik Albritton Jonsson has shown, liberal-capitalist in origin: "Adam Smith and his successors in the classical liberal tradition . . . looked to the natural world for a model of self-regulating balance that justified their faith in market exchange" (125).[57] The key capitalist notion of "exchange," as Theodor Adorno argues in his essay "Progress," is similarly premised on the fantasy of a natural tendency toward homeostasis: "Exchange is the rational form of mythical ever-sameness. In the like-for-like of every act of exchange, the one act revokes the other; the balance of accounts is null. If the exchange was just, then nothing should really have happened, and everything stays the same." And yet, always, "the societally more powerful contracting party receives more than the other" (143).

Morris, like all nineteenth-century Marxists, recognized that the idea of free exchange obscured the market's remainders of profit and surplus value, but he also saw that an ideology of natural balance obscured environmental injury under capitalism. He saw no self-regulating balance in the market's interactions with the natural world and said in his 1883 lecture "Art under Plutocracy," "I tell you the very essence of competitive commerce is waste" (80). This suggests an environmental remainder, "waste," the leftover from the supposed equilibrium of capitalist exchange; in the socialist utopia of *News from Nowhere*, by contrast, "nothing is wasted" (119).[58] In his 1884 lecture "Art and Socialism," Morris reiterated that under "the grasp of inexorable Commerce . . . our green fields and clear waters, nay the very air we breathe are turned . . . to dirt; . . . under the present gospel of Capital not only there is no hope of bettering it, but . . . things grow worse year by year, day by day. Let us eat and drink, for tomorrow we die—choked by filth" (116). The idea of self-regulating balance under capitalism belies the gathering storm of ecological destruction that Morris, like Ruskin before him, saw coming, and their arguments resonate at a moment when the drive for growth at all costs has been responsible for stalling if not outright preventing political checks on the release of greenhouse gases into our atmosphere.

The overhanging threat of climate change bears, as we have seen, a historical relation to the enormous profits generated through the mining of subsurface resources during the industrial extraction boom, and Morris was deeply familiar with this mode of wealth generation since his family's fortune came from shares in the Devon Great Consols copper mine. As Charles Harvey and Jon Press explain in their history of the Morris family's mine holdings, the connection to Devon Great Consols began with Morris's father, who worked in finance, and his uncles, who were both involved in the coal trade. William Morris Sr. bought 272 out of 1,024 shares in Devon Great Consols (36), and one of its mines, Wheal Emma, was named in honor of his wife, William Morris's mother (35). The early Victorian period was "the apogee of copper mining in Devon and Cornwall" when "the problem of drainage had been solved through the application of powerful, fuel-efficient steam engines," and by midcentury there were over one hundred such mines in production in the region (33). Initially Devon Great Consols paid huge dividends, and when Morris Sr. died unexpectedly in 1847, his wife and children became dependent on the mine's fortunes (39). As each of her nine children came of age, Morris's mother gave them thirteen Devon Great Consols shares (Boos, *History* 22); the shares came to her eldest son, William, in 1855, and the wealth made it possible for him to devote himself to books and art (C. Harvey and Press 40).

Morris held shares in Devon Great Consols until 1877 and was a member of the mine's board of directors until 1875. He was also "persuaded to join the board" of British Mining and Smelting, Limited, "a highly speculative concern which failed in 1874" (42). As he moved toward a socialist analysis of society that would fully bloom in the 1880s, Morris withdrew from the Devon Great Consols board—evidently sitting on his top hat to mark the occasion (Thompson 192)—and divested himself of shares in the mine. By this time, however, it was clear that the mine's most profitable days were over. The copper supply at Devon Great Consols began to run out by the late 1860s, and the venture shifted to mining arsenic (a common by-product of copper mining). As Florence Boos and Patrick O'Sullivan recount, "by 1870, Devon Great Consols was already supplying half the world's arsenic," and by 1880 the mine's output of arsenic "overtook [its] copper production"

(15). The *Mining Journal* reported in 1876 that practically "the whole of the arsenic which is produced in the kingdom comes from Devon and Cornwall," and that Devon Great Consols is "the largest arsenic works, not only in England, but in the world." That it was a near-exhausted copper mine goes unremarked in this rosy account: "the fact that the two counties can furnish what is practically an inexhaustible supply [of arsenic] is one of the most cheering features of the year" (1 January 1876, 11).[59]

As a result of his long involvement in Devon Great Consols, Morris had firsthand experience of mining as what Boos and O'Sullivan call "an especially exploitative as well as extractive" form of capitalism (25). One of the singular aspects of the mining industry, stressed throughout this book, is that underground mineral resources do not regenerate, at least not on human timescales. A table from C. Harvey and Press charts the decreasing output of Devon Great Consols over the years (see figure 3.4). The exhaustive dynamic of mining, which greatly accelerated under industrial pressures, would have been obvious to Morris, primed as he was to see it thanks to his growing antagonism to capitalism, and so would the wide social, cultural, and environmental reverberations of this dynamic. Given that the mining industry was capable of generating enormous surplus value for owners and shareholders, that it was based on an extractive and depleting process, and that (in the case of arsenic mining) its effect was to bring poisonous material to the surface of the earth, Morris could hardly have had better preparation to write a post-extractive energy utopia like *News from Nowhere*. His early education in extraction capitalism, coupled with his political awakening, enabled a prescient attention to the damaging environmental effects of extractivism across his body of work.

News from Nowhere has been Morris's most lasting legacy to environmental discourse, but arguably his lectures and essays were more influential in his own day, and they are filled with condemnations of industrial pollution. In "The Lesser Arts," a lecture that was first presented in 1877, the year he divested from Devon Great Consols, Morris thundered: "Is money to be gathered? Cut down the pleasant trees . . . blacken rivers, hide the sun and poison the air with smoke and worse, and it's nobody's business to see to it or mend it: that is all that modern commerce . . . will do for us" (53). A few years later, Morris officially identified himself as "one of the people called Socialists" in his 1883 lecture "Art under Plutocracy," and again, pollution figured prominently in his case against capitalism: "To keep the air pure and the rivers clean . . . is it too much to ask civilization?" (63). "Whole counties of England, and the heavens that hang over them," he says, have "disappeared beneath a crust of unutterable grime" (64).

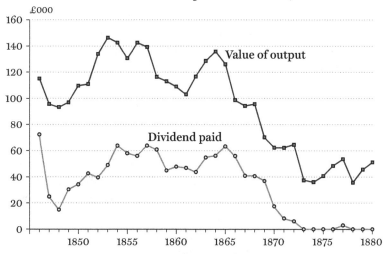

Devon Great Consols: Output and Dividends, 1845–80

FIGURE 3.4. "Devon Great Consols: Outputs and Dividends, 1845–80."
From *Art, Enterprise and Ethics: The Life and Works of William Morris* by
Charles Harvey and Jon Press (London: Frank Cass, 1996), 37.
Used by permission of the authors.

This unutterable grime came, of course, from beneath the surface of the earth, from the stored-up solar power of the Carboniferous era, from the extraction and combustion of coal. Morris had a great deal of fellow feeling for coal miners, whose hardships epitomized capitalist labor exploitation, but his aversion to extraction-based life was strong enough that he once told a group of Scottish miners, while on a propaganda visit to Coatbridge, "For myself, I should be glad if we could do without coal, and indeed without burrowing like worms and moles in the earth altogether; and I am not sure but we could do without it if we wished to live pleasant lives, and did not want to produce all manner of mere mechanism chiefly for multiplying our own servitude and misery" (qtd. in Glasier 81). His stance toward coal, and toward mining more generally, provides an environmental-political context for what we might call Morris's surface aesthetic: the oft-noted tendency toward exteriority and ornamentation in his writing, which Julie Carr describes as "Morris's depth problem" (151). Such an aesthetic prioritizes environmental materiality: in *News from Nowhere*, as noted above, Hammond tells Guest that the spirit of the age is "overweening love of the very skin and surface of the earth on which man dwells." He adds, tellingly, "all other moods save this had been exhausted" (174). *News* offers a juxtaposition between a society premised on the exhaustive process of mining and a society premised on a quasi-erotic—or, at any rate, fleshly and sensuous—relation to the earth's surface.[60] As Morris put it in his 1881 lecture "Art and the Beauty of the Earth," "surely there is no square mile of earth's inhabitable surface that is not beautiful in its own way, if we men will only abstain from wilfully destroying that beauty" (24).

Elsewhere we have seen the steam economy's easy absorption into narrative architecture, such that for Marlow in *Heart of Darkness*, the leaky pipes of the steamboat's engine register as mere "surface-truth," one of those "incidents of the surface" that distract us from confronting a deeper reality (44–45). But Morris, as a thinker, rejects the conceptual stratification of ideal and material that structures Marlow's account of reality here. *News* offers instead a surface world where all reality is spread before us in what we might call, to borrow Marlow's phrase, "surface-truth." Benjamin Morgan has argued that in Morris's work, "the plane of meaning recurrently collapses into the plane of materiality. This should be understood not as the elimination of higher-order faculties of thought or conceptual reflection but rather as a dialectical reinvestment of matter itself with conceptual force" (*Outward* 212). Such a collapse offers a particularly congenial representational environment for a post-extractive energy utopia—a utopia of earthly surfaces that relies on no subterranean energy base. We see a similar collapse of hierarchies of being in Hammond's comment that the "way of looking at life" in Nowhere is akin to "the spirit of the Middle Ages, to whom heaven and the life of the next world was such a reality, that it became to them a part of the life upon the earth" (175). Such a collapse is evident, too, in Clara's observation that what distinguishes Nowhere from the capitalist past is that humans no longer make the "mistake" of "looking upon everything, except mankind, animate and inanimate—'nature,' as people used to call it—as one thing, and mankind as another" (219).

Nowhere has reached, at once, the end of nature and the end of history. The planes of "nature" and "society" have collapsed together in the glorious aftermath of capital's terminal crisis. Less remarked, however, is the novel's depiction of the concomitant collapse of the extraction economy and the split-level world it created, a world where life above ground was enabled by deposits from below. The novel's climactic moment is Ellen's erotic cry to the surface of the earth: "O me! O me! How I love the earth, and the seasons, and weather, and all things that deal with it, and all that grows out of it. . . . The earth and the growth of it and the life of it! If I could but say or show how I love it!" (241). The shattering quality of her declaration can be gauged by the response it provoked from an early reviewer, who objected to such "a strong utterance" being made merely about the "luxuriant development of matter" (Hewlett 822). In his 1891 review, "A Materialist's Paradise," Maurice Hewlett griped that Morris "exaggerated the dependence of human nature upon its environment," and that "he is convinced that the conditions of human welfare are physical" (819). Morris's "cherished remedies" for social ills, in Hewlett's view, "are but skin-deep" (825). The review is clarifying in its very obtuseness, for "skin-deep" is exactly the point: the utopian remedy Morris prescribes is the pursuit of a single-plane society, a society of the surface built around our inescapable interdependence with the natural world that sustains us. *News from Nowhere*, as Hewlett says, "points to a deep joy in mere sensation, and to a deeper, vaster ignorance of what underlies it" (822). Read in light of the history of fossil capitalism, this "ignorance" of what underlies the surface seems more like a disentanglement from it: Morris's aesthetic realization of post-extractivism.

"Unpleasant Creatures from Below":
The Time Machine

The Time Machine, like *News from Nowhere* and other speculative texts of its era, begins with an anonymous narrator and a cumbrous frame narrative, estranging readers from the fantastic material of the main story to make it more fictionally credible through distance. Like *News from Nowhere*, too, *The Time Machine* is set mainly in the future, in Greater London, and concerns the social-natural relation. The parallels are no accident. Wells wrote his novel in direct riposte to *News to Nowhere*, and its opening lines even compare the Time Traveller to William Morris, designer of the famous Morris chair: the Time Traveller, too, has patented his own chairs, which "embraced and caressed us rather than submitted to be sat upon" (59).[61] Like Morris, the Time Traveller seems capable of imagining the betterment of human life through every conceivable means, from chairs to social organization. Such are the motivations behind the time machine he has invented. Like Guest in *News from Nowhere*, he hopes to see a future of social improvement: "What strange developments of humanity, what wonderful advances upon our rudimentary civilization, I thought, might not appear" (78).

The Time Traveller's friend, the Very Young Man, imagines more profitable uses for the time machine—"Just think! One might invest all one's money, leave it to accumulate at interest, and hurry on ahead!" (63)—but it turns out that speculative capital has disappeared somewhere between the late Victorian society of the Time Traveller and the future social world of the Eloi and Morlocks, and so, apparently, has extractive industry. When he visits the palace of green porcelain, the Time Traveller encounters a mining museum, which positions the work of industrial extraction in a historical past that the future will one day leave behind: "I went through gallery after gallery. . . . In one place I suddenly found myself near the model of a tin mine, and then by the nearest accident I discovered, in an air-tight case, two dynamite cartridges!" (132–33). The cartridges are dummies, meant, like the model tin mine, to evoke the past era of industrial extraction for the museum's erstwhile visitors.

The Time Traveller hails from a time of robust global extractivism, as is evident in the materials that make up the time machine: "Parts were of nickel, parts of ivory, parts had certainly been filed or sawn out of rock crystal" (68).[62] But the future world to which the Time Traveller travels initially appears to be a garden utopia not unlike that of "Sultana's Dream" or *News from Nowhere*—all surface and no sublayer. Those parts of the future world that are made from extracted materials, like the bronze pedestal and the marble sphinx, are greatly aged—"thick with verdigris" and "weather-worn" (80). The floor of the Elois' great hall "was made up of huge blocks of some very hard white metal . . . and it was so much worn . . . by the going to and fro of past generations, as to be deeply channelled" (85). A "vast structure" made up of "a great heap of granite, bound together by masses of aluminium," was now reduced to "derelict remains" (88). Aluminum was not produced on an industrial scale until the late Victorian period and was a cutting-edge material at the time of Wells's writing, but whatever novel forms of extraction and metallurgy were originally required to build this world of the future, they are now long over, and a "condition of ruinous splendour" holds sway (87).[63]

Of course, this surface view of a future without sublayer proves to be mistaken. An infrastructure of wells—or what seem to be wells—reveals the first glimpses of the world underground that subtends and supports this garden utopia: "a peculiar feature, which presently attracted my attention, was the presence of certain circular wells . . . of a very great depth" (100). Meanwhile, the disappearance of the Traveller's time machine similarly conveys a "sense of some hitherto unsuspected power" (95) at work in this world. What is the underground, invisible source of power that is holding Eloi society together? It is not coal, though the structural parallels with Wells's own coal-based society are evident. "Putting things together," the Time Traveller determines there must be "an extensive system of subterranean ventilation" beneath him (100); this is not it, but he is moving closer to a revelation of the hidden power underground. The narrative progresses in the manner of the scientific method, with the Traveller formulating a hypothesis to explain the future world, which is subsequently proven or disproven by further experience, exemplifying Clarke's argument about the parallels between scientific speculation and the work of allegorical interpretation that speculative fiction often demands. That the Time Traveller's speculation here concerns the presence of a subterranean social layer, subtending and supporting this future society, provides through the medium of the novel's setting a spatial figuration of the multilayered narrative of speculative fiction.

Eventually the Time Traveller realizes that the future world is inhabited not only by the Eloi but by a second species, also descended from humans, who live underground—the Morlocks.[64] Looking like something out of Dr. E. H.'s report "Subterranean Caverns, and Their Inhabitants," discussed earlier, these "unpleasant creatures from below" (113) have "the bleached look common in most animals that live largely in the dark—the white fish of the Kentucky caves, for instance," and their "large eyes" show the "extreme sensitiveness of the retina" that comes from living without light (108–9). That the Morlocks are figured as the descendants of miners has been noted by many a reader, perhaps starting with an early review in the 13 July 1895 issue of the *Spectator*: "the race of the underworld,— *the race which has originally sprung from the mining population,*—has developed a great dread of light, and a power of vision which can work and carry on all its great engineering operations with a minimum of light" (Hutton 263, my emphasis).[65] The Time Traveller comes to the conclusion that "the earth must be tunneled enormously" beneath him and that "it was in [the Morlocks'] artificial Underworld that such work as was necessary to the comfort of the daylight race was done" (109), parroting a common industrial-era discourse about coal mining.

Determining to descend one of the shafts to visit the world below, the Time Traveller lowers himself "in the throat of the well." Looking up, he sees his Eloi friend Weena's "agonized face over the parapet," looking down at him (114). The moment reverses a visual motif we have observed in such texts as *Hard Times* and *Jane Rutherford*, where the focal character looks down the mining shaft at a person or people who lie below; here, the Time Traveller looks up from the interior of the shaft to see Weena's face framed by the firmament above: "Glancing upward, I saw the aperture, a small blue disk, in which a star was visible, while little Weena's head showed as a round black projection" (115). The moment echoes the experience of Axel, narrator of *Journey to the Centre of the Earth*, in making his initial descent down a volcanic shaft—"Lying on my back, I opened my eyes and caught sight of a brilliant object at the other end of the 3,000-foot-long tube, converted into a giant telescope. It was a star" (89–90)—indicating *The Time Machine*'s debt to that novel's hollow earth imaginary.

Above ground is Weena, a star, and the firmament; underground there is "the thudding sound of a machine below" (115). Wells's figuration of a split-level world allegorizes energy as well as labor, for the Time Traveller's recognition of the crucial work that takes place below ground parallels such revelations of the miner's labor in texts like George Orwell's *The Road to Wigan Pier*: "The machines that keep us alive, and the machines that make the machines, are all directly or indirectly dependent upon coal. In the metabolism of the Western world the coal-miner is second in importance only to the man who ploughs the soil. He is a sort of grimy caryatid upon whose shoulders nearly everything that is *not* grimy is supported" (Orwell 18). In Wells's novel, it seems the ugly Morlocks serve as "grimy caryatid" for the "*not* grimy" Eloi. A caryatid is a female figure used as architectural column or support. Underground mining labor was an entirely male occupation in Britain at this time, as discussed above in relation to "Sultana's Dream," but the image of the miner as caryatid conveys the kind of spatial relation often evoked in accounts of mining labor: a relation where the invisible work of the "unpleasant creatures from below" provides the necessary base for the carefree pleasures of life aboveground.

While underground, the Time Traveller describes feeling "as a man might feel who had fallen into a pit" (119)—much like Stephen Blackpool, perhaps—indicating, again, how Wells's novel figures this underworld as a mine or pit, a space of extractive labor. Taken out of context, Orwell's description of the coal mines of northern England and their relation to the surface world above could easily describe the relation between the Morlocks and the Eloi: "Watching coal-miners at work, you realise momentarily what different universes people inhabit. Down there where the coal is dug it is a sort of world apart. . . . Yet it is the absolutely necessary counterpart of our world above" (29). The parallel between Orwell's and Wells's texts suggest how extractivism shapes discourse and literary form, which in turn shape extractivism by changing the way we think and talk about it, even across genres. Undergirding Wells's secondary world, then, is a conception of subterranean settings as wholly separate sites of socially necessary labor and the fuel on which society runs; Wells goes on to upend this relation, however, as he traces the Time Traveller's efforts to grasp the future energy system he encounters. Struggling to understand what he sees below ground, the Time Traveller reflects on the "tendency to utilize underground space for the less ornamental purposes of civilization. . . . Evidently, I thought, this tendency had increased till Industry had gradually lost its birthright in the sky. . . . It had gone deeper and deeper into larger and ever larger underground factories. . . . Even now, does not an East-end worker live in such artificial conditions as practically to be cut off from the natural surface of the earth?" (109–10).[66] Contributing to this dynamic, the Time Traveller presumes, is the long history of enclosure and the privatization of the commons, "the closing," in the interest of "richer people," of "considerable portions of the surface of the land" (110).[67] Here the dynamic between miner-like underground workers and surface enclosure of land produces an almost perfect bilevel allegory of capitalism: "in the end, above ground you must have the Haves, pursuing pleasure and comfort and beauty, and below ground the Have-nots; the Workers getting continually adapted to the conditions of their labour" (110).

This turns out, however, not to be the whole story, for the Morlocks' subterranean society will eventually prove to be in parasitic relation to the surface world above. The fuel is above-ground, not below: the Eloi are food for the Morlocks, providing energy in the form of calories. Underground, the Time Traveller sees not only "great shapes like big machines," and "dim spectral Morlocks," but also "a little table of white metal" laid with "a red joint" (116). The meat gives him "a vague sense of something familiar" (120), and finally he realizes that the "Eloi were mere fatted cattle, which the ant-like Morlocks preserved and preyed upon" (125). In a strange evolutionary twist, the descendants of miners have become herders of cattle—a grotesque pastoral. These cattle, or Eloi, live simple lives above ground, unaware of the deeper relations that structure their existence. As the Time Traveller reflects, "I understood now what all the beauty of the over-world people covered. Very pleasant was their day, as pleasant as the day of the cattle in the field. . . . And their end was the same" (141). Here, Wells's countering of Morris's "epoch of rest" is evident enough: "There is no intelligence where there is no change and no need of change" (141). The unenterprising Eloi have become mere biomass.

Wells's Morlocks, who have adapted to life below ground, are thus figurative descendants of industrial-era coal miners as well as evolutionary victims of extraction ecology. Recalling the discussion in chapter 1 of miners as "a race apart"—a people with their own language and even, it was thought, their own forms of physical inheritance—Wells need not go far from the conventional discussion of nineteenth-century miners to a speculative narrative about the evolutionary effects of the "conditions of underground life," conditions that have here stunted human stature as well as human virtue (110).[68] Like the dwarves in *The Hobbit*, the Morlocks could be said to descend from ancient folk types of small chthonic creatures, but both Wells and Tolkien reimagine such types in light of evolutionary theory and industrial extraction. In his essay "The Mesh," Timothy Morton describes the consequences of evolution for the idea of species in ways that are helpful here: "when you take an evolutionary view of Earth, an astonishing reversal takes place. Suddenly, things that you think of as real—this cat over here, my cat . . . —become the abstraction, an approximation of flowing, metaphorical processes" (19–20). It turns out that "the real thing is the evolutionary process—the cat is just an abstraction" (20). *The Time Machine* effects this kind of evolutionary paradigm shift with respect to the human.[69] Humans are but a temporal instantiation of a process that may eventually lead to Morlocks and Eloi, and Morlocks and Eloi, too, are just temporal instantiations of an evolutionary process that extends before and beyond them. Evolution, as Morton says, "undermines the idea . . . [of] rigid boundaries between species" (20), which is part of the point of Wells's book, as is the point that there is no correct temporal standpoint from which to define a species: "Wherever we look, up close or far away, over very short or very long timespans, we fail to find a point, goal, origin, or terminus in the process of evolution" (Morton 22).

We might say that Wells's novel encourages us to think about species and speciation in four dimensions: to imagine species as dynamic and continually evolving through time rather than as fixed forms. But the novel also encourages us to think about the impacts of energy on human development, for the Morlocks' adaptations are the direct result of their underground labors, echoing the kind of biopolitical speculation about mining communities and reproduction that I discuss in chapter 1. A species exists but for a moment in the long stretches of time that form the durational structure of Wells's narrative, but his novel suggests that the impacts of industrial mining may last for hundreds of thousands of years in the evolutionary record, not to mention the geological record. What are the long-term impacts of mining, not just in generational time (e.g., exhaustion) but in evolutionary time? Wells offers a nightmarish response to this question by imagining miners evolving into a predatory subterranean species that preys on the consumers of their labor who inhabit the surface above.[70]

The future impacts of extraction-based society were widely discussed in Wells's time, as we have seen, but typically these were only imaginatively projected for a few hundred years out. How will extraction-based society have fared by AD 802,701, a date that, as Aaron Rosenberg points out, "is unfathomably distant, some 160 times longer than the roughly five thousand years of recorded human history" (79)? Southwest suburban London, in Wells's future imaginary, will have a much warmer climate (106), for one thing, and will be subject to forest fires due to dry foliage and a lack of rain (136). This sounds comparable to aspects of life under climate change, and indeed MacDuffie has elaborated on the ways that *The Time Machine* anticipates the forms of environmental repression that climate change demands of us ("Charles" 556–58). Discussing Wells's depiction of "the inevitable extinction of the species, life, the planet itself" (558), MacDuffie zeroes in on the problem the novel poses of how to live a meaningful life when one is witnessing, as *The Time Machine* puts it, "the life of the old earth ebb away" (147). Within the narrative scale of evolutionary time rather than planetary time, however, the novel also poses nearer-term questions about the social organization of energy and about how to live life on the surface once one is aware of the necessary support provided by the invisible caryatids below. Extraction ecologies demanded acrobatics of denial, in other words, even before rising awareness of the climate crisis: denial of the labor of others and of the ruin of sacrifice zones.[71]

At the end of Wells's narrative, the Time Traveller invites his listeners to hear his story as they will: "Take it as a lie—or a prophecy. Say I dreamed it in the workshop. Consider I have been speculating upon the destinies of our race, until I have hatched this fiction" (151). Articulating the connections between dreams and speculative fictions with which this chapter began, the Time Traveller initially seems to call into question the reality of his experience, but one look at his time machine, "squat, ugly, and askew, a thing of brass, ebony, ivory, and translucent glimmering quartz" is enough to remind him that "the story I told you was true" (152). The materiality of the time machine, especially its basis in extracted materials, counterbalances the dreamlike, prophetic qualities of the narrative, and the interpenetration of the material and the prophetic is at the heart of the story the narrative tells, for the Time Traveller, as the anonymous narrator explains in the epilogue, "saw in the growing pile of civilization only a foolish heaping that must inevitably fall back upon and destroy its makers in the end" (155).

The Time Machine offers a pessimistic prognosis for the distant future in a universe of diminishing energy, as a number of critics have discussed, but also a nearer-term pessimistic prognosis for an extraction-based society premised on oppressive underground labor. Wells's first big hit, the novel is an early expression of what would emerge in his career as an enduring preoccupation with resource exhaustion. Three years later, in *The War of the Worlds*, for example, he would depict a Martian species forced to search beyond the frontier of its planet for a means of life: "That last stage of exhaustion, which to us is still incredibly remote, has become a present-day problem for the inhabitants of Mars. . . . To carry warfare sunward is, indeed, their only escape from the destruction that generation after generation creeps upon them" (8–9). The Martians' attack on Earth is carried out with energetic and extractive technologies such as a heat ray and an automatic machine for mining and processing aluminum (132–33).[72] Later, in *The World Set Free* (1914), Wells would draw on the new imaginative resource of radioactivity to envision what MacDuffie describes as "a future of plenitude" enabled by the discovery of radium and the new frontier of nuclear energy (*Victorian* 247).[73]

The Wells novel that is most relevant to my reading of *The Time Machine*, however, is *Tono-Bungay* (1909), a semiautobiographical hodgepodge that draws on elements of all the genres featured in this book—provincial realism, adventure narrative, and speculative fiction—and, indeed, exemplifies the arguments I have made about all three genres.[74] *Tono-Bungay* positions extractivism at the heart of the capitalist project, but also moves the energy extraction narrative out of the fossil fuel underworld and into the terrain of radioactive energy. As Michael Martel has described, key discoveries around radium in the first decade of the twentieth century suggested the possibility of an infinite source of self-generating energy, and hence a possible solution to fossil fuel exhaustion (155); the radioactive narratives discussed by Martel deploy such possibilities to challenge the fin de siècle entropy narrative and its scientific bases, but *Tono-Bungay* instead positions radioactive energy firmly within the exhaustion narratives of fossil capital, emphasizing the socio-environmental liabilities of extraction more generally—not just fossil fuel extraction.

Early in his career, *Tono-Bungay*'s protagonist, George Ponderevo, observes with "young, receptive, wide-open eyes" the filth of the coal economy when he sees a valley "all horrible with cement works and foully smoking chimneys" (47) and "barges and ships" that fuel the works (47–48): "When I saw colliers unloading, watched the workers in the hold filling up silly little sacks and the succession of blackened, half-naked men that ran to and fro . . . the stupid ugliness of all this waste of muscle and endurance came home to me" (48). His skepticism provoked by this recognition of the intensive labor and infrastructure of the coal economy, George will

grow up to engage, ultimately, in a new kind of energy extraction adventure: a journey to Africa to steal a pile of a fictional radioactive substance
called quap. The energy source may be new, but the means of obtaining
it is not. Mirroring the imperial extraction narratives discussed in chapter 2, George and his uncle establish "the London and African Investment
Company" (221) with the goal of procuring a "great heap of quap that lay
abandoned or undiscovered on the beach behind Mordet's Island among
white dead mangroves and the black ooze of brackish water" (223). The
dead trees and black ooze recall the blighted landscape of the Azuera Peninsula in Conrad's *Nostromo*, but here they speak to the possibly "*cancer-
ous*" effects of radioactive material (329, Wells's italics). Quap is, George
says, "the most radioactive stuff in the world," "a festering mass of earths
and heavy metals, polonium, radium, ytterbium, thorium, cerium and
new things too. There's a stuff called Xk—provisionally" (224). Translat-
ing the advanced energy frontier of radioactivity into the extractive colonial adventure tale, Wells's quap comes not from a laboratory but from a
remote beach in Africa where it lies in heaps "for the taking!" (225).[75] In
an astonishing chapter titled "How I Stole Heaps of Quap from Mordet
Island," George makes off with these phosphorescent heaps that glimmer
with a "faint quivering luminosity" (327) and kills an African man in the
process. The radioactive quap burns right through his boat on the way
home, however, and sinks, with the ship, to the bottom of the Atlantic,
leaving George and his uncle to bankruptcy.

Tono-Bungay speaks to Wells's continuing interest after *The Time
Machine* in energy systems, energy exhaustion, and the networks of
finance and empire in which they are strung. It speaks, too, to the growing
impact of radioactive and protonuclear fictions in early twentieth-century
speculative fiction—a tendency that we will also see in J.R.R. Tolkien's *The
Hobbit* (1937). Jessica Hurley, in a recent discussion of nuclear waste and
realist narrative, argues that since "the threats of radioactive materials are
impossible to contain in specific times and places," "radioactive materials
push the probabilistic epistemology" of realism "beyond the limits of its
reach" (783). Although focusing on the specific environmental calamities
of uranium mining and nuclear waste in post-1945 United States, Hurley's
comments on radioactivity and the antirealist imagination are relevant to
early twentieth-century speculative fiction too. For Tolkien, a veteran of
World War I who wrote the earliest parts of his legendarium while conva-
lescing from the Battle of the Somme, was well attuned, like Wells, to the
apocalypticisms of the age.

"Riddles in the Dark": The Hobbit

In assessing the work of a writer like J.R.R. Tolkien, it can be difficult to separate out a thematics of extractive exhaustion and energy depletion from a broader attraction to eschatology and timescapes of diminishment and decay. As Anna Vaninskaya notes, "time-as-process in Tolkien is a winding down and a running out" (14), and as Michael Saler writes, Tolkien "believed modernity represented the nadir of [an] ongoing process of historical decline" (165). But three key features of *The Hobbit* (1937) suggest its specific affective and epistemological debt to energy extractivism, in tune with other speculative fictions from the post-1870 period: the prominence of underground settings; the depiction of creatures adapted, like miners, to subsurface life; and the narrative role of object agency, channeling new understandings of matter that were emerging with the study of atomic energy as a potential post-extractive power source.[76] Tolkien himself disliked allegorical interpretations of his work and used an extractive metaphor to describe the folly of critics who use fairy stories "not as they were meant to be used, but as a quarry from which to dig evidence" often leading to "strange judgements" ("On Fairy-Stories" 119). To read *The Hobbit* as part of a cultural matrix dedicated to mediating the impacts of quarries, mines, and large-scale extraction would perhaps seem a strange judgment from this perspective, and yet Tolkien also said that "Fantasy is made out of the Primary World" and that "a good craftsman loves his material." For this reason he first encountered through fairy stories "the wonder of the things, such as stone, and wood, and iron" ("On Fairy-Stories" 147). The wonder that accrues to underground things in *The Hobbit* powerfully exemplifies literature's energy imaginaries and extractivist forms in the industrial era.[77]

Like William Morris, H. Rider Haggard, and other authors in this study, Tolkien's family and financial history were tied up with the industrial extraction boom. He was born in 1892 in Bloemfontein, South Africa, where his father had moved from Birmingham, England, to take a position in the Bank of Africa. At this time "the Orange Free State of which Bloemfontein was the capital was becoming an important mining region with new gold and diamond discoveries encouraging investment from European and American venture capitalists" (White 11), and "the British-controlled Bank of Africa had been deeply involved" in funding the diamond and gold mines (Gorelik 4). At age three, Tolkien left South Africa with his mother, Mabel, to visit England, and during this visit his father died suddenly back in South Africa at age forty, leaving the young family in difficult straits: his father's "capital had been invested in Bonanza Mines but the dividend provided Mabel the sum of only thirty shillings a week, which in 1896 was barely sufficient to eke out a modest existence" (White 18). Tolkien never returned to South Africa, but stories from his early childhood there were often recounted in the family circle, providing, perhaps, a sense of how our destinies are shaped by the pursuit of extractible wealth.[78] Certainly such a sense pervades *The Hobbit*.

As a fantasy novel, *The Hobbit* differs in some ways from the other speculative works in this chapter, and it may be helpful to identify these differences before positioning the story within the critical framework I have proposed. Vaninskaya describes "fantasy" as an ambiguous term that did not emerge as a common generic designation until the 1960s, but Tolkien uses it more or less interchangeably with "fairy-stories" in his influential 1939 lecture/essay "On Fairy-Stories."[79] Like the other texts discussed in this chapter, *The Hobbit* depicts what "On Fairy-Stories" calls a "Secondary World"—thereby coining a phrase—but its secondary world is more firmly detached from the "Primary World" than others we have seen.[80] In *The Coming Race*, the underground realm of the Vril-ya is spatially distant from but still continuous with the primary world above; Ladyland in "Sultana's Dream" exists in a dream, yet seems to envision what the primary world could look like in the future; *News from Nowhere* and *The Time Machine* are both clearly set in the future, in secondary worlds meant to be seen as temporally continuous with the primary world. Middle-earth, by contrast, is "outside Time itself" ("On Fairy-Stories" 129). A novel like *The Time Machine*, Tolkien says, exists in part "to survey the depths of space and time" (116); such magnitudinous aims motivate fantasy too but cannot be satisfied by the device of a machine because of the more complete autonomy of the secondary world. Tolkien also excludes from fantasy "any story that uses the machinery of Dream . . . to explain the apparent occurrence of its marvels" (116). The dream device is deployed or partially deployed in three of the texts discussed in this chapter, but fantasy, Tolkien says, aspires to a state of "Enchantment" rather than dream, to a secondary world with no explicable connection to the primary world (143).

Despite fantasy's more scrupulous separation of worlds, it has served, as Saler argues, a similar "social function" as do genres like science fiction and utopia of "challeng[ing] normative outlooks and advanc[ing] alternate possibilities to the status quo" (192), thus I group all these worldbuilding genres together as speculative fiction. All secondary worlds provide the grounds for reimagining energy, environment, and society, but one distinction between science fiction and fantasy often emphasized by critics who want to disambiguate their cognitive effects is the presence of super-natural phenomenon in fantasy's secondary worlds. Jameson, for example, calls magic "the fundamental motif of fantasy" and says that "if SF is the exploration of all the constraints thrown up by history . . . then fantasy is . . . a celebration of human creative power . . . [and] the omission of . . . those material and historical constraints" (66). This might suggest that fantasy can offer only energy mystification, but in my view the supernatu-ral elements in *The Hobbit* such as Smaug the dragon or the ring of power contribute to energy and extractive dreamwork in ways that correspond with vril, solar power, or force-barges, and they prompt similar recogni-tion of energy and extraction as nature-culture systems.[81]

We can complicate the SF/fantasy binary with a third term, too, in that *The Hobbit* is also a work of children's literature, and the line between possible and impossible has always been more permeable in literature for children. Tolkien originally wrote *The Hobbit* for his children, and though he claimed that as a child reader *Treasure Island* had "left [him] cool," in *The Hobbit* we can see the influence of Stevenson's treasure-hunting adventure novel, even down to the inclusion of a treasure map ("On Fairy-Stories" 134).[82] We can also see the direct influence of Morris, a writer whose impact on Tolkien was immense.[83] The effects of *The Time Machine* on Tolkien's writerly imagination can be seen in his frequent references to it in "On Fairy-Stories," and another key influence was Edward Augus-tine Wyke-Smith, a British mining engineer whose *The Marvellous Land of the Snergs* (1927) featured a race of small people. As these examples suggest, extractive literature and Medieval literature together produced *The Hobbit*: Tolkien took the name "Middle-earth" from an Old English poem (Anderson 4), and it conveys the layered conception of reality that underlay heaven-earth-hell cosmogonies but was also repurposed in the era of industrial extraction, as we have seen, to figure a bifurcated reality where life in the critical zone is supported by necessary stocks from the subsurface.

The prevalence of underground settings in speculative fiction from the industrial era, including *The Hobbit*, suggests how that subsurface had come to rival the forest in the role of menacing space at the edge of the known, just as the underground had supplanted the forest as the primary site of energy stores. Going into the forest for fuel was a fraught undertak-ing in folktales and fairy tales of the past: in Jacob and Wilhelm Grimm's version of "Hansel and Gretel," for example, the children's stepmother first lures them into the woods by taking them to the forest to collect firewood (Grimm 143). With extraction-based life, the underground becomes a similarly eerie setting stocked with the fuel to power human life—fuel that is dangerous to gather—and the world becomes divided along a binary not of forest versus civilization but of underground versus surface.[84] In *The Hobbit*, such a bifurcated spatial imaginary has even been absorbed into idiomatic language, as when Thorin the dwarf asks, "Now what on earth or under it has happened?" (295).

The significant role of the underground in *The Hobbit* is evident from the first line of the novel: "In a hole in the ground there lived a hobbit" (29).[85] Bilbo Baggins's hole is "not a nasty, dirty, wet hole," but "a very comfortable tunnel without smoke" (29). *The Hobbit* was published just six months after Orwell's *The Road to Wigan Pier*, and if the prominence of subsurface settings in both works reflects the preoccupations of an extraction-based society, the pointed lack of smoke in Bilbo's hole only underscores this vexed correspondence between subterranean Middle-earth and the real-world fossil economy. Bilbo's hole is one of a variety of underground settings in the novel that run the gamut from cozy to ter-rifying: the troll cave, the goblin tunnels under the mountain, Gollum's underground lake, the wood elves' underground palace, and Smaug's lair. In many of these settings dwell creatures specially adapted to life under-ground, like the Morlocks in *The Time Machine* or the creatures described in the *Mining Journal* article "Subterranean Caverns, and Their Inhabit-ants," discussed above. In Gollum's underground lake, for example, live "fish whose fathers swam in, goodness knows how many years ago, and never swam out again, while their eyes grew bigger and bigger and bigger from trying to see in the blackness" (118). Gollum is also "a small slimy creature" with "two big round pale eyes" who "lurk[s] down there, down at the very roots of the mountain" (118–19). The dwarves, who live to mine, are well-adapted to underground life: "We like the dark" (46); and the trolls, whom Bilbo and the dwarves encounter early on, are creatures of the earth who "must be underground before dawn, or they go back to the stuff of the mountains they are made of" (80). The goblins, too, live in the "deep, deep, dark, such as only goblins that have taken to living in the heart of the mountains can see through" (107). Peopled with all these chthonic beings, *The Hobbit* generates a subterranean evolutionary imagi-nary like that of *The Coming Race* or *The Time Machine*, raising similar questions about biological adaptation and the fossil economy.

In its various subterranean scenes, *The Hobbit* cultivates the feeling of underground life where one is alienated from the primary organiza-tional structure for human time—the solar day. As the wizard Gandalf says, "You lose track of time inside goblin-tunnels" (142). In Smaug's cave,

Bilbo gets lost in the long night of underground time: "two nights and the day between had gone by . . . but Bilbo had quite lost count, and it might have been one night or a week of nights for all he could tell" (299). Not just time, but perception and experience change underground. In the "Riddles in the Dark" chapter, just before Bilbo happens upon the ring of power, he "opened his eyes, [and] wondered if he had; for it was just as dark as with them shut. . . . He could hear nothing, see nothing, and he could feel nothing except the stone of the floor." Deprived of his senses, he will soon make an amazing transformation, with the help of the ring, from blindness to invisibility: "suddenly his hand met what felt like a tiny ring of cold metal lying on the floor of the tunnel. It was a turning point in his career, but he did not know it" (115). Bilbo's unknowing appropriation of an underground resource with environment-changing capacities—a golden ring capable of producing the sensory condition of the underworld (invisibility) in the sunlit world above—mirrors the extractive and energetic dimensions of other speculative narratives in this chapter: the underground setting as the site of finite extractible resources and the speculative search for new sources of illimitable energy. The moment of dramatic irony when Bilbo finds the ring ("it was a turning point in his career, but he did not know it") recalls, too, the epistemological conditions of adventure fiction, on which Tolkien draws in imagining Bilbo's quest.

Bilbo is a well-to-do hobbit, but the opening of the novel specifies that his wealth comes from inherited land, not extractive adventure schemes: "The Bagginses had lived in the neighbourhood of The Hill for time out of mind, and people considered them very respectable, not only because most of them were rich, but also because they never had any adventures" (30). Perhaps because Bilbo inherited "something a bit queer" from his mother's side (31), he nevertheless finds himself setting off with a group of dwarves as their designated "burglar" or "Expert Treasure-hunter" (49)—the narrative equivalence of the two terms is itself suggestive—on a treasure-seeking adventure that Gandalf says will be "profitable" (36).[86] Though the dwarves are skilled miners, the object of the quest is premined and preprocessed treasure, as in the treasure stories in chapter 2. This hoard the dwarves lost long ago when attacked by Smaug, the dragon who still guards it. Deprived of their ancestral riches, the dwarves in the intervening years "had to earn our livings as best we could . . . sinking as low as blacksmith-work or even coalmining" (56). Their roots in the Lonely Mountain go back to Thorin's grandfather, Thror, under whose rule the dwarves mined the mountain, made halls and workshops inside it, "found a good deal of gold and a great many jewels," and "grew immensely rich" (54).[87]

The dwarves' fondness for mining, crafting, and all things metallurgical and mineral is a crucial aspect of Tolkien's worldbuilding and at the heart of the central quest of *The Hobbit*. The dwarves are an extraction-based people, who, as Thorin explains to Bilbo, "never bothered to grow or find" food for themselves, instead trading smith work for provisions (55). They possess "a fierce and a jealous love" for "beautiful things made by hands and by cunning and by magic" (45).[88] When the dwarves show up at Bilbo's hole unannounced in the opening chapter, they sit around his table "talk[ing] about mines and gold" (39); when they spend a night with Beorn the skin-changer, they speak "most of gold and silver and jewels and the making of things by smith-craft, [though] Beorn did not appear to care for such things: there were no things of gold and silver in his hall, and few save the knives were made of metal at all" (177). A song encoding the dwarves' cultural memories describes "the deep places of their ancient homes" where the "dwarves of yore . . . shaped and wrought" "many a gleaming golden hoard." It is to this ancient home that they intend to return on a quest "to claim our long-forgotten gold" (44). More than gold, Thorin seeks "the great white gem, which the dwarves had found beneath the roots of the Mountain, the Heart of the Mountain, the Arkenstone of Thrain" (287). In the course of reclaiming the treasure, however, Thorin will die, and the Arkenstone will be buried with him, to "lie till the Mountain falls" (350).

That the Arkenstone will persist below and outlast the mountain, like the diamonds left underground at the end of *King Solomon's Mines*, brings an aspect of deep time and posthuman (or postdwarf) perspective into *The Hobbit*'s narrative imaginary. Several critics have written on Haggard's novel as a key influence for Tolkien, and while both texts end with unextractable precious stones, a key difference between them is that the setting of *The Hobbit* is not the longue durée of imperialist capitalism, except by analogy.[89] Its setting is preindustrial, at times even pastoral: "long ago in the quiet of the world, when there was less noise and more green" (31).[90] Middle-earth is a dynamic environment, however, and seems to be a world on the verge of industrial modernity, driven by two factors: the rise of men, and the greed for subsurface wealth. In the course of Bilbo's extractive quest, for example, the adventurers travel through "wild" lands where "bold men" are making inroads, "cutting down trees, and building themselves places to live" (148). Middle-earth is transforming from a world that had been, in Ursula Le Guin's phrasing, "not just pre-industrial" but "prehuman and non-human" (86).

The goblins are the most industrialized creatures in Middle-earth, and they resemble Wells's Morlocks in their adaptation to life underground and their cruel talent for mechanism and biopolitical control: "They make no beautiful things, but they make many clever ones. They can tunnel and mine as well as any but the most skilled dwarves. . . . Hammers, axes, swords, daggers, pickaxes, tongs, and also instruments of torture, they make very well, or get other people to make to their design, prisoners and slaves that have to work till they die for want of air and light" (108–9). The goblins' subsurface world is the world of industrial extraction, marked by terraforming and labor exploitation; from this world come weapons of mass destruction, a detail that places the novel in the context of the 1930s and the rise of new war technologies and atomic power: "It is not unlikely that [the goblins] invented some of the machines that have since troubled the world, especially the ingenious devices for killing large numbers of people at once, for wheels and engines and explosions always delighted them . . . but in those days and those wild parts they had not advanced (as it is called) so far" (109). Here the narrator, pointedly transgressing the line between primary and secondary worlds, ironizes ideologies of modern progress and industrialism that have produced only a more efficient necropolitics.

Much of the rapid environmental change that Middle-earth is experiencing seems to emanate from the dragon Smaug, a dragon for the industrial era. In the region of the Forest River, weather and Smaug are together producing transformative effects: "Those lands had changed much since the days when dwarves dwelt in the Mountain. . . . Great floods and rains had swollen the waters . . . and there had been an earthquake or two (which some were inclined to attribute to the dragon)" (242). The reference to earthquakes associates Smaug with extractive industry, for a connection between mining and seismicity has been recognized since the nineteenth century, and in fact the "first studies of this phenomenon began after a series of earthquakes in 1894 near Johannesburg, South Africa were attributed to gold production in the nearby Witwatersrand fields" (Cook).[91] Middle-earth's landscape registers Smaug's extractivism in other ways too: near the Lonely Mountain, where Smaug dwells, "the land grew bleak and barren, though once . . . it had been green and fair. There was little grass, and before long there was neither bush nor tree, and only broken and blackened stumps to speak of ones long vanished" (257). The description resembles the extractive zones of realist novels like *John Caldigate*, discussed earlier: "There was not a blade of grass to be seen," only "ghost-like skeletons of trees" (Trollope 74).

Smaug is, undoubtedly, a curiously modern dragon, which lends support to his connection with industrial extraction. Displaying the ancient dragon tendency to "steal gold and jewels" and "guard [his] plunder," he also has "a good notion of the current market value" of his treasure, transposing this premodern underworld fantasy into the discursive context of the modern extraction economy (*Hobbit* 55). Smaug's name is often pronounced as a homophone for "smog," and though Tolkien's appendixes would later specify a different pronunciation, the scent of industrial pollution still wafts around the dragon.[92] Smog and foul air are figured, for example, in Smaug's original means of defeating the dwarves, long ago, by creating a fog of smoke from the river with his hot breath: "The river rushed up in steam and a fog fell on the Dale, and in the fog the dragon came on them and destroyed most of the warriors" (56). In Bilbo's quest, when he and the dwarves finally reach the Lonely Mountain, "steam and a dark smoke" issue from within. Balin says that he expected "smokes and steams would come out of the gates: all the halls within must be filled with [Smaug's] foul reek" (258). Inside, the dragon lies amid "countless piles of precious things, gold wrought and unwrought, gems and jewels, and silver" (270). At the sight of it, even Bilbo's heart is "pierced with enchantment and with the desire of dwarves" (271). The fogs and smoke that Smaug produces connect this desire for underground commodities with the environmental hazards of industrialism.

Far from the smoke of the Lonely Mountain is Bilbo's home, the pastoral region of the Shire. Pockmarked with hobbit holes as it is, it is manifestly not an extraction-based community: illustrations of Hobbiton feature a prominent water mill, as in the frontispiece illustration to the 1937 first edition (see figure 3.5), which Tolkien drew himself. By placing the water mill front and center, Tolkien relies on energy to establish the chronotope of the Shire, and he positions the hobbits in a presteam economy with water power as their social base. Middle-earth's various regions are unevenly aligned with industrial modernity, and on his way to join the dwarves on their treasure hunt, Bilbo runs "past the great Mill, across The Water" (64), a detail that stresses his departure from the water-powered pastoral in the direction of the extractivist underworld. On his way back, at the end of the novel, he and Gandalf again "crossed the bridge and passed the mill by the river and came right back to Bilbo's own door" (360). Water power as presteam energy regime marks the limit of the Shire on the journey out and on the return.

After leaving the Shire, Bilbo's first adventure beyond its limit features not only trolls and treasure, but also enchanted objects. Inside the trolls' cave, Bilbo and the dwarves find their first haul of underground treasure as well as swords with "jeweled hilts" and a knife that becomes

The hill : hobbiton-across-the Water

FIGURE 3.5. "The Hill: Hobbiton-across-the Water" by J.R.R. Tolkien.
From the first edition of *The Hobbit* (1937). Bodleian Library, Oxford.
Used by permission of the Tolkien Estate. Bodleian Libraries, MS.
Tolkien Drawings 26,(c) Tolkien Estate Limited 1937.

Bilbo's sword, Sting (82). The swords are ancient and celebrated blades
of elvish make—objects so agential that they have proper names, Orcrist
and Glamdring (94)—and the ostensibly independent agency of metal
and minerals in *The Hobbit* takes on a heightened significance given the
novel's prominent focus on underground extraction. When the goblins

encounter Orcrist, they cower, for "it had killed hundreds of goblins in its time" (110), the sword itself apparently responsible. The swords glow in the presence of the goblins, signaling, again, their energetic agencies: Glamdring "flashed in the dark by itself. It burned with a rage that made it gleam if goblins were about; now it was bright as blue flame for delight in the killing" (112). Here the sword not only is agential ("by itself"); it possesses feelings of rage and delight. The Arkenstone is another energetic object of extraction in the text that shines "of its own inner light" (293), a premodern lightbulb.

Of all the object agencies in *The Hobbit*, the ring of power most closely recalls the era's new developments in atomic theory that were transforming ideas of matter and energy, as discussed in relation to Wells's fiction above.[93] The fact that it is called "the ring of power," for one thing, recalls the early promise of radioactive material as a potentially inexhaustible post-extractive energy source.[94] "Power" had referred to a "source of energy or force available for application to work" since at least 1700, according to the *Oxford English Dictionary*, and although *The Hobbit* predates the beginning of the Manhattan Project by two years, popular fiction had been incorporating "the nuclear muse," to use John Canaday's expression, for decades. From 1900 to 1942 "knowledge of atomic structure was growing at an overwhelming rate" (Canaday 25), and the "public imagination [was] inflamed with all manner of wild fancies in reaction to the discoveries of . . . radioactivity in uranium by Becquerel in 1896, of radium and polonium by the Curies in 1898, and of the possibility of converting matter into energy according to Einstein's relativity theory of 1905" (Brians 4). All these developments were making their way into speculative fiction, as Paul Brians has shown. Wells's *The World Set Free* (1914), discussed earlier, fictionalized the idea "that atomic energy would be derived from the annihilation of matter" (5) and even imagined the first atomic weapons. Tolkien's ring of power emerges from long-standing fantastic tropes but also engages new speculations about the transmutation of matter and energy, which would, it was hoped, offer a solution to the finitude of underground fossil energy stores. Such new conceptions of matter and energy were frequently expressed in metaphors that recast extractive imagery, such as the metaphor of the "treasure-house of energy deep down in the atom" that Canaday discusses at length (32).

Tolkien made numerous revisions to *The Hobbit*'s account of the ring of power between the first edition in 1937 and the revised edition in 1951, and his revisions tend to heighten these associations. Beyond its own agential qualities, the ring of power also saps the agencies of its owner, suggesting how energy sources—even seemingly free energy sources—have a way of overmastering those who wield them, belying any notion of cheap or free nature. This aspect of the ring is further developed in *The Lord of the Rings*, but even as early as the 1951 version of *The Hobbit*, Tolkien stressed the ring's powerful capacity for diminishing its user: "Gollum used to wear [the ring] at first, till it tired him; and then he kept it in a pouch next his skin, till it galled him; and now usually he hid it in a hole in the rock on his island, and was always going back to look at it. And still sometimes he put it on, when he could not bear to be parted from it any longer" (128). The first time Bilbo wears the ring, it seems rather the ring's choice to wear him: "He put his left hand in his pocket. The ring felt very cold as it quietly slipped on to his groping finger" (130). *The Hobbit* affords grammatical agency to the ring; and its effect, when worn, is to reproduce the darkness of underground life: "it made you invisible!" (132).[95] The ring of power could be said to reverse the most familiar use for electricity in Tolkien's day and ours—to illuminate and make visible—suggesting, again, how energy sources have a way of turning back on their users in unintended or even contrary ways. In Tolkien's world, the treasures of the subsurface never come without cost, and the environmental manipulations of the ring convey new energetic agencies that are likewise presented as far more costly, and far less governable, than they seem.

CONCLUSION

If literature is a privileged site for identifying our myths of energy and our patterns of environmental meaning making, we can see in this era's speculative worldbuilding an effort both to expand our stories of what is possible and to stimulate new ways of thinking about energy—and dreaming about energy—as a nature-culture formation. The dream structure is a common device in speculative fiction and appears in several texts I discuss in this chapter. "Sultana's Dream" ends with the narrator falling out of a solar-powered flying car and thus out of her utopian dream. *News from Nowhere* ends with Guest asking, "*was* it a dream? If so, why was I so conscious all along that I was really seeing all that new life from the outside?" He concludes, "if others can see it as I have seen it, then it may be called a vision rather than a dream" (249). Finally, the Time Traveller of *The Time Machine* asks, "Did I ever make a Time Machine . . . ? Or is it all only a dream?" And if it is a dream, "where did the dream come from?" (152). In all these cases, dreaming emerges as a figure for speculation, for imaginative reflection on material relations and how to change them.

The energy humanities, as Axel Goodbody and Bradon Smith say, investigate "the historical and cultural record of our relationship with energy" (11), but they also have a speculative function in and of themselves, one that relates to the mental practice of dreaming and the formal practice of worldbuilding. The historical energy humanities look back to ask what kind of worldbuilding has accompanied the rise of fossil fuels, in terms of imaginative and discursive constructions as well as real infrastructures and built environments, and they speculate on, or dream about, what kind of imaginative and real worldbuilding can contribute to energy transition and the creation of a postcarbon future. Speculative fictions of the past offer a particular angle of estrangement that makes visible possible futures, roads not taken, and moments of inflection before the hardening of historical pathways. These historical speculations correspond with the task of imaginative literature as Ursula Le Guin has defined it: to "restore the sense . . . that there is somewhere else, anywhere else, where other people may live another kind of life. The literature of imagination . . . offers a world large enough to contain alternatives, and therefore offers hope" (87).

Conclusion

The Puritan wanted to work in a calling; we are forced to do so. For when asceticism was carried out of monastic cells into everyday life, and began to dominate worldly morality, it did its part in building the tremendous cosmos of the modern economic order. This order is now bound to the technical and economic conditions of machine production which to-day determine the lives of all the individuals who are born into this mechanism, not only those directly concerned with economic acquisition, with irresistible force. Perhaps it will so determine them until the last ton of fossilized coal is burnt.

MAX WEBER, *THE PROTESTANT ETHIC AND THE SPIRIT OF CAPITALISM* (1905), 123

AS THE ABOVE epigraph from Max Weber suggests, mineral resource extraction during the first century of the industrial era posed an apparent end point for capitalist growth at an unspecified point in the future and provided, or seemed to provide, a temporal structure and a material limit beyond which the existing system could not survive. Although the realities of extraction-based life did not bear out according to such a vision, this book has tried to reenter the exhaustive imaginary of the nineteenth and early twentieth centuries to understand its correspondences with our current environmental moment and to recognize how profoundly it has shaped our modes of understanding and representing the natural world. From the 1830s, the decade when steam power achieved definitive sway within British industrial production, to the 1930s, when the early glimmers of nuclear power suggested the promise of new illimitable forms of energy, narratives of mineral resource extraction permeated literature and culture, carrying stories of exhaustive futures, movement outward and downward, energy existentialism, and energy dreams. These narratives shaped literary genre in profound ways, as the previous chapters have explored, and they tell us much about the epistemological challenges of our present moment: that we have come to understand time, history, and the movement toward futurity as a depletionary process; that we have come to understand global space in terms of the resource frontier and the sacrifice zone; and that we have come to understand energy as a medium of human freedom capable of effecting full-scale social transformation. We will need to work through these frameworks of understanding as we push for a just transition to a decarbonized energy regime and a new order of relations with and within the environment that sustains us.

In examining the rise of industrial extraction, we behold a social and environmental transformation of tremendous force, one that left neither culture nor nature untouched, and one that underscores the inseparability of these two domains in the new epoch known as the Anthropocene. "Capitalist modernity," Jeremy Davies writes, "is a strange-looking geological phenomenon, but it is nonetheless a real one" (192). Extractivism's material trace is visible in the geological record as in the cultural one, but only the cultural record can provide more than a trace of this material realness, for literature also offers the trace of resistance and lament, of adaptation and explanation, of imaginative response and incorporation. This work of mediation continues in the art and literature of today, which inherits the environmental narratives of the past and transmutes them for the present with lasting impacts—an ongoing exchange between culture and its subtending environment. In her 2019 poem "Wagers," for example, Kristin George Bagdanov asks, "When I die where will my body go / where will the oil flow / will it become a new body / will it peat into coal[?]" (*Fossils* 51–52). At stake is a new environmental understanding of ourselves, one that is also evident in Lorine Niedecker's 1968 poem *Lake Superior*, which begins, "In every part of every living thing / is stuff that once was rock // In blood the minerals / of the rock // Iron the common element of earth / in rocks and freighters // Sault Sainte Marie—big boats / coalblack and iron-ore-red" (1). Extraction literature of the long exhaustion narrates the rise of the mass mineral trade Niedecker describes here, a global transformation that, as these two poems express, has bound us to the subsurface indefinitely.

While the previous chapters have emphasized the consequences of my argument for literary studies, I want to conclude *Extraction Ecologies* by asking what we can take from the literary past in imagining a post-extractivist environmental politics. Here I take up the call of Ursula K. Heise and other environmental humanists to formulate "affirmative visions of the future . . . visions that are neither returns to an imagined pastoral past nor nightmares of future devastation" that may provide "the foundation for a new kind of environmentalism" (12). The humanities are sometimes charged with offering all critique and no solutions, and certainly, mining is one area of human endeavor that is easy to denounce yet hard to replace.[1] No one wants their own backyard filled with the wastewater or tailings of an extractive operation, and yet the prospect of no more MRI machines, wind turbines, or electric vehicle batteries is doubtless unappealing as well. There are no obvious answers to this conundrum, and precisely for that reason it feels necessary in closing my argument to wade into the contentious question of mining and environmentalism with some provisional conclusions that emerge from the archive I have explored.

First, by reading these narratives produced amid the rise of industrial extraction, we learn that it is crucial to disambiguate two terms that have often worked in tandem in this study: extraction and extractivism. The activity of mineral resource extraction accelerated in the industrial era to the point of becoming the economic and social basis of a way of life—extractivism. But an activity need not necessarily become a way of life, and if a crucial aspect of extractivism is a depletionary idea of the earth that disregards the ethic of stewardship, we may conclude that it is possible to imagine extraction outside of, or beyond, extractivism. The story of mining will never be a pastoral; to the contrary, as a group of environmental engineers recently wrote, "Mining is inherently an environmentally detrimental process that only becomes less efficient as resources are depleted and ore grades worsen" (Leader, Gaustad, and Babbitt 12). But there are steps we can take toward mitigating our reliance on this detrimental process. These would involve, for starters, a vast reduction in unnecessary extraction—dematerialization, to use the language of economics. Identifying substitute materials to supersede particularly egregious types of mining operations is also crucial. Perhaps most obviously, post-extractivist extraction will require a real commitment to pursuing metal and mineral recycling on a much wider scale, processes for which there are far too many disincentives in the current global market.

Bearing in mind the lessons of chapters 1 and 2 of this book, another crucial goal for a post-extractivist extraction is to overcome the long-standing spatial dynamic of mining—its provincial or frontier setting, its foundation in the idea of the extractive zone, to use Macarena Gómez-Barris's phrasing. The location of mines in places deemed invisible or out of the way, or, rather, the creation of places deemed invisible and out of the way by the siting of mines, fosters the exploitation of labor and land now as it did in the past. From the 1830s to the 1930s, the great fear that attended the rise of extraction-based life was the possible exhaustion of underground resources, and in Britain, as we have seen, such fears were an encouragement to range far afield in extractive expedition. The situation is different in energy and extraction debates today, when extreme extraction techniques such as deep-water drilling, mountaintop removal, fracking, and the mining and processing of tar sands have enabled humans to get more from less, sometimes more locally, often with disastrous environmental consequences. While exhaustion of resources remains an issue for some raw materials, geological distribution now counts alongside factors like labor markets and environmental regulations in the siting of extractive operations.

To take one example of an extracted commodity that exemplifies this dynamic, consider rare earth metals, which despite their name are actually "widely distributed geographically" and "not particularly rare" (Goonan 1). They are a necessary component of technologies we need for decarbonization, such as wind turbines and electric vehicles, and thus are sometimes held up as examples of how green technology relies on environmentally detrimental mining. Extracting and processing rare earth metals, also known as lanthanides or lanthanoids, is a complex process that produces radioactive waste among other hazards.[2] As a result, 84 percent of rare earths are currently mined in China, which has proven more willing than Western countries, for a host of historical reasons, to bear the toxicities associated with them (Van Gosen, Verplanck, and Emsbo 4). In fact, however, technologies to recycle rare earth metals are available and have been unpursued mainly because it remains cheaper to mine the metals than to recycle them (Goonan 9). As of 2016, only 1.5 percent of global supply came from recycled sources (Lee and Wen 1278). It is not difficult to imagine how incentives for recycling or taxations on mining could reverse this situation in a political field where global environmental values were prioritized. Moreover, rare earth minerals are geologically widespread and could be sourced in places with stricter environmental regulations where the impacts of such mining could be lessened—again, in a political field where global environmental values were prioritized.[3]

The literary works treated in the previous three chapters suggest that one of our great inheritances from industrial-era extractivism is the tendency for mining to gravitate to the least protected environments and the least protected workers. To imagine post-extractivist extraction, then, is to envision a world where the benefits and costs of underground mining are borne by the same people, instead of the situation we have now where one group reaps the benefits of extracted materials and one group bears the brunt of the pollution and labor hazards that come with extractive processes. The literary archive from the age of empire forcefully reminds us, in other words, that a post-extractivist model of extraction would need to confront the legacy of colonialism that laid the extractive path dependencies in which we are mired today. It would need to insist that where extraction remains necessary, labor and environmental safety are secured, and that if the toxicities associated with a particular form of extraction and production prove unbearable in one place, they are unbearable everywhere. Such points are obvious enough at face value, and yet, as is equally obvious, politically difficult to effect—which is exactly why they must occupy a more central place in an environmental platform dedicated to post-extractivism. It may be helpful here to think of extraterritorial mining as a form, one that took hold of the world with the Spanish colonization of the Americas and remained central to global extractivism as it developed through the Industrial Revolution, when steam-powered mining equipment and steam-based transportation networks transformed forever the scale of what was possible in the way of overseas extraction.[4] It is a form connected to that of the sacrifice zone, though sacrifice zones are not always extraterritorial as we see in the recent history of US uranium mines, mountaintop removal, and fracking on Indigenous lands or in impoverished communities. Extraterritorial mining and the sacrifice zone are forms of the past and of the present, and they exemplify why it has been a central goal of this book to see extractivism as both cultural and infrastructural, discursive and material.

If many of the changes suggested above—decolonization of extraction, prioritization of work and environmental safety, dematerialization—seem incompatible with global capitalism, this accords with some of the general lessons of the literary archive I have explored: that the advance of industrial capitalism fostered a massive acceleration of extraction, and that capitalism and extractivism do rise and fall together to some extent. This does not mean that communist governments have not pursued detrimental extraction projects (they have), nor that reforms under the current dispensation would not be beneficial for mine workers and for the habitability of the planet (they would); it simply means that there is a fundamental incompatibility between prioritizing profit and prioritizing environmental stewardship, an incompatibility that is clearly visible in the history of industrial extraction and that was already grasped by many in the historical era covered by this book. In 1892 political economist

Charles S. Devas said that whatever "industrial progress" had been made since 1750 must be weighed against the "drawbacks incidental to the technical revolution" including "pollution of streams by manufacturing refuse" and "pollution of air by vapours, gas, and smoke" (75). William Morris put it more fervidly in his 1884 lecture "Art and Socialism": "Less lucky than King Midas, our green fields and clear waters, nay the very air we breathe are turned not to gold (which might please some of us for an hour may be) but to dirt" (116).

But if capitalism fosters extractivism, it does not follow that getting rid of capitalism would, at this juncture, eliminate the need for extraction. With 7.5 billion people on Earth, even were we to pass through a revolution to a green, livable, postcapitalist future made up of walkable human settlements at a scale that would not require motorized transportation, we would still need a means for food to be brought to these settlements, given the large amounts of land necessary to grow food for 7.5 billion people.[5] At this juncture, even a revolution against capitalism would not eliminate the need for decarbonization—for solar panels, wind power, and green batteries all require extraction.[6] Perhaps our only choice is to dig ourselves out of the hole we have made, at least to an extent. We will need to continue extracting underground minerals even as we fight to keep all hydrocarbon reserves buried in the ground, for the former is necessary for the latter, and the latter is necessary to prevent runaway global warming. To decarbonize we will have to extract, with an eye to reducing extraction wherever possible, mitigating its harms, and distributing the harms and the rewards equitably.

As Debeir, Deléage, and Hémery put it in 1986, in one of the first major contributions to what we would now call the energy humanities, "no social alternative can be conceived today that would not establish a new energy system" (xv). Renewables (solar power, wind power, wave power, hydropower) with battery storage offer the best possibility for that new energy system, but the challenge of building a real "social alternative" to accompany a transition to decarbonized energy remains the major political challenge for environmentalists today. It is a challenge of understanding, a challenge for literature, art, and criticism as well as every other domain of knowledge and creativity. The culture of the past is a resource on which we can draw in taking up this work, for it provides an archive of forms to mirror and to unmake. We "who are born into this mechanism," to return to Weber's epigraph, can learn much from listening to the voices of those who witnessed the mechanism being born.

NOTES

Introduction

1. Benjamin Morgan, for example, has called for the expansion of historicist interpretive practices in the age of climate change: "we need to think more broadly about the kinds of practices that count as interpretation . . . not only the restoration of economic or social contexts to the surface of the text, but of geological and climatic contexts as well" ("After" 22). Tobias Menely and Jesse Oak Taylor have suggested, meanwhile, that geohistorical reading practices are central to stratigraphy itself and that "the literary dimensions of geology—a practice of *reading* stratigraphic inscriptions and *narrating* evocative, if improbable, stories—[have] become even more pronounced in the Anthropocene" (2). See Dipesh Chakrabarty ("Climate" and "Planet") for foundational thinking regarding climate change and humanist inquiry.

2. Here I understand form as a category that crosses the literary and the environmental; for work along these lines, see Nathan Hensley and Philip Steer ("Introduction"), and Morgan ("Scale as Form"). Morgan says that "the resonances between interpreting literary forms and reading geological formations have been recognized at least since the nineteenth century" (133), and, recounting the rise of geology as a discipline, Adelene Buckland describes how early geologists studied "the structures and forms of the natural world" while "grappling for the material, visual, and verbal forms by which to instantiate them," concluding that "it was in a combination of each of these forms that 'geology' could be said to exist" (17–18). Recent work by Caroline Levine and Anna Kornbluh has influenced such approaches by modeling new ways to think form and politics together, and, by extension, to think form and ecologies together.

3. The time line originates with Britain but is intended to express the global reach of Britain's imperial-industrial complex.

4. Orthodox definitions of industrialism identify it with a set of labor practices rather than energy and resource practices, but I am influenced here by the work of critics like Andreas Malm who conceptualize labor, energy, and material resources together.

5. Political ecology is an important guide for establishing the historical scope of this book, but the period from the 1830s to the 1930s also makes sense from the angle of literary history. This era saw the consolidation of a particular form of realism with the industrial novel and the work of such writers as Charles Dickens and Elizabeth Gaskell at its earlier edge and saw the proliferation of nonrealist genres such as adventure narrative, science fiction, and fantasy later on. The second half of the 1930s makes for a convenient place to end since it saw the beginning of a tradition of proletarian novels written by miners, including Walter Brierley's *Means-Test Man* (1935) and Lewis Jones's *Cwmardy* (1937) and *We Live* (1939); this development transformed literature about mining, as Holderness discusses, in important ways that I do not take on here.

6. Dynamite was patented in 1867 and manufactured beginning in 1870; TNT, or trinitrotoluene, began to be used as an explosive for mining in the late nineteenth century. See Simon Quellen Field for more on the history of explosives.

7. Robert C. Stauffer's translation (140). For more on Haeckel and his place in evolutionary theory, see Robert J. Richards. For more on the emergence of this term in the nineteenth century and its place in literary studies today, see my essay "Ecology."

8. See Allen MacDuffie's related argument "that the literature of the [post-Darwin] period . . . reveals a culture simultaneously struggling to come to terms with, and struggling to avoid, its relationship to the natural world, and thus its situation on this planet" ("Charles" 545).

9. This is not to say that mining before the industrial era is an unimportant subject, but because of its different scale, it carries different meanings. For a study of mining in early modern Europe, see Phillip John Usher's *Exterranean: Extraction in the Humanist Anthropocene*.

10. Throughout this book, I use "mineral" in its conventional rather than technical sense. As Bob Johnson notes, coal is not technically mineral but "organic matter composed of plants . . . that became fossilized (or mineral-like) under geological pressure in the absence of oxygen," but the term has long been used to describe fossil fuels as in the US Mineral Leasing Act of 1920, "which lumped fossil fuels into the category of the mineral . . . alongside other inorganic resources" (10).

11. As Elizabeth Povinelli and others have noted, this revolution in coal mining was closely connected to the birth of geology as a discipline: "the Industrial Revolution massively expanded the Lancashire, Somerset, and Northumberland coalfields. . . . But the exploitation of the coalfields also uncovered large stratified fossil beds that helped spur the foundation of modern geologic chronology: the earth as a set of stratified levels of being and time. In other words, the concept of the Anthropocene is as much a product of the coalfields as an analysis of their formation" (Povinelli 10). Geographer Bruce Braun puts it thusly, articulating an observation that is crucial to my project in this book: "nature's ordering in and through modern forms of knowledge is related to, and in part constitutive of, the ways in which nature is integrated into forms of economic and political rationality" (13–14).

12. James Winter claims that "only indirectly was steam technology a crucial factor in the huge nineteenth century expansion of the mining and quarrying industries" since "traditional tools, skills, and muscle power" were still required at the face or wall. He acknowledges, however, that "rapid improvement in pumping and hauling technology . . . allowed mines and quarries to reach deep below the water tables" and that "the application of steam energy to industry and transportation caused an enormous outpouring of coal, iron ore, copper, stone, clay, tin, lead, zinc, salt, sand, and shale" (134–35). To my mind, steam is in no way an "indirect" factor simply because its major impacts were felt first in mine draining and transport rather than in the actual work of extraction.

13. An advertisement from the 11 April 1868 issue lists some of these works: *Joint Stock Companies, and How to Form Them*; *Miners' Manual of Arithmetic and Surveying*; *Ventilation of Coal Mines*; and *Conversations on Mines, &c., between "A Father and Son."* Many other British mining periodicals circulated in this period including *Iron Trade Exchange* (launched 1847), *Miners' Weekly News* (launched 1873 in Coventry), and *Mining World* (launched 1871).

14. On coal and industrial Britain, see, for example, Debeir, Deléage, and Hémery; Malm; Pomeranz; Sieferle; and Wrigley (*Energy* and *Path*).

15. Klein came to this understanding partly through the work of Indigenous feminist authors such as Leanne Betasamosoke Simpson (see Klein, "Dancing").

16. Technically coal is not a mineral but dead living matter and, thus, is theoretically regenerative, but not on anything close to a human timescale. Louis Simonin in *Mines and Miners* (1868) describes "the eventual exhaustion of the coal-fields (an exhaustion which, from the data we possess, can be calculated with some approximation to truth)" as "one of the gravest and most important of subjects" (265), and he refers to a possible solution that had evidently been floated: the planting of forests to (eventually) create more coal. He disregards the idea owing to the impossible timescales involved: "Some recommend

planting fresh forests to replace coal at some future day . . . but the world never goes backward" (269). The Carboniferous period, when much coal formed, was approximately three
hundred million years ago, and though nineteenth-century thinkers still debated the age
and origins of coal, they understood that vast stretches of deep time separated coal's formation from its industrial use. P. W. Sheafer, a geologist and mining engineer, expressed
in 1881 that the "ingenuity of scientists is taxed to account for this wonderful accumulation of fuel, once vegetable, now mineral" and asked, "How many ages were consumed in
the process . . . who can tell? . . . It may well be said that all the years since the creation of
man would be too short a time to produce a bed of coal" (4). Human history may be but
a glimmer in the timescale of underground mineral resources, but within a few human
generations, it was understood, these vast, sublime geological resources could be used up.

17. For a discussion of concerns about soil fertility in the nineteenth century, see Mark
Larabee. As his narration of the rise and fall of the guano trade (roughly 1840–90) shows,
guano exhaustion was a sped-up version of what many feared would happen with coal.
Guano, a fertilizer that comes from bird excrement, is potentially regenerative but takes so
long to form that, like coal, it is essentially a nonrenewable resource. For more on guano
see A. J. Duffield.

18. By the late nineteenth century, certain animal populations such as seals and elephants had been hunted to the brink of extinction. On seals, see John Miller ("Fiction");
on elephants, see chapter 2.

19. Jeremy Davies helpfully reminds us that "limits to growth are *produced* through
economic and ecological relations," which "does not make it any less true that infinite consumption growth is impossible on a finite planet," but does mean that "invoking biophysi-
cal limits to growth cannot substitute for analysis of [those] relations" (199). Ted Benton
provides an example of such analysis: "if we think in terms . . . of nonrenewable resources
as a natural limit, a society that shifts its resource base, or builds resource recycling into
the intentional structure of its labor processes, may effectively transcend what previously
were encountered as limits" (173).

20. Sandro Mezzadra and Brett Neilson argue that extraction's material impacts have
been felt especially since the rise of neoliberalism in the 1970s: "While these activities are
by no means historically new, they have reached unprecedented and critical levels as the
rush to convert materials, both organic and inorganic, into value has escalated" (2).

21. See Deanna K. Kreisel, "Sustainability," for a discussion of this term in the context
of the period, emphasizing how "sustainability" functions as an economic concept rather
than an environmental one.

22. My use of the phrase "long exhaustion" is meant to draw a connection between the
depletion-based society that emerged in the nineteenth century, premised on what were
understood to be finite underground resources, and the depletion-based society we inhabit
today, premised on the use of fossil fuels that are generating unsustainable levels of greenhouse gases and producing an inhospitable climate for future generations. A limitation
of this formulation ("long exhaustion") is that my book does not address the important
developments in exhaustion, depletion, finitude, and extraction that occurred between
the 1930s and today, e.g., the Great Acceleration, the revolution in plastics, or even, really,
the rise of oil and petroculture, which began before the 1930s and which this book only
touches on. Limits are a feature of book writing as well as world ecology, but despite the
necessary limits that constitute the boundaries of this study, it is my sense that a case
could be made for an interconnected "long exhaustion" extending from the industrial era
to today. Thus I use the phrase to illustrate the ideological continuities that connect the
first century of industrialism to our own century, passing over the intervening chapters
by necessity.

23. I take the "borrowed time" formulation from Rob Nixon (69) and discuss it further in chapter 1.

24. In his preface to *The Gold Mines of the Rand* (1895), Frederick H. Hatch is hesitant to engage the question at all, owing to the complexities of calculating mine exhaustion: "With regard to the duration or life of the Witwatersrand mines, either collectively or individually, we refrain from speculating, as we consider such estimates worse than useless, being dependent on so many indeterminate factors. It is evident that in any given mining area there is only a certain limited quantity of the noble metal, and exactly what proportion of it will be extracted must depend on the particular circumstances that govern the cost of obtaining it" (7).

25. Promotional prose for mining speculations often emphasized short-term gains with an urgent disregard for future returns. An advertisement in the 5 January 1884 issue of the *Mining Journal*, for example, offers shares in two mines, one in Cornwall and one in Venezuela, both "SPECIALLY RECOMMENDED FOR A GREAT RISE. SHARES in THE FOLLOWING MINES if purchased NOW will probably yield a HANDSOME PROFIT in a short time" (6).

26. For more on mining and nineteenth-century political economy, see Antoine Missemer, who argues that the dominant Ricardian theory of land was agricultural in nature and suffered from a theoretical deficiency with respect to mining and extraction. The marginalist movement, of which Jevons was a leader, emerged partly to remedy this deficiency: "After Jevons' seminal contribution, which led to the constitution of an autonomous economic analysis of fossil fuels, British economists continued to create new concepts and tools (e.g. mining rent) to provide a better understanding of the exhaustion of natural resources. This reveals a persistent concern with coal depletion [from] the 1860s to the 1890s" (22).

27. I discuss this topic in an earlier note in this chapter, citing primary sources by Simonin and Sheafer. For recent scholarship on the subject, see also Jesse Oak Taylor (*Sky*, especially 127–28) and Naomi Yuval-Naeh.

28. Sheafer provides data to show how the "modern growth and ultimate decadence of this great empire may be calculated from the statistics of her coal mines. In 1800 her coal product was about 10,000,000 tons; in 1854 it was 64,661,401 tons; and in 1877 it swelled to 136,179,968 tons. This period was a time of continued prosperity, when England ruled the world, financially and commercially. In the twenty-three years from 1854 to 1876, inclusive, she produced the enormous quantity of 2,210,710,091 tons of coal; and, more wonderful still, exported only 222,196,109 tons . . . consuming the rest within her own borders." And yet, Britain's coal output has been decreasing of late, and Sheafer asks, "Is this the beginning of the inevitable decline?" (11).

29. In the latter part of the period covered in this book, oil appeared to some to promise a new fuel frontier, or perhaps just a stopgap measure, for what seemed to be a rapidly dwindling supply of coal. Keizaburo Hashimoto wrote in 1930, for example, that "the reserves of coal in the crust of the earth are not inexhaustible and they are distributed in such a way, that almost all the developed industrial countries in the world are anticipating a more or less nearly and complete exhaustion of their coal resources," and he found "the advantages of oil over coal [to be] enormous" in terms of cost, labor, and supply (21). Michael Tondre discusses the horizon of oil as it emerged in early twentieth-century British literature, and critics such as Stephanie LeMenager (*Living*), Timothy Mitchell, Jesse Oak Taylor (in chapter 5 of *The Sky of Our Manufacture*), Jennifer Wenzel, and the contributors to Ross Barrett and Daniel Worden's volume have all treated the materialities of oil and their analogues in literature and culture.

30. On the feedback loop as an environmental form that also models the mutually determinative relationship between discourse and environmental materiality, see Jesse Oak Taylor (*Sky* 40) and Devin Griffiths ("Petrodrama" 621).

31. Much poetry about extraction was also produced in this era, often focusing on extractive labor and the biopolitical demands of extraction, as in Elizabeth Barrett Browning's group-voiced poem "The Cry of the Children" (1843) or Joseph Skipsey's group elegy "The Hartley Calamity" (1862), written for the victims of the Hartley Colliery Disaster. Regrettably, I have not the space to make a real case for extraction and poetry within this volume, apart from a few asides here and there, but I hope that future scholarship will take up such an inquiry.

32. As is now widely known, earth systems scientists put forward the term *Anthropocene* to describe a new geological epoch following the Holocene and marked by irreversible human impacts. As Davies explains, "in geology the word *epoch* has a specific technical meaning. A geological epoch is a midsize section of the planet's history." Thus, "the idea of the Anthropocene epoch lets us understand the ecological crisis of the present day in the context of the distant past" (2). Studies of the Anthropocene—which has been variously termed the Capitalocene, Plantationocene, or Chthulucene by critics who seek to emphasize capitalism, racial capitalism, or species extinction, respectively (see chapter 2 of Haraway)—now constitute a major subfield in the environmental humanities, one that has called into question the periods and scales at which the humanities tend to operate. (See Chakrabarty, "Climate" and "Planet.") As to whether and when there *was* a nature that existed outside of human impacts, I leave this question to scholars focusing on earlier periods than my own, noting here only that the subject is one of debate, with various scholars settling on the human use of fire (one to two million years ago), agriculture (approximately ten thousand years ago), or early capitalism (approximately five hundred years ago) as possible origin points for a fully anthropogenic nature. For more on the Anthropocene and its consequences for literary study, see, for example, Timothy Clark (*Ecocriticism*), A. Tsing et al., Menely and Taylor, and Davies.

33. See Tina Young Choi and Barbara Leckie for a reading of Babbage's image (573). See Dana Luciano ("Romancing") for a discussion of Babbage and "matter's memory" (106) and the "more or less straightforward guarantee of the permanence of human action" (107).

34. Nuclear tests from the 1940s to the 1960s dispersed radioactive isotopes, with plutonium as "the most significant for stratigraphic purposes" since "the half-life of plutonium-239 is 24,110 years; it binds tightly to soil particles, and its preanthropogenic concentration in the earth's crust was extremely low" (Davies 105). Plutonium-239 is made from uranium (in a nuclear reactor by bombarding the uranium with neutrons and then allowing the resulting material to undergo beta decay), and thus, as Debeir et al. have argued, "nuclear energy is still based on the mobilization of a stock, limited by definition, of fossil 'fuel', namely uranium" (167).

35. Davies and Menely and Taylor offer useful overviews of the various arguments on this score. See also Jason Moore and Malm on the alternative framework of the Capitalocene and Haraway on the Plantationocene. Kathryn Yusoff, drawing on the work of Sylvia Wynter and other critical race theorists, argues for identifying the origin of the Anthropocene with colonial genocide in the Americas and the beginning of the Atlantic slave trade in the sixteenth century, since this is the point at which "Blackness becomes characterized through its ledger of matter" (Yusoff 35). She makes a case against "the 1800 Industrial Revolution" as the "natal moment" of the Anthropocene (39), since, in her view, the apparently new modes of material relation that characterize the Industrial Revolution actually stem from "slavery and its organization of human property as extractable energy properties" (40). Although Wynter does not engage explicitly with the idea of the Anthropocene, she anticipates Yusoff's argument in claiming that "the Negro, the Native, the Colonial Questions . . . can be clearly seen to be the issue, not of our present mode of economic production, but rather of the ongoing production and reproduction of . . . our present biocentric ethnoclass genre of the human, of which our present techno-industrial, capitalist

mode of production is an indispensable and irreplaceable, but only a proximate function" (Wynter, "Unsettling" 317). Elsewhere Wynter discusses "the West's epochal historical rupture, effected from the Renaissance onward," in its thinking about the natural world and humanity's place in it—the "de-supernaturalization of the physical cosmos" ("On How" 159)—in ways that would support Yusoff's case for the sixteenth century as a definitive moment in human-environmental relations.

36. For more on the lifespan of CO_2 in the air, see Carbon Brief and Duncan Clark, Zeke Hausfather, and Lisa Moore.

37. In the period covered by this book, some political and economic thinkers came to attribute Britain's rise as an imperial power to geological accident. George Bernard Shaw wrote in "Common Sense about the War" (1914), for example, "The general truth of the situation is . . . that we are cursed with a fatal intellectual laziness, an evil inheritance from the time when our monopoly of coal and iron made it possible for us to become rich and powerful without thinking or knowing how: a laziness which is becoming highly dangerous to us now that our monopoly is gone or superseded by new sources of mechanical energy" (35).

38. In a piece that appeared as *Extraction Ecologies* was nearing completion, Alok Amatya and Ashley Dawson maintain, not unlike Ghosh, that "all too often . . . resource extraction and the colonial location of such extractions are rendered invisible in novels written during Britain's Age of Empire" and that "in the nineteenth century, the novel comes to be characterized by a constitutive absence that centers on empire and resource extraction" (9). For this reason they focus their own analysis on late twentieth- and early twenty-first-century novels. My book argues, instead, that extractivist forms crucially structure narrative genres in nineteenth- and early twentieth-century literature of the British Empire, though critics have tended to overlook their significance.

39. Alongside Chang (*Novel*), see, for example, Jesse Oak Taylor (*Sky* 15) and recent work by John MacNeill Miller ("Ecological," "Weird," and "Mischaracterizing").

40. See David Coombs and Danielle Coriale for more on strategic presentism.

Chapter One. Drill, Baby, Drill: Extraction Ecologies, Futurity, and the Provincial Realist Novel

1. As discussed in the introduction, my use of "extraction" is specific to the mining of underground commodities, and my use of "extractivism" refers to the cluster of socio-cultural and socio-environmental conditions that attended the rise of industrial mining.

2. Central to my notion of the provincial realist novel is the idea of "semi-detachment" articulated by John Plotz and the "linkages to a greater world beyond" that are at the heart of this genre ("Provincial Novel" 362). For more on the provincial or regional novel as it developed in English, see Raymond Williams (chapter 5 of *Writing in Society*), Ian Duncan ("Provincial or Regional Novel"), and Plotz (chapters 4 and 5 of *Portable Property*, chapter 4 of *Semi-detached*, and "Provincial Novel").

3. Just how uninhabitable? David Wallace-Wells's *The Uninhabitable Earth* walks us through various warming scenarios and some of the catastrophes we can expect.

4. Tina Young Choi and Barbara Leckie suggest that long nineteenth-century novels "not only describe slow causality . . . but also offer it as an experience" (582).

5. On chrononormativity, see Freeman's *Time Binds*; also relevant here is Dana Luciano's term "chronobiopolitics, or the sexual arrangement of the time of life" (*Arranging* 9). I am indebted to Freeman's and Luciano's elaboration of these temporal formations, which I seek to situate in an environmental as well as a social and reproductive context.

6. See also Alexis Lothian's *Old Futures*, which argues against the idea that "reproduction and heterofuturity are . . . easily equated" and aims to develop "new ways to think

reproduction and futurity together without . . . submitting to the clichés of a singular reproductive futurism" (9).

7. Raymond Williams has noted that "the available *forms* of the novel" in the nineteenth century "centred predominantly on problems of the inheritance of property and of propertied marriage" (*Writing* 233–34). For new work on the marriage plot, see Jill Galvan and Elsie Michie's recent collection, which warns against critical assumptions of "unity and teleology" that often color approaches to the marriage plot (4), and yet shores up the marriage plot as the structural backbone of nineteenth-century fiction, a "familiar narrative structure [that] serves many different purposes" (9).

8. Indigenous people now often make up the front lines of protest against extraction capitalism, calling to mind a history that goes back centuries, for miners were often the leading edge of white settlement in the Americas, and British-backed extraction projects in nineteenth-century Latin America wrought environmental disruptions hostile to Indigenous lifeways. See Rebecca Solnit; Courtney W. Mason; and Mark David Spence. On extraction regions as imperialist profit centers, see Svampa (66).

9. Most of Latin America was not formally colonized by Britain, but Britain was by far the biggest investor in nineteenth-century Latin America, and British firms were responsible for extraction and infrastructure projects that transformed the region, as I discuss in chapter 2.

10. See Fredrik Albritton Jonsson for an account of Parliament testimony on coal exhaustion in the 1830s.

11. For more on Jevons's time in Australia, see Philip Steer ("Gold" and chapter 2 of *Settler*).

12. See Paul Warde for a fascinating account of the history of forest management and its key role in the rise of a notion of sustainability—the idea that "intergenerational justice" must be a factor in the legal management of resources, such that the resources are understood to "belong to posterity" (160, 161).

13. Here Jevons's structure of thinking resembles Malthusian pessimism in the earlier part of the century, even though the nineteenth-century agricultural revolution, and especially advancements in fertilizer, had mitigated Malthusian fears concerning population and food supply. For more on Malthus and his influence on nineteenth-century understandings of natural resource limits, see MacDuffie (*Victorian*, especially 3–12).

14. Edward Hull in *The Coal Fields of Great Britain* was somewhat more optimistic than Jevons on the question of Britain's coalfields, estimating a total duration of over six hundred years, but being a geologist rather than a political economist, he made this calculation on the assumption of a steady rate of consumption, whereas Jevons calculated increasing annual consumption on the basis of historical data. And even so, Hull thought that Britain would reach peak coal within a few decades, entailing a period of drawdown from that point forward, and projected a long, slow exhaustion that would begin very soon. From surveys of British coal mine personnel he concluded that "the present output could be maintained, and somewhat increased for a period of about 20 or 25 years, after which a gradual diminution might be expected; in other words, that the maximum production of coal would be reached." Within a few decades he presumed that "a large number of collieries would necessarily be 'worked out,' and consequently closed, and also, that the better, and thicker, seams of coal would be exhausted" (435). Most salient in Hull's projections is his account of the "downward curve" by which exhaustion would happen (436). Exhaustion would be gradual, not sudden, but the gradual decline would begin very soon.

15. See Andrea Wulf for a discussion of the wide influence of *Political Essay on the Kingdom of New Spain* (152).

16. The story of Real del Monte Mine is worth repeating briefly here. After its founding, ships were sent out full of equipment and a battery of miners, engineers, and machinists,

mainly from Cornwall. The "three ships filled with one thousand tons of machinery, pumps, etc." finally arrived at the mine "after untold trials in transportation and erection" (Romero 74). A long struggle with water ensued; the mine was rich but proved difficult to drain. Finally in 1848, "by order of the bondholders," the struggling concern was turned over to Mexican owners (75). After the transfer, Real del Monte "hit a bonanza by exploring veins . . . that the previous managers had overlooked" (Tenenbaum and McElveen 63).

17. The industry press mystified the language of exhaustibility, as I also discuss in the introduction. Russell Gold Mining Company, for example, recounted in the *Mining Journal* a "big lode" that "will supply ore . . . for years to come—in other words, inexhaustible— of a high grade, being free gold, heavy and easily saved" (6 November 1886, 1278). Here "inexhaustible" means, at best, "years to come."

18. Perhaps more than that of any author in the period, Conrad's oeuvre explores the dynamics of extraction and exhaustion in the world ecology of industrialized imperialism. Focusing on *Nostromo* in this chapter and *Heart of Darkness* in the next, I take up two of his extraction narratives, about silver and ivory, respectively, highlighting his broad interest in extractivism. Elsewhere, in his narratives "Youth" and *Victory*, Conrad focuses more squarely on coal; for illuminating accounts of these texts, see Taylor ("While") and Samuel Perks.

19. Similarly, we must go to the sea, to stories like Conrad's "Youth," to find accounts of the global transportation network that connected extraction to consumption.

20. Amitav Ghosh asks whether the realist novel may be unfit for the representative tasks that climate change presents since "the very gestures with which it conjures up reality are actually a concealment of the real" (23), and Timothy Clark argues that "the main artistic implication of trying to represent the Anthropocene must be a deep suspicion of any traditionally realist aesthetic" ("Nature" 81). Although not specifically focused on environmental questions, Anna Kornbluh has recently suggested to the contrary that the realist novel is itself speculative and has called for readings of realism "not as symptom but as mediation" (55). Such a vision of realism aligns with my argument that extraction's remainders leach into the marriage and inheritance plots of provincial realism.

21. Robert Higney, for example, reads the mine as "only one among many" imperialist institutions in the novel (86). Huei-Ju Wang brings welcome attention to the Indigenous labor on which the mine depends but has little to say about silver extraction, viewing the laborers themselves "as the material base of the Sulaco mine" (8). J. Hillis Miller instead sees a "meaningless material base" in the novel: "the phrase 'material interests'" is a contradiction, because "matter has no 'interest' in itself" but rather "becomes interesting" through "the circuits of exchange and substitution" (170). See Hensley and Steer ("Signatures") for a reading of coal as the unacknowledged material base of *Nostromo*.

22. The use of so-called conflict minerals in the tech industry today has helped fund deathly conflicts, as in the Democratic Republic of Congo. In the United States, the 2010 Dodd-Frank financial-reform bill included a section on conflict minerals, requiring US companies to disclose their use of minerals from mines controlled by armed groups in the Congo, but many companies have not complied with the rules on reporting.

23. See Luz Elena Ramirez for a discussion of the sources and personages on whom Conrad draws in creating the fictional history of Costaguana (80–83).

24. Schmitt explains that the term "informal empire" was first used by Ronald Robinson and John Gallagher and suggests that British imperialism in Latin America, with its emphasis on commercial and economic interests rather than land and sovereignty, was "a precursor to what has been considered a decidedly later phase of imperialism" (*Darwin* 98–99). See Robert Aguirre (*Informal*) for a discussion of the term in relation to literary and cultural studies. Nasser Mufti says that with *Nostromo* Conrad "deterritorializes

Britain's civilizing mission" and "conceives of imperial interventionism as taking place *any-where* in the peripheries, even those sites previously untouched by Britain" (114). This is a compelling argument but must be considered alongside Britain's specific history of extractive imperialism in Latin America.

25. Eloise Knapp Hay has argued, contra most of the novel's critics, that imperialism is not a central concern of *Nostromo*, and that the novel's only imperialists are Holroyd and Sir John, the railway manager. Charles Gould, she says, "is insistently presented as a native Costaguanero, his family having been rooted in the region for generations" (83), and is "innocent of 'imperialist' designs" (84). This ignores that Gould is of English descent, is educated in England, spends his youth living in Europe, and is called "the Englishman" or the "Inglez" by other characters over a dozen times in the novel. Conrad's novel actually shows how colonial settlers expand the project of British imperialism by providing local allies to facilitate extraction and investment, as is the case with Gould and his British railway allies and American investors.

26. We find a similar formulation in Richard F. Burton's chapter "The Birth of the Babe" in volume 1 of *The Highlands of the Brazil* (1869). Here Burton describes the processing of gold in a Brazilian gold mine with an extended birth metaphor. As the gold is extracted from the ore, "the golden fluid is poured like a bar of soap into an oblong mould of cast iron previously warmed" and "thus the babe is born and cradled" (260).

27. There are key differences between the narrative mode of *Heart of Darkness*, a novel I discuss in the next chapter, and *Nostromo*, as J. A. Bernstein notes: in *Heart of Darkness* "one character narrates from a more or less consistent vantage point," while "*Nostromo* presents an ambiguous narrator who manages to slip inside characters' heads at times, at other times offering a semi-omniscient viewpoint" (204). These differences relate to the genre distinctions between the two texts, as will, I hope, become clear in chapter 2.

28. Conrad's story "Youth" represents extractive temporalities in similarly anachronous terms, with multiple characters who violate chrononormativity: Marlow is second mate on a voyage to ship coal from northern England to Bangkok, but he is only twenty years old; the captain of his ship, Beard, is sixty but is on his first command; the captain's wife, Mrs. Beard, is "an old woman with . . . the figure of a young girl" (71).

29. Citations for *The Mill on the Floss* refer to the 2007 Broadview edition.

30. See Seamus O'Malley for a reading of *Nostromo* as historical novel and "study of modernist historiography" (52).

31. See Griffiths for more on Eliot's comparative historicism (*Age of Analogy* 166–210). On Eliot, history, and epochal change, see Mary Wilson Carpenter; Suzanne Graver; and Hensley.

32. The narrator claims, for example, to "possess several manuscript versions" of the life of St. Ogg, the town's patron saint (154).

33. Joseph Wiesenfarth quotes an 1869 letter from Emily Davies that notes that Eliot "read about the Trent to make sure that the physical conditions of some English rivers were such as to make the inundation possible," and he conjectures that the great flood referred to by the novel's old men refers to an actual November 1771 event (Eliot, *Writer's* 168). As Oliver Lovesey notes in his introduction to the novel, "Eliot wanted to be scrupulously accurate both in her novel's account of natural history from thirty years before and in her references to water rights" and researched both extensively (32).

34. The "shifting baseline syndrome" is a term coined in 1995 by Daniel Pauley in a discussion of fisheries science: "each generation . . . accepts as a baseline the stock size and species composition that occurred at the beginning of their careers, and uses this to evaluate changes. When the next generation starts its career, the stocks have further declined, but it is the stocks at that time that serve as a new baseline." To counteract this tendency,

he called for "developing frameworks for incorporation of earlier knowledge . . . into the present models . . . adding history to a discipline that has suffered from lack of historical reflection" (430)—in other words, the fisheries science version of a historical novel.

35. See Shuttleworth, Jonathan Smith, and Buckland for various points of view on how catastrophism and uniformitarianism, divergent nineteenth-century geological theories, inform Eliot's novel.

36. For a longer discussion of water power and steam power in Eliot's novel, and the variant temporalities of these two energy regimes, see my article "Fixed Capital and the Flow." Tamara Ketabgian also discusses water power and steam power in *The Mill on the Floss*, focusing on how steam technology serves in the novel as a model for organic life.

37. See Kreisel ("Superfluity") for a full discussion of saving, investment, and political economy in the novel.

38. Poor ventilation in weaving sheds eventually led to a provision in the Factory and Workshop Act of 1901 that "the arrangements for ventilation shall be such that during working hours in no part of the Cotton Cloth Factory shall the proportion of carbonic acid (carbon dioxide) in the air be greater than nine volumes of carbonic acid to every ten thousand volumes of air" (W. Williams 196).

39. Jules Law examines the confused legal situation around water rights in the context of Eliot's novel, at which time the laws governing irrigation technology were in their infancy.

40. On Lincolnshire as the model for the novel's setting, see Birch (ix). On ironstone mining in Lincolnshire, see Sympson (chapter 14, "Mines and Minerals") and Richard Meade (chapter 8, "Lincolnshire Iron Industries").

41. The other thwarted marriage plot concerns Stephen Guest; I will not discuss this plot at length but would note that Maggie's decision not to elope with Stephen is another means by which the novel resists conventional narrative means of establishing futurity and continuity.

42. In her influential study *Darwin's Plots*, Gillian Beer observes that "the novel as a form is particularly dependent on the future for its pleasures" (172), and that in most of George Eliot's novels "the future is suggested through progeny" (173); the two exceptions are *The Mill on the Floss*, which Beer does not discuss at length, and *Daniel Deronda*, where, she says, "succession and progeny" are achieved through "race and culture" (173).

43. See Clare Walker Gore for a reading of Philip and Maggie's relationship as a rewriting of the "redemptive disability plot," an unusual case where it is the disabled character "rather than his beloved, who lives on into the future beyond the last page of the novel" (174–75). Nevertheless Wakem's story, Gore says, is not full enough to compensate for the ending's "overwhelming sense of fruitlessness" (182).

44. Reading Maggie and Philip's thwarted marriage plot in terms of environmental circumstances diverges from the approach of recent critics like Talia Schaffer, who positions the novel in the context of a historical shift from familiar to romantic marriage; and yet the historical shift in marriage that Schaffer describes tracks with underlying changes in social organization such as the shift to extraction-based life. In Schaffer's reading, *The Mill on the Floss* "takes the forms of marriage that allowed people to express their yearning for a larger, richer sociality and argues that modernity is transforming them into individualizing, isolating experiences" (195). Such a transformation is related, I would suggest, to the shift toward an extraction-based society, which was premised on a break with the interests of future generations, or, to use Schaffer's terms, a break in the bonds of kinship as they extend into the future.

45. *Germinal* was translated to English in 1894 by Havelock Ellis and was widely discussed among British socialists, as I describe in my book *Slow Print*. Although its naturalist

mode distinguishes it significantly from the novels I discuss in this chapter, *Germinal*'s climactic sex scene between two characters trapped in a mine, one of whom dies shortly after consummation, treads some of the same ground treated here: "This at last was their wedding night, in this tomb, on this bed of mud. . . . They loved each other in despair, in death itself. There was no sequel. . . . She was dead" (486).

46. Joseph Kestner has also discussed *Jane Rutherford*'s domestic approach to the mining question, arguing that the novel conveys how the "microcosm of the patriarchal family, with its unjust distribution of power, reflects the situation of the industrial world itself" (379).

47. I am grateful to Tom Randall, trustee of Somerset Coalfield Life at the Radstock Museum, for biographical information on Frances (Fanny) Mayne (1811–55). For more on Charles Otway Mayne (1808–67) and the Wellsway Disaster, see "Diabolical Plot."

48. There is some discrepancy in the ages of the victims as reported in the press, but here I follow the information provided by the Northern Mine Research Society (see "Wellsway Colliery").

49. Abolitionist in its politics, *True Briton* ran an illustrated serial version of *Uncle Tom's Cabin* in 1852, which was early among the various British serializations.

50. See Laura Marcus's third chapter for more on railway novels. We are still waiting (as far as I am aware) for a critic to bring an environmental humanities approach to this body of fiction.

51. Kestner discusses the importance of the 1842 act in terms of the novel's historical setting.

52. MacDuffie has written that "the idea that a parent's professional skills would . . . directly encode themselves onto the bodies of offspring was . . . widely held" in the Victorian period, a holdover from Lamarckian theories of inheritance ("*Jungle*" 25). For more on discussions of miners' bodies in this period, see Cregan-Reid.

53. Timothy Mitchell, in *Carbon Democracy*, criticizes "the old argument" to explain coal miners' militancy, "that mining communities enjoyed a special isolation compared with other industrial workers." He emphasizes instead the particular autonomy enjoyed by mine workers underground as well as the "flow and concentration of energy [that] made it possible to connect the demands of miners to those of others. . . . Strikes became effective, not because of mining's isolation, but on the contrary because of the flows of carbon that connected chambers beneath the ground to every factory, office, home or means of transportation that depended on steam or electric power" (20–21).

54. A poem that ran in the 30 June 1853 issue of *True Briton*, just after the fourth installment of *Jane Rutherford*, made use of pit talk, suggesting that it was not unfamiliar to the journal's readers. The poem, called "The Miners' Complaint," is a reprint from *Traveller's Library* and includes such words as "marrows," "jud," and "jenkin." A glossary appears at the end of the poem, telling us that "marrows" means "companions" and "jud" and "jenkin" are technical terms used in hewing coal; clearly the journal did not presume all readers' knowledge of pit talk.

55. Relevant here is Rebekah Sheldon's account of reproductive futurism as "a causal structure that seeks to secure the self-same" (40), an assurance "that procreation will guarantee reproduction of the past into the future" (57), giving life to patriarchy and capitalism with the birth of each child.

56. The risks of mining accidents and the risks of exhaustion were sometimes discussed together in terms of their generational inequities, as we see in *The Report of the South Shields Committee, Appointed to Investigate the Causes of Accidents in Coal Mines*, produced in the wake of the St. Hilda's Colliery Disaster, which killed fifty-one men and boys: "It is, then, important, not only for the sake of humanity, but for the continued prosperity of Britain, that the hope of the future be not sacrificed to the interest of the present,

and that a SAFE, ECONOMICAL, AND WELL-ARRANGED SYSTEM of working those mines be established" (6).

57. The Wellsway Disaster was ruled at the time to be the result of foul play: "some diabolical malice had caused this calamity wilfully and premeditately [*sic*]," "the rope having been wantonly injured" ("Diabolical Plot"). No suspect was identified, and at the inquest the verdict was returned of "Wilful Murder against some person or persons unknown." Given the lack of evidence there is reason to doubt whether the rope was not really just faulty or worn rather than purposefully damaged: "In all probability there was no culprit, but the merest suspicion that one existed would prevent any prosecution for manslaughter due to negligence" ("Wellsway Colliery").

58. Mayne gestures at such rhetoric earlier in the novel: "Part of the miner's misery is unavoidable—much more is occasioned by his own reckless conduct in the mine and improvidence out of it" (99). For an example of this argument outside the novel, see the *Mining Journal*'s 1882 discussion of accidents and fatalities in Australian mines: "No amount of supervision will remove the natural carelessness of men's natures, and the familiarity with danger appears to have such an effect upon many, as to induce them to display greater heedlessness in pursuing their work" ("Mine Inspection").

59. The 3 June 1876 issue of the *Mining Journal* reprinted statistics on the number of mining fatalities that happened per year in Britain, taken from the President's Address at a meeting of the North of England Institute of Mining Engineers: "the total number of deaths which occur in and about mines in Great Britain and Ireland still average upwards of 1000 per annum," the president says. Around 1852 "it was as high as one death for every 226 persons employed," but in 1874 it had fallen to one death "in every 510 persons employed" per year—still a significant number (10). According to Sidney Webb, writing in 1921, "in Great Britain alone, possibly as many as a hundred thousand miners' lives have . . . been sacrificed" in colliery accidents (2). This estimate does not include, of course, the many miners that died in British mines overseas, where native or Indigenous miners' deaths may not have been properly counted. When they were, their fatalities were sure to exceed those of white workers, as in an 1888 accident at the De Beers Mine in South Africa, where a fire in one of the shafts killed "24 white men and 178 [Black Africans]" (*Reunert's* 28).

60. Durham Records Online provides a list of the victims, with names and ages, and accounts for each family group (Whitehead).

61. For more on the Hartley Colliery Disaster, see McCutcheon; Thesing and Wojtasik; "Frightful"; "Dreadful Colliery Accident"; and "Hartley Colliery Disaster." See also, on other mining disasters of the period, Marjorie Levine-Clark and Martha Vicinus.

62. Mayne's fictional plaque is based on a stone commemorating the Wellsway Disaster in Midsomer Norton churchyard. (I am grateful to Tom Randall for this information.) The original stone was replaced by a new one in 1965 ("Wellsway Colliery").

63. See Gallagher for an influential reading of this first sentence (70) and a full discussion of *Hard Times*'s relation to utilitarian thought. See also Courtemanche on *Hard Times* and "the utilitarian-industrial complex" (399).

64. For one example of the tendency to overlook *Hard Times*'s rural settings, see Patricia Johnson: "*Hard Times* takes place in Coketown, a milltown" (129). See John Parham ("Bleak") for a discussion of the city and country settings in another Dickens novel, *Bleak House*.

65. The airs *were* actually "killing airs," as Brimblecombe explains: pollution helps fogs form, and in the fog of 1873, for example, "there were 700 more deaths than normally expected in London at that time of year." In such "great fogs," "stagnant conditions allowed the pollutant concentrations to build up to high concentrations. These periods of continuous fog were followed by increased rates of mortality" (123).

66. John Tyndall published "Note on the Transmission of Radiant Heat through Gaseous Bodies" in 1859 and is generally credited with discovering "the absorption of radiant heat by gases" though he does not mention the implications for climate (Jackson 114). He was apparently unaware of the 1856 publication "Circumstances Affecting the Heat of the Sun's Rays" by the American woman Eunice Foote, a publication that did reference climate but which seems to have been largely ignored (Jackson 114). Swedish chemist Svante Arrhenius was a key early theorist of the greenhouse gas effect and calculated in 1896 that "if the quantity of carbonic acid increases in geometric progression, the augmentation of the temperature will increase nearly in arithmetic progression" (267). See Jesse Oak Taylor (*Sky*) and Heidi C. M. Scott (*Chaos*) for more on the greenhouse gas effect in nineteenth-century literature and thought.

67. Jesse Oak Taylor has interpreted fog and smoke in the Victorian novel as a measure of how humans have coproduced the modern climate by way of fossil fuel combustion (*Sky*); Heidi C. M. Scott has described the moral misgivings that attended such imagery from the start: "New symbols arose that Victorians struggled to comprehend: chiefly, the moral implications of the beleaguering smoke. . . . They perceived in it not just ecological impacts, but moral culpability too" ("Industrial" 589).

68. In a discussion of early Dickens novels that does not include *Hard Times*, Boone suggests that "if we attend to the non-urban settings" in Dickens's work "without dismissing them as mere pastoral interludes, we can see Dickens offering ethically important depictions of the nonhuman world" ("Early Dickens" 100). *Hard Times*, however, never directs us to read its nonurban settings as pastoral in the first place; these abandoned mine shafts clearly show that the countryside is itself industrialized.

69. Such an analysis might rhyme, in some ways, with those critics who have read the novel as opposing "rationally knowing" with "subjectively sensing" (see Lupton 151), but what I am more directly drawing on here is Stacy Alaimo's influential materialist notion of trans-corporeality: "Imagining human corporeality as trans-corporeality, in which the human is always intermeshed with the more-than-human world, underlines the extent to which the substance of the human is ultimately inseparable from the 'environment'" (2).

70. Starr, for example, reads such figuration as a comment on the industrialization of literature and the emergence of the novel as a manufactured good; Dugger reads the novel's manufacturing world as "modeled on the publishing world which produced [the] novel" (151), such that the "capitalist factory owners" are "analogous to authorship" and "the consumer market" is "analogous to readership" (164).

71. In a related reading, Ketabgian suggests that the automaton-like characters of *Hard Times* represent an "industrial metaphorics" meant to convey modernity's "mechanical forms of feeling" (70).

72. Louisa has a habit of staring into the hearth fire, and Patricia Johnson has written that Louisa herself is "fuel for the system of marriage and reproduction" that the novel depicts (134).

73. In Kearns's view, Dickens "loses [Blackpool] on the road for so long" because Blackpool is reduced to "impotence" and "canceled out as a character" (860).

74. Leitch reads *Hard Times* as "end-oriented" in an apocalyptic or eschatological sense (149).

75. "Gin-pit" employs a now-obsolete use of "gin," defined in the *Oxford English Dictionary* as a "mechanical apparatus used to draw up ore, water, etc., from a mine shaft; *esp.* one employing a windlass powered by draught animals, wind, or water."

76. This impossibility is evident, too, in Gertrude's wry comment about Walter's drinking habits: "the after-life would hold nothing in store for her husband: he rose from the lower world into purgatory, when he came home from the pit, and passed on to heaven in the Palmerston Arms" (30).

77. This despite the fact that a massive, monthlong national coal strike in 1912, as noted earlier, had ground the industry to a halt just before this novel's appearance.

78. Granofsky has read Paul's difficult relationship with his father in terms of Paul's attraction to and repulsion against his father's dirtiness, noting that "the emotional freighting of dirt with so much disgust . . . is a bit odd in the context of the culture described in *Sons and Lovers* because all the miners emerge from the pit coal-blackened every working day of their lives" (247).

79. The use of the shift system in mines, to maximize productivity after the passage of laws limiting miners' workday, exacerbated this underground detachment from natural rhythms, since the miners might be working hours at odds with daylight. Sidney Webb offers an account of how hard the shift system was on miners' wives, in particular, since they were apt to have more than one miner in the house (father and son) for whom they were expected to wash and cook, and there was no guarantee that these miners would not be on different shifts (71–72).

80. On *Sons and Lovers* as bildungsroman, unconventional bildungsroman, or incongruent bildungsroman, see Beards, Gillis, Sheehan, and Wiedenfeld. Weidenfeld argues that "Lawrence is trying to write a development narrative in which development is conceived as a socially embedded process" (311), and that critics who read *Sons and Lovers* as straightforward bildungsroman must "avoid or downplay large chunks of the narrative, such as Paul's relationship with his father and the family history that begins the narrative" (305). Both elements of the text mentioned here serve to locate Paul's development in the socio-environmental context of extractivism.

81. See Duncan for an account of the origin of the genre and its relation to the marriage plot ("Bildungsroman" 17).

82. Raymond Williams sees *Sons and Lovers* as introducing into English fiction both coal mining as "a central sector of working-class life" and the rejection of coal mining, for it exemplifies "the novel of working-class childhood, and of the move away from it," which "from *Sons and Lovers* on . . . is a significant English form. The working-class childhood is strongly written; the move away from it is given equal force" ("Working-Class" 115, 119).

83. Such belatedness is also the case for Catherine, a young worker in the coal mine in Zola's *Germinal*: "Since she was ten she had been earning her living in the pit . . . and if she had reached the age of fifteen untouched by man, it was because she was late in developing, and indeed the decisive change into womanhood was yet to come" (131).

84. See Doherty for a longer analysis of the role of sacrifice in *Sons and Lovers*. Doherty does not take up the idea of the sacrifice zone, despite the novel's extractive landscape, but his reading of sacrifice as the grounds for subjectivity in this novel could be usefully connected to its extractive milieu.

85. As Valeur and Berberan-Santos explain, phosphorescence was a subject of scientific debate in the years leading up to Lawrence's novel: "In the late 19th century, the question arose whether cold light violates the second law of thermodynamics, in that heat cannot flow from a colder body to a warmer body" (735). Experiments in fluorescent and phosphorescent lighting were underway at this time, as reported in W. S. Andrews's 1911 article for the *Engineering Review*.

Chapter Two. Down and Out: Adventure Narrative, Extraction, and the Resource Frontier

1. The "topdress" passage also evokes Britain's reliance on fertilizers mined overseas. Guano from Peru was often compared to gold—A. J. Duffield's *Peru in the Guano Age* (1877) makes an explicit comparison of "the Age of Guano with the Age of Gold" (9)—and in an 1890 letter to the London *Times*, Henry Kains-Jackson alleged that "the earth is

declining from its primal fertility" and that British agriculture relied overmuch on "fresh discoveries in the sub-strata of the earth's unknown mines of fertilizers that can be transmuted into agricultural gold." Kains-Jackson warned that these "mines are giving out," and "the balance is on the wrong side of the agricultural Budget, and is not passing on as father to son" (Kains-Jackson).

2. In a realist novel like Trollope's *John Caldigate* or Conrad's *Nostromo*, the colonial setting serves as a brief interlude in the main trajectory of the plot or as a place of settlement; in adventure narrative, however, the narrative action involves a journey outward to an extractive setting that remains a frontier from which one must return, preferably enriched by treasure.

3. John Coatsworth offers examples of such practices: "Europeans transformed the natural and human resource base of the entire New World" through "technologies and organizational forms" such as "deep-shaft mining" and "commercial credit" (27).

4. On mining and native dispossession, see, for example, Mark David Spence: "The first sustained contact between the Yosemite and whites took place in the midst of the Gold Rush, as thousands of Forty-niners invaded the central Sierra Nevada. In their feverish quest for some trace of the Mother Lode, miners brought epidemic diseases to native communities and destroyed carefully tended ecosystems. Moreover, the growth of mining camps and settlements also spawned a series of violent conflicts" while "the environmental destruction wrought by mining practices undermined seasonal hunting and gathering cycles" (102, 104). On the resource imbalance, see, for example, Svampa: "in 2011 agricultural, mineral, and commodity raw materials represented 76 percent of the exports of the countries of the Union of South American Nations, compared to only 34 percent for the world as a whole" (65).

5. Jason Moore's work builds on that of critical geographers like David Harvey, who argued in 1985 that "capitalism can open up considerable breathing-space for its own survival through pursuit of the 'spatial fix.' . . . It is rather as if, having sought to annihilate space with time, capitalism buys time for itself out of the space it conquers" ("Geopolitics" 338). Moore's analysis focuses on the ecological relations in which this spatial dynamic unfolds.

6. Similarly, in Benjamin Disraeli's *Sybil* (1845), the narrative perspective begins where the mine ends: "They come forth: the mine delivers its gang and the pit its bondsmen" (178).

7. D. H. Lawrence was actually from a mining family and, as noted in the previous chapter, described his uncle's death in a pit accident in his short story "Odour of Chrysanthemums," but even here the story focalizes the perspective of the miner's wife, Elizabeth Bates, waiting at home for her husband. Readers learn of the accident as she does, secondhand.

8. This is not to suggest that prior to Orwell no writer had taken readers down a mine. Émile Zola's *Germinal* (1885) was a breakthrough text that was, as noted in the previous chapter, influential within British socialism—Orwell's political and intellectual milieu. Of all the books discussed in *Extraction Ecologies*, *The Road to Wigan Pier* is probably the closest literary descendant of Zola's novel, even though it is not a novel, for Orwell's revelatory documentary style shares much with Zola's naturalism. For an example of a firsthand account written in English of a descent down a mineshaft that predates Orwell's, see Richard Burton's chapter "Down a Mine," in *The Highlands of the Brazil* (volume 1).

9. Long journeys to the coal face were a source of complaint for nineteenth-century miners. The working-class newspaper *Glasgow Sentinel* published a series of letters in 1864 from a Scottish miner, Thomas Gemmell, who emigrated to the United States and found "that the coal seams in his mines were . . . much thicker than those in most of the older Scottish pits" and "could be worked through level openings in the sides of the hills" so that "miners could usually walk to their workplaces in the mine through a relatively short

tunnel dug into the side of the hill" (Laslett 528). By the 1870s, however, US mining had also transitioned to the "deeper, shaft-operated coalpits" used in Britain (535), which Gemmell found so burdensome.

10. For primary and secondary sources that offer background on this Latin American context, see Aguirre, Bulmer-Thomas, Bunster, Burton, Coatsworth, Dahlgren, Deckard, Duffield, G. H., Hanley, Heath, Humphreys, *Observations*, Ramirez, Rippy, Romero, Tenenbaum and McElveen, United Mexican Mining Association, Ward.

11. For primary and secondary background on the South African context, see Churchill, Goldmann, Hatch and Chalmers, Huggan and Tiffin, *Industrial Prospects*, Kaufman, Maxwell, McCarthy, John Miller ("Environmental"), Munich, *Reunert's*, and Wybergh. For a recent discussion of British expeditionary literature about Africa starting with David Livingstone's *Missionary Travels and Researches in South Africa* (1857), see Adrian S. Wisnicki.

12. As mentioned in an earlier note attached to chapter 1, Schmitt (*Darwin* 98–99) as well as Aguirre (*Informal*) discuss the term "informal empire" in relation to literary and cultural studies.

13. As Bulmer-Thomas explains, "the period from the mid-nineteenth century to the First World War witnessed the rise of new export products throughout Latin America in response to the demands created by the Industrial Revolution." Precious metals were partially eclipsed as "new mineral products emerged . . . and rapidly came to prominence in the export structure" including copper, tin, and nitrates (63).

14. For a discussion of how the earliest efforts at rail transport were closely tied up with the colliery industry, see Dunn (52–55) and Sieferle (131). Stockton and Darlington Railway, the world's first steam railway, opened in September 1825 in northeast England, built to move coal from the mines to the ships.

15. The Congo Free State was ruled by King Leopold II of Belgium, who had been granted personal control over it at the Berlin Conference of 1884–85. As Huggan and Tiffin describe, Leopold's interest "was purely one of asset stripping," achieved through forced labor and other brutalities (159). In 1908, faced with international outcry, the Belgian government took over administration of the region and renamed it the Belgian Congo. Wisnicki says that *Heart of Darkness* "contributed to these developments" (135).

16. *Reunert's* tracks the diamond mines' increasing depths: "By the end of 1882 the deepest open workings in the blue ground of Kimberley Mine were over 400 feet beneath the surface" (27); at the De Beers Mine, "by 1886 the open workings there were 400 feet below surface and the underground mining 650 feet below" (30); by 1888 an "800-foot shaft was sunk in De Beers Mine," while at the Kimberley mine "from 1889 the new shaft was sunk to 1200 feet below surface" (48).

17. Racial capitalism understands slave labor and racialized labor markets as constitutive to the development of capitalism as a world system, not incidental to it. Some parts of the world—most notably, Brazil—still relied on slave labor in the mines during the period covered by this book. (Brazil agreed to end the slave trade in 1830, after importing more enslaved Africans than any other country during the era of the Atlantic slave trade, but did not enforce this ban until 1850 and did not abolish slavery until 1888 [Hanley 26].) Burton's chapter "The Black Miner" in volume 1 of *The Highlands of the Brazil* (1869) offers a rosy account of "the exceptionally humane treatment of the slave in the Brazil" (270), intended to encourage British investment in Brazilian mines despite the persistence of slavery there. The 15 July 1882 issue of the *Mining Journal*, however, includes a report on a parliamentary inquiry concerning the use of enslaved workers in the St. John Del Rey Mining Company, a British-owned outfit in Brazil, as late as the 1870s: "In the House of Commons on Tuesday Mr. Pease asked the Attorney-General whether, during the administration of

the late Government, steps were taken in view of a criminal prosecution of the directors of the St. John del Rey Mining Company for the working of a large number of slaves in their mines." It was argued that while "the conduct of the directors was very reprehensible . . . as regarded a criminal prosecution, great difficulties presented themselves" (851).

18. Although antiracist capitalization practices are currently a matter of debate and best practices may well change, I have opted in this book to capitalize the terms "Black" and "Indigenous" in reference to peoples, but not "white," since the capitalization of "white" is a practice now associated with white supremacist groups. Here I am following the most recent guidelines of the Associated Press, who consulted widely in reaching this decision (see Daniszewski, "Why" and "Decision").

19. Gold was discovered in New South Wales and Victoria in 1851, setting off the Australian gold rushes. As Philip Steer writes in his recent book *Settler Colonialism in Victorian Literature*, "between 1852 and 1861, the colony of Victoria alone received almost 300,000 emigrants from Britain; during that time, more than 28,500,000 ounces of gold were extracted from the continent." And yet, Steer continues, "the sensational discoveries at the antipodes stand out, paradoxically, for their seemingly negligible impact on Victorian literature." Steer shows that the rushes did influence some writers, including Trollope and William Stanley Jevons (80), and helped "catalyze the revaluation of mid-Victorian narrative models" (108), "silently naturalizing a settler-colonial hermeneutic" (115). But Australia is not a prominent setting in extractive adventure narrative in the same way as Latin America and Africa.

20. Travel narratives and novels about the Australian gold rushes do exist, but many are now hard to track down, and none are widely known. The most prominent is William Howitt's memoir *Land, Labour, and Gold* (1855), which documents the author's two years of traveling through central Victoria during the height of the gold rushes. For more on Australian adventure narrative, see Robert Dixon. For a more general discussion of the subject, see Steer (*Settler*).

21. Svampa characterizes "neoextractivism" as "large-scale enterprises, a focus on exportation, and a tendency for monoproduction or monoculture" (66). Her focus is South America, but the inequalities she describes also bedevil contemporary Africa.

22. Simonin, in *Mines and Miners*, uses similar language of visuality and concealment: "The patient and enduring labors of the coal-miner are deserving of serious examination on the part of all. If the multitude could but see them, they would be full of curiosity and eagerness; but the ground covers the greater number of these bold and grand colliery undertakings, and the world at large knows little or nothing about them" (251).

23. On Enlightenment taxonomies of living species, see Heather Swanson et al. as well as Andreas Hejnol.

24. After the burst of public recognition that greeted her memoir, Seacole was forgotten, as shown in a 24 December 1954 article from the *Times*, "Mrs. Seacole: A West Indian Nurse in the Crimea," which I found pasted into the Cambridge University Library's first edition of *Wonderful Adventures*. The author, "A Correspondent," regrets that Seacole "finds no place in recent biographies of Florence Nightingale," and says, "when I sought her own autobiography, *Wonderful Adventures in Many Lands*, from the shelves of the University Library at Cambridge I found the leaves unopened." The leaves of this particular volume have, fortunately, been opened many times since.

25. This raises the question of generic forerunners to the treasure-hunt adventure narrative that emerged in the nineteenth century with the rise of industrial extraction; travel writing is one key antecedent, as suggested above, and pirate stories are another. Monica F. Cohen argues that "nineteenth-century pirate fiction begins with Defoe" and that, not unlike treasure-hunt stories, such fiction "provided a means for expressing ambivalence

about an emergent consumer society in which new economies of exchange seemed to unmoor wealth from . . . the ownership of land" (19). Daniel Defoe's *Captain Singleton* (1720) is an important precursor here, given Singleton's acquisition of gold and ivory on his roving global travels, and yet Defoe's novel is also "anomalous," as Ian Newman argues, in that "it devotes most of its energies to depicting a character that shows no interest in the ontologies typically associated with modern individualism" including its property relations (571).

26. In the Penguin edition of *Wonderful Adventures*, Sara Salih summarizes the case against the idea that *Times* correspondent W. H. Russell, a friend and supporter of Seacole, was the editor and maintains that "the identity of Seacole's editor is elusive" (185).

27. Critical work on Seacole has emphasized how she "rewrites the conventions of women's autobiography," as Jessica Howell puts it, "by stressing her profitable profession and public identity" (33).

28. Citations of Seacole refer to the 2005 Penguin edition.

29. Both of these events were figured as the "second discovery of the New World," as H. G. Ward described Britain's move into Latin America (481). The California Gold Rush was similarly styled in Henry Vizetelly's memoir *Four Months among the Gold-Finders in California*: "The El Dorado of the early voyagers to America has really been discovered" (13).

30. Britain abolished the slave trade in 1807 but did not pass the Emancipation Act until 1833 and did not effect full emancipation until 1838. As Tricia Lootens describes, Britain made a rapid about-face in transitioning from a slave-holding empire to one that touted the moral superiority of its antislavery stance compared to the United States. *Wonderful Adventures* bolsters this notion of Britain by presenting the American gold seekers encountered by Mrs. Seacole as viciously racist.

31. Although the Panama Canal did not open until 1914, construction was well under-way in 1904, when Hudson's novel appeared.

32. See Aguirre (*Mobility*) for a detailed account of the building of the Panama Railroad.

33. Take, for example, Dunlop's account of Panama, published five years before Sea-cole's and directed to British capitalists: "The physical geography of the chain of the Andes, from Guatemala to the Isthmus . . . and the geological formations throughout induce me to believe that the spurs of the mountains and transverse valleys, between the chain and the sea, are still full of rich mining districts of virgin ores, which will well repay working out and careful search." Here Britain's extractive advantages in relation to a debt colony are already appreciated, and readers are encouraged to exploit them: "Some of these South American States, which owe large sums of debts to England, would surely offer every facil-ity to European enterprise, which might enable their soil to assist in repaying their debts" (31–32). Though the primary lender is now China rather than Britain, the situation Dunlop describes could equally be applied to Latin America today (see, for example, David Dollar).

34. Jim Hawkins does not provide us with the exact location of Treasure Island, but several clues indicate that it is a Caribbean island, and Stevenson himself wrote that it is "not in the Pacific" and is "supposed to have been in the West Indies" (see note 214n76 of Oxford edition).

35. *Treasure Island* not only was widely read at the time of its publication but was sub-sequently incorporated into school curricula and read by generations of students, as Bris-tow has discussed (30). Its impact on the extractive imagination can hardly be overstated.

36. Citations of *Treasure Island* refer to the 2011 Oxford University Press edition.

37. Here I am leaning on Naomi J. Wood's compelling reading of the map as an exem-plum of "the perils of paper [money]," given that the map is seen to represent the treasure and yet leads only to a rifled cache (77). See Sally Bushell, too, for a detailed discussion of Stevenson's map, which, she notes, is also a chart (617).

38. As mentioned in an earlier note attached to the introduction, sources on coal and British industrialization include Debeir, Deléage, and Hémery; Malm; Pomeranz; Sieferle; and Wrigley (*Energy* and *Path*). Extraction in Britain was not limited to coal, of course; as discussed elsewhere in this book, metals and rocks were mined and quarried there, including tin, copper, iron, and limestone. The British Geological Survey's 2014 list of active mineral workings in Britain includes fifty-six different extracted commodities (Cameron et al. xvi).

39. According to Heidi V. Scott, such arguments were also made in the context of Spanish imperialism: "the very existence of abundant mineral deposits below the surface was interpreted by some Spaniards as clear evidence that God had intended Peru's territories to be possessed by the Spanish" (1860).

40. For this reason adventure narrative is closely connected to the speculative subgenre of hollow earth fiction, which I discuss in chapter 3.

41. For helpful discussions of Marx's account of resource scarcity and natural limits, see Perelman, and Benton, as well as Moore.

42. Another cache with bar silver is left unexcavated on the island, and the last line of the novel ("Pieces of eight!"—spoken by Silver's parrot) likewise refers to a silver coin. Wood has read such references, and especially the pirate antihero, Long John Silver, in light of nineteenth-century debates about the gold standard and silver standard. It is a brilliant reading, but I would suggest that the novel's apparent ambivalence toward silver also evokes the great colonial silver mines of Latin America—storied extraction sites such as Potosí, where the forced labor of the *mita* nearly wiped out the Indigenous peoples compelled to work the mines.

43. This structure of thinking persisted as a framework for understanding famine in the Indian colonial context, as Kathleen Frederickson writes: "a brute Malthusianism . . . inheres in many British analyses of famine through the 1870s."

44. Marah Gubar has recently argued, in a book that takes up *Treasure Island* among other works, that children in "Victorian and Edwardian children's books . . . resist the Child of Nature paradigm . . . in favor of the idea that young people have the capacity to exploit and capitalize on the resources of adult culture" (4–5). My aim here, by contrast, is to position children's literature and its historical relation to child labor in the context of a world ecology of capitalism, which requires grappling with how the child is often regarded in such texts *as* a resource of adult culture.

45. It is a key tenet of children's literature criticism that the genre of children's literature emerged in tandem with the movement against child labor. A new ideal of childhood as a special time of development, to be spent learning and playing rather than working, helped fuel a vision for children's literature as well as the movement to restrict child labor. Christopher Parkes argues that children's literature at this time worked to redefine "the relationship of the child to the marketplace in order to accommodate the child within capitalist society" (1). This was a long process that extended through the nineteenth century and beyond; as Gubar writes, "well into the 1920s, activists such as Margaret McMillan were still fighting to ensure that poor children [had] the opportunity to experience childhood as a protected period" (204–5). Child labor continues to persist in many parts of the world today, of course, including in the extraction industry; see, for example, Annie Kelly on child labor and cobalt mining.

46. On this topic see also Peter Newell; and see Bob Johnson (61–84), who discusses an 1827 hearing in British Guiana where a planter was asked whether "steam is now employed in Berbice in substitution of labour formerly executed by slaves" (66). The hearing offers, Johnson says, "a glimpse of the emerging equivalence between carbon technologies and human labor as it first took shape" (65). For a discussion of *Treasure Island*'s veiled evocations of slavery and the slave trade, see Loman.

47. Wally Seccombe's study of working-class family life during the Industrial Revolution suggests that while "industrial capitalism did not inaugurate the use of child labour" it did rely on "a more voracious consumption" of it, such that children were working longer hours with fewer breaks at a greater intensity outside the home under the supervision of employers rather than kin. Some industries, such as the textile mill, were particularly reliant on child labor. Seccombe notes that "over half the English cotton factory labour force in 1816 were under eighteen" (35–36).

48. Despite their antinationalist tendencies, the pirates are clearly aligned with England: Captain Flint is explicitly identified as "English," and prior to sailing with him, Long John Silver worked under another pirate named Captain England.

49. Haggard's novel was one of the most popular among a host of novels set in Mexico around this time: Aguirre writes that "between 1857 and 1900, at least thirty British and American novels were set in either contemporary or pre-conquest Mexico or in the Mexican days of the Southwest" (*Informal* 138). Haggard's *Heart of the World* (1895), published two years after *Montezuma's Daughter*, is also an extractive adventure romance set in this part of the world.

50. "Not more than a tenth of the many mining enterprises organized by Englishmen for the purpose of operating in Mexico . . . were notable for big dividends" (Rippy 99), but the period of Mexican history during which Haggard's novel was published (the Porfiriato, roughly 1884 to 1910) saw a surge in British mine speculation on the heels of legal changes. Santiago explains that until 1884, "the mining laws derived from colonial Spanish law, which maintained that 'the juices of the earth' (*los jugos de la tierra*) belonged to the Crown" (62); to mine, concessions from the government had to be obtained. "Between 1884 and 1909," however, "the Mexican government passed laws asserting the principle that whoever owned the surface of the earth owned whatever existed below it," and "the owner was free to exploit . . . without government permits" (63). Initially these laws applied only to mining, but in December 1901 "the first law specific to petroleum appeared" (63), marking the beginning of a new stage in the exploitation of Mexico's underground resources that would last until the nationalization of oil in the Mexican Revolution (see Santiago).

51. The *Mining Journal* described the contents of one such bonanza in a 12 January 1884 report on the United Mexican Mining Company: "La Luz [broke] into bonanza in 1843, and continued to give profits up to 1856. The amount of silver extracted from the mine was fabulous. . . . Upwards of $7,000,000 (after payment of expenses) were divided amongst the owners" (36).

52. As discussed in an earlier note attached to this chapter, the 15 July 1882 issue of the *Mining Journal* includes a report on a parliamentary inquiry into the use of enslaved workers in the St. John Del Rey Mining Company, a British company in Brazil. The report notes that "it was by no means unusual for British subjects in slave-owning countries to have slaves" (851).

53. This recalls the "industrial revery" trope Mary Louise Pratt identifies in the writings of "the capitalist vanguard" in early nineteenth-century Latin America (150).

54. As the United Mexican Mining Association wrote in its 1827 report to the directors, several of their mines lack "a sufficient supply of wood for steam engines" (United xxv).

55. See Ramirez for a discussion of *Montezuma's Daughter* in the context of a longer history of literary depictions of Montezuma's first contact with Cortés (37–38).

56. See Burnett for a discussion of Sir Walter Ralegh's quest for El Dorado and the significance of this mythic city for British imperialism in Guyana, the purported site of the legendary city. British Guiana, founded in 1831, was Britain's only formal colony in South America, and long-standing British interest in the region dates back to Ralegh's expedition. As Burnett explains, "El Dorado and Ralegh's routes mattered to those who set about defining what British Guiana was in the nineteenth century" (47).

57. Guatemoc's name is more commonly rendered as Cuauhtémoc.

58. Pearson details Haggard's main historical sources for this narrative: William J. Prescott's *History of the Conquest of Mexico* (1843), Bernal Díaz del Castillo's *The Conquest of New Spain* (1632), and Jourdanet's French translation of Fray Bernardino de Sahagún's *General History of the Things of New Spain* (1569) (33).

59. See Hultgren for an overview of postcolonial criticism on Haggard, a thoroughgoing imperialist who had himself labored for the empire in the Transvaal.

60. This connects to John Miller's argument that Haggard's work is marked by a paradoxical "support of and retreat from the culture of Victorian capitalism" ("Environmental" 158).

61. On Haggard's trip to Mexico he "secured permission from the Mexican authorities to dig for antiquities and search for Montezuma's gold" (Deckard 40), but dazed with grief, he abandoned this plan and accompanied Jebb on a journey to his silver mine in Chiapas instead.

62. References to *King Solomon's Mines* refer to the 2002 Broadview edition unless otherwise noted.

63. Just as *Treasure Island* opens with a map, the first edition of *King Solomon's Mines* includes a fold-out frontispiece captioned thusly: "Fac-simile of the Map of Route to King Solomon's Mines, now in the Possession of Allan Quatermain, Esq., drawn by the Don José da Silvestra, in his own Blood, upon a Fragment of Linen, in the Year 1590" (36). As with *Montezuma's Daughter*, extractive desire is the hook from the outset.

64. Of the novel's publication history, Bristow writes, "Having read through the manuscript, [Andrew] Lang informed Haggard that it might be possible to place the novel with Harper's *Boys' Magazine*. Instead, it went to Cassell's who performed a successful advertising campaign that guaranteed the book a wide number of reviews. . . . At a time when the size and pricing of novels was changing, Haggard's single-volume work reached a larger public than the three-decker novels that were becoming expensive by comparison. 31,000 copies were sold within a year, and so its distribution extended well beyond the intended juvenile readership" (135).

65. In his 1877 text *The Gold Mines of South Africa*, Thomas Baines attributed the rush in Africa to the exhaustion of other mining regions: "rainbow hues of hope attracted to our shores a host of adventurers from well-nigh exhausted auriferous regions in California, Australia, and New Zealand" (qtd. in Stiebel 124–25). Haggard was familiar with Baines's work (Stiebel 126).

66. Monsman asserts that "everyone knows that the trio . . . went to Kukuanaland for its diamonds; the rescue of Sir Henry's brother served only as a convenient pretext for recovering Solomon's treasures, the objective of the lost George Curtis in the first place" (280).

67. Haggard's novel attributes the mine's construction to ancient biblical peoples, not to the Black Africans who live in the region; as many critics note, this aligns with the false, racist belief at the time that Great Zimbabwe was not built by Black Africans. See Stiebel.

68. John Miller notes that "the idea of nature's breasts" was "a favorite metaphor of Haggard's" ("Environmental" 168), one that appeared in other writings beyond these two novels. Not for nothing does Hensley call Haggard "Victorian literature's clumsiest stylist" (196), but a paucity of metaphors did not preclude dogged commitment to those he favored.

69. Laura Chrisman discusses Haggard's account of these ancient mines in terms of the 1870s archaeological investigations into the ruins of Great Zimbabwe. Conflating Great Zimbabwe with King Solomon's mines offers Haggard, she says, "a mythical/historical sanction for the present practice of South African mining" (37).

70. As Levine-Clark describes, "'a fall of coal' at the Homer Hill Colliery in the West Midlands town of Cradley trapped the miner Thomas Shaw underground." Local men tried but failed to disinter his body, and local officials decided to seal the mine to prevent

the release of dangerous gases (22). Levine-Clark says the "historical record suggests that Shaw's body remains in the mine" (25).

71. Likewise, it was Kurtz's "impatience of comparative poverty," and the effect of that comparative poverty on his marriage suit, that "drove him out there" to the Congo (Conrad, *Heart of Darkness* 120). Uneven economic gains and class inequality in Britain and Europe played a key role in motivating the workers of empire, both novels suggest.

72. Dryden says that Conrad "reviled Haggard's romances" (48) and argues that his work "subvert[s] the myth of imperial adventure in such fiction as Haggard's" (197), an interpretation that does not contradict the idea that Conrad was indebted to the forms Haggard used.

73. That blank space on the map, and the invitation it represents, has a long history: Denis Cosgrove describes how "Herodotus, recorder of the global knowledge produced by Alexander's [fourth-century BC] expedition" replaced linear boundaries with open spaces "implying a more diffuse, open frontier zone, an 'empty space' or uninhabited waste potentially available for further human expansion. This represents a critical shift toward a more imperialist vision" (44).

74. Delayed decoding is the technique by which Marlow, as narrator, provides us first with initial sensory impressions of an event, only later explaining their context or cause. For example, when Marlow's steamer is attacked with arrows at the inner station, the first sign of the attack is when Marlow "saw my poleman give up the business suddenly, and stretch himself flat on the deck." Then Marlow sees "sticks, little sticks, were flying about—thick: they were whizzing before my nose." Finally, a half-page later, he reveals the cause: "Arrows, by Jove! We were being shot at!" (55). A famous example is when Marlow looks through his binoculars and sees the "round carved balls" (64) outside Kurtz's lair; they are not identified as severed heads for eight more pages. For an ecological reading of how delayed decoding functions in Conrad's "Youth," see Taylor ("While"). On delayed decoding, environment, and "the foundational identification between character and setting," see McCarthy (64).

75. MacDuffie's comments on how and why *Heart of Darkness* disturbs the conventions of its genre are relevant here, though my focus is adventure narrative rather than nineteenth-century realism: "For Conrad, it's not just that such conventions are not up to the task of representing nature, but also that they have suppressed its reality to such an extent that the only way to access it now is through strategic, ironic excursions into alternative and unreal modes of discourse and perception" ("Charles" 560).

76. Conrad had spent time in the Congo Free State during his maritime career. While *Heart of Darkness* was published just in time to be read in light of Britain's conflict with the Boers in South Africa, the novel also responds more generally to the late nineteenth-century "scramble for Africa" and especially to Belgium's role in that historical travesty. Huggan and Tiffin write that "Conrad's portrait of the Company is based, at least in part, on Leopold II's *Association Internationale Africaine* and its successor association . . . 'fronts' for the amassing of Leopold's private wealth" (178), but as Ryan Francis Murphy notes, Conrad is also "careful not to single out Leopold directly, as so doing might work to absolve the remainder of Europe of its crimes" in the region (7). On this subject see also Wisnicki, and see this chapter's earlier note on the topic.

77. Here I am drawing on Fred Moten's reevaluation of the object as category in "Resistance of the Object: Aunt Hester's Scream": "The history of blackness is testament to the fact that objects can and do resist. . . . While subjectivity is defined by the subject's possession of itself and its objects, it is troubled by a dispossessive force objects exert such that the subject seems to be possessed—infused, deformed—by the object it possesses" (1).

78. See Allewaert for a discussion of the "ideological commodity fetish" at stake in Marx's *Capital* in relation to the "decidedly material fetishes that circulated in [the] Atlantic world" (205). Despite the sublimated connections that Allewaert identifies, some critics

have argued that Marx's theory of commodity fetishism assiduously evades the question of slavery and the context of the Black Atlantic, as in Walter Johnson's reading: "The magnificent critique of the commodity form with which Marx began *Capital* . . . unfolds from a detailed consideration of the nature of a bolt of linen. . . . But wait: *a bolt of linen?* . . . Marx's substitution of (British) flax for (American) cotton as the emblematic raw material of English capitalism enabled him to tell what in essence was a story of the commodity form artificially hedged in by British national boundaries" (301–2). Other critics argue, to the contrary, that Marx's analysis of US slavery and the Civil War was actually a crucial underpinning for the ideas he elaborated in *Capital*, as in Tom Laughlin's recent analysis of Marx and the Lancashire "Cotton Famine." Regardless of how we situate Marx himself, it remains the case that nineteenth-century political economy did generally tend to ignore slavery as a backward-looking, precapitalist formation, despite the massive amounts of profit achieved through its barbarities. Lisa Lowe argues that Victorian fiction (e.g., Thackeray's *Vanity Fair*) could be more attentive than political economic writings to "the intimacies between colonial commodities . . . and the imperial relations to Africa, Asia, and the Americas to which they were inextricably tied" (83), which suggests "the role of literature as a particular mediation of the global social and economic forces" (73–74). Lowe's account of literature's mediating capacities, which both recognize and suppress these colonial intimacies, is an inspiration for my approach in this book to the literature of global extraction.

79. Rubber can be sustainably grown and harvested (which is part of why it would not have worked for Conrad's symbolic purposes), but in the Congo Free State, it was an extraction industry. As McCarthy writes, "Congo rubber was wild. Unlike sustainable plantation rubber, its gathering generally involved the killing of the rubber vine and thus occasioned an expanding circle of jungle searching. The manner of this prospecting was yet more inhumane than the ivory trade, and Conrad would have known all about the Belgian system of quotas and hostage taking where men and women were forced to collect rubber at the threat of their children's limbs" (43). See also Elizabeth Chang, who notes that Conrad "revised the desired commodity in the Congo from the then-ascendant milky sap of the rubber tree to the already exhausted material of ivory" (*Novel* 127), because "the disavowal of a renewable and manageable plant life" was "a necessity of Conrad's fictional world" (128).

80. See Huggan and Tiffin (162), McCarthy (42), and Marzec (72). Ryan Francis Murphy's compelling reading is that the novel's "tendency to detach the substance from its animal source and imbue it with symbolic meaning" is meant as an ironic echo of the attitude of European consumers, who used ivory piano keys and billiard balls without paying attention to their material origins (10). Jesse Oak Taylor notes that two months before the first installment of *Heart of Darkness* ran in *Blackwood's*, the magazine ran an article by Alfred Sharpe "on the question of whether the African elephant was fated with extinction" ("Wilderness" 37–38), which recommended regulation of the ivory trade to forestall extinction. Ivory was a novelistic symbol freighted with exhaustion, in other words, when Conrad chose it.

81. Fossil ivory, as Conrad is using the term, is not "fossil" in the same sense as fossil fuel, e.g., preserved remains from an earlier geological period; fossil ivory is merely buried ivory, whether buried by nature (when an elephant dies) or by design. Some was evidently buried for safekeeping: in a discussion of Herbert Ward's *A Voice from the Congo* (1910), Huggan and Tiffin describe elephant tusks being buried by Bangala villagers "in case they are needed to redeem villagers captured by Arab slave traders or neighbouring hostile groups." This suggests how "the trades in ivory and slaves had long been intrinsically interwoven" (164). Ryan Francis Murphy describes the rise of the ivory trade in nineteenth-century Africa, controlled first by Arabs and then by Europeans, and notes that for the Indigenous tribes of central Africa, ivory had use value, for fencing and door posts and the like, but not exchange value—until the involvement of foreign powers (1).

82. I am grateful to Deanna K. Kreisel for this observation.

83. This feature of the novel has been discussed by critics interested in its intertextual references to Dante's *Inferno* and other narratives depicting journeys to the underworld. See Robert Evans, and Thomas Cleary and Terry Sherwood.

84. Conrad's *Victory* likewise shows, as Samuel Perks writes, that "capital moves to new and prospectively more profitable sites of extraction" and "leaves if a more profitable site is brought into the colonial capitalist world-ecology" (257)

85. I am grateful to Jesse Oak Taylor for this observation. Taylor has suggested that "elements of Marlow's account read like . . . a foray into planetary deep time, in which the human intrusion has proven only a momentary disruption" ("Wilderness" 21).

86. One must note here, as Troy Boone writes, that "Conrad's professional life as a sea-man (1879–93) coincides with one of the most profound technological transformations of the industrial era, the displacement of sail by steam in commercial and military vessels in the nineteenth century." Conrad consistently depicts the shift to steam as "a degradation of the seafaring life," one that alienated the mariner from the natural world (Boone, "Dirty Weather" 101–2).

Chapter Three. Worldbuilding Meets Terraforming: Energy, Extraction, and Speculative Fiction

1. For recent ecocritical accounts of Ruskin's lecture, see Vicky Albritton and Fredrik Albritton Jonsson (especially 34–42), Allen MacDuffie (chapter 5 of *Victorian*), Heidi C. M. Scott ("Industrial"), Jesse Oak Taylor ("Storm-Clouds"), and Daniel Williams.

2. Elizabeth Chang offers a useful overview of the hollow earth subgenre, tracing back to the early nineteenth-century hollow earth theories of John Cleves Symmes ("Hollow Earth"), and notes that in one of the earliest such narratives, the American utopia *Symzonia* (1820), the journey underground is prompted by a supposed exhaustion of surface world resources: "the resources of the known world have been exhausted by research, its wealth monopolized" (Seaborn qtd. in Chang, "Hollow Earth" 387). For a wider-ranging take on imaginary underworlds, see Rosalind Williams.

3. Wrigley's work (*Energy* and *Path*) has established the key role of energy in the Industrial Revolution, though some recent thinkers have taken issue with the Malthusian logic of his arguments. In this chapter I follow recent critics like Nicola Whyte and Axel Goodbody who conceptualize the Industrial Revolution as an energy revolution—"the Industrial Revolution was driven by the most significant change in the generation, distribution, and consumption of energy in modern times before the decarbonization of the economy and society on which we have recently embarked—namely, the shift from wood, water, wind, and muscle power to coal and its combustion in steam engines" (26)—while also acknowledging that energy is a nature-culture formation and that social factors like land enclosure and the factory system played a key role in industrialization.

4. The catch-all category of speculative literature contravenes a tendency among critics of science fiction, utopia, and fantasy to emphasize the divisions among these genres. In using it, I draw on the work of Uncertain Commons and scholars like Alexis Lothian, who prefers the term "speculative fiction" because it emphasizes "imagining things otherwise than they are, and . . . creating stories from that impulse" (15). Recent pieces by Margaret Ronda ("Social Forms") and Jayna Brown and Alexis Lothian emphasize features of speculative literature, such as worldbuilding and secondary worlds, that enable "imaginative nonalignments" with a given social reality (Ronda "Social Forms"). We can also find uses of the term in the period discussed in this chapter: H. G. Wells refers to "a certain speculative writer of quasi-scientific repute" (meaning himself) in his 1898 novel *The War of the Worlds* (127).

5. I borrow the "interregnum" formulation from Rob Nixon (69), "dithering" from Kim Stanley Robinson, and "derangement" from Amitav Ghosh.

6. The genealogy for all these genres is complex and contested, but I rely here primarily on the work of Darko Suvin for science fiction, Matthew Beaumont for utopian fiction, and Anna Vaninskaya for fantasy. If we agree, as I do, to call Bulwer's and Wells's novels science fiction, they are science fiction *avant la lettre*; Wells himself used the term "scientific romance." (See Stableford for a discussion of this term.) But Suvin, channeling Virginia Woolf, baldly claims of science fiction in Britain, "if ever there was in the history of a literary genre one day when it can be said to have begun, it is May Day, 1871" ("Victorian" 148), the date that saw the publication of *The Coming Race*; he calls *The Time Machine*, moreover, one of "the basic historical models for the structuring of subsequent SF" (*Metamorphoses* 222). On utopia, Beaumont describes the turn-of-the-century moment when the genre proliferated across English-language literature with "hundreds" of works "prophesying a future society from whose perspective the present state of affairs seemed manifestly unsatisfactory" ("Party of Utopia" 166; see also his *Utopia Ltd.*). "Fantasy," like "science fiction," is a later generic descriptor, as Vaninskaya explains, and she calls the classification of Tolkien as a fantasy writer "entirely *ex post facto*." She traces the designation to Ballantine's 1960s "Adult Fantasy" series (1) but notes that "fantasy" was used earlier in the "small world of British pulp genre magazines (a magazine called *Fantasy* appeared briefly in 1938)," at which time "its differentiation from science fiction was . . . minimal" (17). In the United States, the related genre of "sword and sorcery" also emerged in pulp magazines in the 1930s, as Mark Jerng describes (103), but its secondary worlds drew more closely on "the annals of human civilizations" than did Tolkien's "cosmology around elves" (109). Suvin's foundational criticism maintained a strict separation between SF (including the subcategory of utopia) and the lesser (in his view) genre of fantasy, but more recent critics have emphasized their shared speculative project of building secondary worlds and the shared habits of mind they produce. China Miéville, for example, notes that "if the predicates for a fantasy are clearly never-possible *but are treated systematically and coherently within the fantastic work*, then its cognition effect is precisely that normally associated with sf" (45).

7. Suvin, for example, points to the "immediate stimuli" provided by the Franco-Prussian War and the Paris Commune, along with other political and economic factors ("Victorian" 148). Michael Robertson similarly positions the utopian turn within an economic downturn from 1873 to 1896 (17). The Uncertain Commons suggests a connection between the fictional work of speculation and the rise of finance capitalism, a connection I discuss later in the chapter. Saler offers a cultural explanation, describing a "fin-de-siècle turn toward the ironic imagination, animistic reason, and imaginary worlds" that was "historically specific," drawing on aestheticism and its doctrine of artistic detachment to generate "new ideas about the imagination, imaginary worlds, and the search for specifically modern forms of enchantment" (160).

8. See Wrigley (*Path*) for details on how much energy was generated in England's preindustrial organic economy, which relied on the flow of solar energy, versus the industrial economy, which relied on coal stocks. See Rabinbach for a discussion of the nineteenth-century rise of productivism, "the belief that human society and nature are linked by the primacy and identity of all productive activity" (3).

9. See, for example, two recent journal special issues on the energy humanities, one edited by Jeff Diamanti and Brent Ryan Bellamy and one by Axel Goodbody and Bradon Smith.

10. See Underwood for a longer discussion of the term *energy* as it arose in the eighteenth and nineteenth centuries (33–43) and "the troubled relation" between ideas of energy and industry in this period (54).

11. Metabolism, to use Povinelli's simple definition, refers to "the full range of chemical and mechanical processes that all organisms (all life) use to grow, reproduce, and maintain

their integrity" (39), and it is a crucial concept for understanding Marx's ecological thought. John Bellamy Foster defines Marx's "metabolic relations" as "evolving material interrelations" (10), and Paul Burkett writes that "Marx always insists on [capitalism's] necessary material basis and substance in an interchange or metabolism between people and nature" (38). For a helpful digest of Marx's ecological thought and metabolic analyses, see Ronda ("Nature").

12. The case that I am making for speculative fiction's capacity to attend to the work of nature rhymes with Benjamin Morgan's recent argument that utopia is "the literary and aesthetic mode in which what Moore describes as world-ecology becomes available to representation" ("How We Might" 156).

13. How many years? The answer is complex, but at least a portion of those emissions will likely persist for millennia. See Carbon Brief and Duncan Clark, Zeke Hausfather, and Lisa Moore.

14. This is why I focus on the discourse around resource exhaustion rather than entropy: the time line of heat death was so extended that its implications for the social were limited compared to the generational timescales of mineral exhaustion. Much work on literature and thermodynamic law, including entropy, has already been written; see, for example, Gillian Beer (*Open*), Bruce Clarke, Barri Gold, Ted Underwood, and Tina Young Choi. In his treatment of the subject, MacDuffie interprets concern about entropy not only as "a signifier for cosmic anxieties" but as "an expression of unease about resource use, energy expenditure, and widespread environmental damage" (*Victorian* 66).

15. The term *speculation* was originally visual, and when it evolved to take on an economic significance it came to express "a reach toward those futures that are already latent in the present" (Uncertain Commons 14). In their introduction to a special issue devoted to the connections between speculative fiction and speculative finance, Higgins and O'Connell note that "speculative fiction, as a particularly modern form, has often (if not always) been attuned to the speculative and fantastical nature of capitalist economics" (2). Critics in the special issue (apart from Steve Asselin, who discusses nineteenth-century science fiction) mainly focus on recent contexts such as venture capital and neoliberal financialization, but there is a longer history here: Philip Steer, for example, has discussed "the blurring of generic boundaries between financial debates about colonial investment and utopian narratives of settlement" in the Victorian period (*Settler* 128).

16. On the debate over the Carmichael Mine and the use of public subsidies to fund what promises to be a losing investment, see John Quiggin; Ben Smee; and Peter Hannam.

17. Suvin has argued that science fiction "is not—by definition cannot be—an orthodox allegory with any one-to-one correspondence of its elements to elements in the author's reality" (*Metamorphoses* 71), but Jerng convincingly argues that "the development of secondary worlds invites *the desire for allegorical interpretation*" (107), that "allegory is characterized precisely by the instability of these two poles" (the secondary world and the author's reality), and that "it is this instability that sets off a process of cognition" (108).

18. See again Chang on the hollow earth genre. Suvin has called *The Coming Race* "a monotypic sub-genre unto itself" encompassing various SF subgenres including alternative history, extraordinary voyage, and future war ("Extraordinary" 231).

19. The novel was published anonymously, so its fame cannot be wholly chalked up to the fact that Bulwer was "the second most widely read novelist in Victorian England," outsold only by Dickens (Lane 598). I will refer to the author as Bulwer, following the most common convention, though he is variously referred to as Bulwer Lytton and Bulwer-Lytton. He is listed in the *Oxford Dictionary of National Biography* as Edward George Earle Lytton Bulwer Lytton, formerly Edward George Earle Lytton Bulwer, first Baron Lytton. As Andrew Brown explains in this entry, "His daunting array of names is a source of frequent confusion. . . . For the first forty years of his life his surname was Bulwer though

out of respect for his mother's family, to whose estates he was heir, he often styled himself Edward Lytton Bulwer. When his mother died in 1843 and he came into his inheritance, he changed his surname by royal licence to Bulwer Lytton (without a hyphen, though others sometimes supplied one)."

20. Citations for *The Coming Race* refer to the 2008 Broadview edition.

21. Susan Stone-Blackburn describes how Bulwer drew on scientific and evolutionary discourses as well as occult and paranormal discourses to imagine vril (247).

22. Other underworld peoples who do not use vril are subject to a form of energy racism: those nations that are not "held capable of acquiring the powers over the vril agencies" are "regarded with more disdain than citizens of New York regard the negroes" (88).

23. We are told little about the location of this mine, but we do know that it is not in an English-speaking country (167).

24. With a different time system, Vril-ya drama, too, moves to the beat of a different drum, and "all classic unities are outrageously violated" (103).

25. Judge, for example, sees vril as a figure for "force of habit" (145), and Suvin sees in the close relation between the terms "Vril-ya" and "vril" an "idealist metaphysics" where "language . . . is significant and 'realist'" ("Extraordinary" 243). Clarke, in *Energy Forms*, rightly positions *The Coming Race* in an emerging canon of energy literature, but perhaps because his book was published in 2001, he is more concerned to situate vril in the context of classical thermodynamics than the fossil fuel energy regime that has so spectacularly backfired on us in ways that have become especially apparent since the turn of the millennium. Barri Gold likewise positions *The Coming Race* at the intersection of literature and thermodynamics, exploring how energy concepts such as equilibrium, force, and diffusion govern the Vril-ya's political system.

26. Although vril has brought the Vril-ya beyond the state of competitive capitalism that characterizes life under fossil fuels, it has not had the effect of generating social equality, which Suvin chalks up to Bulwer's conservative politics: "the puzzling retention of degrees of wealth—totally unnecessary with unlimited energy literally at everybody's fingertips" reveals "a tortured transposition of the situation of British bourgeois 'aristocracy' and of capitalist individualism" ("Extraordinary" 243). Bulwer gives the Vril-ya both aristocratic idleness and the individuation through class difference that underlies bourgeois ideology, exemplifying how speculative fiction often remains beholden to the structures of subject formation available in the current system.

27. Here the novel recalls early theorists of global warming who speculated that the greenhouse gas effect, as we now term it, could one day be wielded as a benevolent power for human comfort. Both Joseph Fourier and John Tyndall emphasized the insulating power of the atmosphere as a "cosmically benevolent" force (Heidi C. M. Scott, *Chaos* 46). See also Jesse Oak Taylor (*Sky* 26).

28. See Gibson for a broader discussion of nineteenth-century science fiction (including utopian science fiction) by Indian writers.

29. Nijhawan notes that the story, though written and published in English, was translated into Bengali and other Indian languages soon after its publication (93). In 1908 "Sultana's Dream" was also published as a book by S. K. Lahiri and Company in Calcutta.

30. See Logan for a discussion of the *Indian Ladies' Magazine*, a publication that was founded by Kamala Satthianadhan and offered "a unique platform for Indian women writers" (xxix), though critics have debated whether it promoted a "nationalist identity" or an "unpatriotic adherence to an Anglicized-Christian-humanist value system" (241). Though the magazine published original literature by Indian women, their work appeared in the context of an English canon, perhaps in service of creating a broader, mixed canon. The September 1905 issue of the magazine, for example, which ran "Sultana's Dream," features "The Damsel of the Sangrael" from a Dante Gabriel Rossetti painting as frontispiece, as

well as an article on Charlotte Brontë and Wordsworth's poem "Daffodils" ("I wandered lonely as a cloud"), while the previous issue, from August 1905, includes a frontispiece image from Ravi Varma, a well-known Indian painter.

31. For more background on Hossain, see Bharati Ray, Md. Mahmudul Hasan ("Commemorating" and "Muslim"), Yasmin Hossain, or Mohammad Moniruzzaman Miah. Hasan notes that while Hossain's name is typically rendered as Rokeya Sakhawat Hossain in academic discourse, there is a "lack of clarity" about her name since women's "names were also thought to be *purdanasin* (secluded/private) at that time." Hasan further explains that Hossain used both "Roqiyah Khatun" and "R. S. Hossein" ("Commemorating" 44–45) and that the honorific "begum" is often attached to her name such that she is popularly known as Begum Rokeya ("Muslim" 739). In the *Indian Ladies' Magazine*, "Sultana's Dream" was published under "Mrs. R. S. Hossein."

32. On the novum (the invention or innovation that promotes cognitive estrangement in science fiction), see Suvin, *Metamorphoses* (63–89).

33. Page numbers reference the story's original publication in the *Indian Ladies' Magazine*, though I occasionally draw on paratextual material from the 1988 Feminist Press edition.

34. Sister Sara explains, "Our fields are tilled by means of electricity, which supplies motive power for other hard work as well, and we employ it for our aerial conveyances too. We have no rail road nor any paved streets here" (85).

35. On the Hydrogen Fuel Initiative, see "Hydrogen Fuel Technology." On the Hydrogen Fuel Initiative as disingenuous greenwashing, see David Mazovick.

36. The October 1905 issue of the *Indian Ladies' Magazine*, which followed "Sultana's Dream," included a response titled "An answer to Sultana's Dream" by Padmini (Padma Satthianadhan Sengupta, daughter of Kamala Satthianadhan), which does not entirely backpedal from the feminism of "Sultana's Dream" but softens it considerably. Here the women of Ladyland are again attacked, but this time they are defenseless: "The sun-force machine was out of order; it would not work. The enemy was coming. What were they to do?" (117). After the men save the day—"They rushed upon the enemy; the foe was annihilated. There was no need even of the sun-force"—the men and women agree to form an egalitarian society.

37. This figuration of a female, single-sex utopia as a garden utopia with no mines also appears in Wells's *Tono-Bungay*, published a few years after Hossain's story. Here a character named Ewart envisions "a sort of City of Women" with "a walled enclosure" and "dozens of square miles of garden—trees—fountains" (175) because "women like that sort of thing" (176). No man is allowed inside this "city-garden of women" "except to do rough work, perhaps." The men live instead "in a world where they can hunt and engineer, invent and mine and manufacture" (176). As Frances Bonner has said, "Sultana's Dream" offers an "early instance of the identification of 'bad' male science with weaponry and 'good' female science with gardens" (294). On gardens and gender, see also Merchant.

38. On the development of early solar power, see A.S.E. Ackermann; John Perlin; and Charles Henry Pope.

39. On Archimedes and solar power, see Pope (25) and Ackermann (541).

40. Perlin's take on solar energy goes back six thousand years to when "the Stone Age Chinese built their homes so that every one of them made maximum use of the sun's energy in winter" (xix).

41. See Oliver Meth and "Green Peace Report on Air Pollution in Mpumalanga."

42. See my introduction for a discussion of the rise of the oil in the early twentieth century.

43. Hossain's interest in connecting science to feminism can be gleaned from her Bengali writings as well as from "Sultana's Dream" (which was her only major publication in

English). Miah quotes from Hossain's *Abarodhbasini*, where she compares purdah to "the lethal carbonic acid gas, which being a painless killer, its victims are never alert to its hazards. Women kept confined to the home die a slow death by the effect of this fatal gas known as purdah" (43). Elsewhere, in another essay written in Bengali, Hossain connected extracted mineral wealth with gendered oppression: Yasmin Hossain writes that Hossain's *Stri Jatir Abanati* "identified jewellery as the primary symbol of slavery. Drawing parallels between the iron shackles of slaves and the golden chains with which a woman is ornamented, she wrote 'our precious jewels are but ornamentations of slavery'" (2).

44. "Sultana's Dream" is unusual among feminist utopias for its lack of focus on reproductive labor. Toward the end of the story we see the Queen walking with her four-year-old daughter, but otherwise there is little attention to the labors of rearing or birthing children (though Ladyland does possess a more open kinship structure, with an expanded network of "sacred relations"). Lothian notes that "more than a hundred utopian texts by women were published in Britain and America between the years 1869 and 1920" and that "overwhelmingly, reproductive futurism is the engine through which these . . . feminist futures are imagined" (40). Later feminist utopias would have a noteworthy boom with the emergence of second-wave feminism, and these also center reproductive labor as a primary concern.

45. Rolston's book has perceptive observations to make about gender and coal mining, but it should be noted that climate change and the environmental impacts of coal are woefully neglected in this study. A more comprehensive anthropological account of gender and the mining industry would consider not only women miners, but how women more generally will disproportionately bear (and are disproportionately bearing) the impacts of climate change owing to entrenched gender inequalities.

46. On India's indigo plantations, see Banerjee ("Drama"); on its tea plantations, see Rappaport.

47. The total output for all Great Britain in 1883 was 163,737,327 tons (*Cork Industrial Exhibition* 53).

48. The epigraph to Hasanat's article is from Francis Bacon, "The discipline of scientific knowledge, and the mechanical inventions it leads to, do not merely exert a gentle guidance over nature's course; they have the power to conquer and subdue her" (*The New Organon* 78, qtd. in Hasanat 114), and Hasanat argues that the women of Ladyland have "acquired the Baconian scientific knowledge so well" that they exert similar power over nature and "abus[e] nature's bounty" (115)—a gendered reversal that ultimately does nothing to transform human-natural relations. Hasanat discusses Ladyland's use of a "wonderful balloon" that draws water from the atmosphere to control rainfall and temperature (84), a form of ostensibly benevolent geoengineering that seems, to Hasanat, like hubris. Ladyland's control of the weather enables high agricultural yields: "We are all very busy making nature yield as much as she can" (85); at the same time, the Queen insists, "We enjoy Nature's gifts as much as we can" (86), indicating a tension between nature as worker and nature as giver in Hossain's utopia. See also Maitrayee Chaudhuri on this topic.

49. On the planet romance as a subgenre of science fiction, see Suvin, *Metamorphoses* 103.

50. Apart from the reference to the Kelmscott Press edition above, all other references cite the 2003 Broadview edition.

51. Extracted materials do exist in Nowhere, as evidenced by the bronze doors that Dick has damascened (74) or Bob's silver filigree belt buckle (62). Industrialized extraction, however, is over; these materials seem to be gathered on a much smaller scale, or perhaps even recycled (of the British Museum, for example, which "many people have wanted to pull . . . down and rebuild," Dick says, "there is plenty of labour and material in it" [99]). Other subsurface materials have been superseded in Nowhere: as Guest journeys up the Thames, he sees that "cast-iron bridges had been replaced by handsome oak and stone ones" (201).

52. *News from Nowhere* (like "Sultana's Dream") is a garden utopia where the concept of the "garden," as Arwen Spicer argues, inhabits a "discursive ambiguity" incorporating both "human agency" and "nature's agency," both "the uncultivated (wild) and the cultivated (humanly managed)" (67). When Guest asks Hammond why, if Nowhere is a "garden," they still have "wastes and forests," Hammond replies, "we like these pieces of wild nature . . . so we have them," and "as to the forests, we need a great deal of timber, and suppose that our sons and sons' sons will do the like" (120–21). Such references to both "wild nature" and forests as sustainable plantation speak to the ambiguity Spicer describes. In a recent analysis, Morgan suggests that the figure of the garden in *News* conveys "humanity's total and harmonious control of natural systems" such that "processes within the natural world . . . take the place of 'machinery' within nineteenth-century industrial society" ("How We Might" 147). If Nowhere has achieved such a level of mastery over nature, however, it has not been achieved through recognizably scientific, technological, or industrial means. Unlike "Sultana's Dream," where scientific investigation has produced solar power and weather balloons, in Morris's utopia "the so-called science of the nineteenth century" is dismissed wholesale as "an appendage to the commercial system" (175).

53. Utopian hydropower is both useful and beautiful: as Guest journeys up the Thames, he sees "many strikingly beautiful" mills including one "as beautiful in its way as a Gothic cathedral" (234).

54. Trains are apparently absent from the socialist future, for Guest refers to "the railway having disappeared" (225), though Henry Morsom, a resident of Nowhere, says that in the past during the transition to socialism some trains did run and were "worked in some way," Guest says, "the explanation of which I could not understand" (217).

55. On fogs and coal pollution, see Peter Brimblecombe and Jesse Oak Taylor (*Sky*); see also my notes on the topic attached to chapter 1. It was well understood at the time that "high levels of pollution . . . aid the formation of fog," and that the fogs "were thicker, more frequent and of a different colour from those of the past" (Brimblecombe 109). Meteorological records show that "fog frequency appears to have reached a peak in the 1890s" (111), after which point regulation such as the London Public Health Act of 1891 started to reverse the visible effluvium of combustion (163).

56. Critics such as Raymond Williams have identified Morris as one of the first who "began to unite [the] diverse traditions" of socialism and environmentalism ("Socialism" 46). For approaches to Morris along these lines, see Boos ("Aesthetic"), Gagnier, Gagnier and Delveaux, Gould, Kent, O'Sullivan ("'Morris'"), and Parham ("Sustenance"). See Benton for a discussion of capitalism as an "ecological crisis-generating mechanism," and the extent to which Marx understood it as such (178).

57. For a discussion of the paradigms of equilibrium, balance, and homeostasis in the natural sciences, see Brendan Kavanagh, and Benjamin J. Murphy.

58. Here I draw on Ronda's use of the term *remainder* "as a means of considering the relations between ecology, history, and form as they become newly visible in the devalued remnants of capital's circuits of production, circulation, and consumption" (*Remainders* 13). The lack of waste in Nowhere, however, suggests not just a society with no remainders but also Morris's susceptibility to the common Victorian "fantasy of a self-contained system where surplus is metabolized in such a way as to nourish and maintain that system" (Kreisel, "'Form'" 103).

59. Green pigments containing arsenic were used in the wallpapers produced by Morris's firm, as was common at the time, and an ongoing debate concerns the extent to which Morris knew or cared about the potentially poisonous effects of such pigments. Two unresolved questions are at stake in this debate: whether arsenic pigments in wallpaper actually posed a health risk, and whether Morris's failure to entertain the potential risk

was related to his wealth derived from an arsenic mine. See Andy Meharg and Patrick O'Sullivan ("William").

60. For a discussion of eroticism and reproduction in the novel, see Shea.

61. See Parrinder or Beaumont ("Red Sphinx") for more on *The Time Machine* as a response to Morris's communist utopia. See Godfrey's collection for more on the relation between Wells and Morris.

62. As Cantor and Hufnagel note, "the raw materials of the time machine are the raw materials of empire" (54), and the ivory, they observe, suggests resonances between *The Time Machine* and *Heart of Darkness*, a novel discussed in chapter 2. Caroline Hovanec also considers *The Time Machine* in relation to *Heart of Darkness*, emphasizing, like Cantor and Hufnagel, the question of modernism's origins.

63. Some of the earliest aluminum smelters in Britain were established in the Scottish Highlands, not long before the publication of Wells's novel, because the extraction from ore demanded large amounts of electricity, which could be supplied here through hydropower (Oglethorpe 559).

64. Heidi C. M. Scott notes that the Morlocks and the Eloi exemplify "allopatric speciation, which occurs when two populations of a single species are spatially separated and evolve apart as they adapt to different conditions" (*Chaos* 66). See Redford on subterranean spaces in Wells's fiction and Shackleton on geological discourse in *The Time Machine*.

65. To cite a later example of such a reading, Holderness speculated in 1984 that "the Morlocks represent [a] bourgeois conception of the miners, evolved to an ultimate extremity of estrangement and terror" (21).

66. See M. Williams et al. for a discussion of the "limited subsurface record" of preindustrial civilizations compared to the "complex burrow systems at and below ground level" of nineteenth-century London and subsequent megacities (146).

67. See Carolyn Lesjak for more on the enclosure of common lands in the long nineteenth century and the extended time line of enclosure's impacts. See Gould for a discussion of the commons preservation movement and open-space socialism in Wells's day.

68. See chapter 8 of MacDuffie's *Victorian Literature* for a reading of *The Time Machine* that centers degeneration theory and social class.

69. See Elana Gomel for a longer discussion of evolution in *The Time Machine* and the multiple chronotopes Wells uses to explore evolutionary temporality: "time travel necessitates historical determinism, while contingent evolution denies it" (341).

70. If *The Time Machine* imagines the negative values of extractivism in biological terms, this fits with what Caleb Fridell describes as Wells's general tendency "to associate the idea of a linear, continuous progress with technology, while associating regress with biology" (175).

71. The structural demand for denial within industrial-capitalist life also comes up in *News from Nowhere*: describing the keen satisfactions of the future, when the pleasures of the surface no longer depend on exploitation below, Guest says, "I could enjoy everything without an afterthought of the injustice and miserable toil which made my leisure" (183). As critics such as Emelyne Godfrey and Teresa Trout have discussed, Wells's mother's position as housekeeper at a wealthy estate gave the young Wells a clear view of the upstairs/downstairs dynamics of class hierarchy.

72. See Otjen for a reading of *The War of the Worlds* as "a reaction to the fear of late nineteenth-century fossil fuel overconsumption" (129).

73. See Canaday for a discussion of *The World Set Free*'s influence on Hungarian physicist Leo Szilard, a nuclear pioneer.

74. For example, the marriage plots of *Tono-Bungay*, unsurprisingly given its focus on extraction and exhaustion, end in much the same way as the novels featured in chapter 1.

In George's first ill-fated marriage to Marion, "no children came to save us," for Marion felt "a disgust and dread of maternity" as "the fruition and quintessence of the 'horrid' elements in life" (187). With Beatrice, the love of his life, romance blooms briefly in a chapter titled "Love among the Wreckage," where George describes love as a "mighty passion, that our aimless civilization has fettered and maimed and sterilized," "a fruitless thing" (372). Beatrice declines his proposal because "I couldn't be . . . any sort of wife, any sort of mother. I'm spoilt" (378). Meanwhile, George's aunt and uncle who helped raise him have no children of their own, but when his uncle dies his aunt says of her husband, "he was my child and all my children" (368). Summing up in his final chapter, George says, "I have told of childless Marion, of my childless aunt, of Beatrice wasted and wasteful and futile. What hope is there for a people whose women become fruitless?" Such reproductive plots are, he asserts, a condition of broader economic-environmental forces: "It is all one spectacle of forces running to waste, of people who use and do not replace . . . hectic with a wasting aimless fever of trade and money-making" (381).

75. George suspects that the heaps are "the outcrop of a stratum of nodulated deposits" (328), but for the full particulars he tells readers to consult "the *Geological Magazine* for October, 1905" (329).

76. I focus here on *The Hobbit* with only minimal reference to other parts of Tolkien's legendarium including *The Lord of the Rings*, which, like *The Hobbit*, depicts the Third Age of Middle-earth, and the writings gathered in *The Silmarillion*, which depict the First and Second Age. This is partly because *The Hobbit*, published in 1937, fits the time period of my study, and, more significantly, because of the prominent structural role of the extractive quest within its narrative. *The Lord of the Rings*, published in 1954 at the height of the nuclear age, could be said to depict the opposite of an extractive quest: a quest to dispose of a burdensome golden treasure, to send it back to the bowels of the earth.

77. Chris Brawley has suggested (following Le Guin in "The Critics, the Monsters, and the Fantasists") that an agential shift from human to nonhuman is a special affordance of fantasy, making the genre particularly amenable to ecological thinking (293). Many critics have discussed environmental relations in Middle-earth, and others have addressed the significance of metal and mineral treasure in this world, but I have found none that think the two together in terms of extraction ecologies. On environment, see, for example, Brawley, Flieger ("Taking"), Jeffers, Eleanor R. Simpson, and Thiessen. On treasure, see, for example, Flieger ("Jewels"), Kinane, and Loughlin.

78. Boris Gorelik recounts that in 1920 Tolkien applied for a professorship at the University of Cape Town, a position that was financially sponsored by "the mining empire of De Beers" (8). He was awarded the position but declined for family reasons.

79. "On Fairy-Stories" was originally presented as a lecture in 1939 and first appeared in print in 1947.

80. Tolkien capitalizes "Secondary World" and "Primary World" in "On Fairy-Stories," but in keeping with more recent critical practice I will not capitalize them hereafter.

81. See Tally for more discussion of fantasy as enabling a critical perspective on the present, as with utopia or SF.

82. William H. Green describes the influence of *Treasure Island* and *King Solomon's Mines* on *The Hobbit*, calling Tolkien "the most influential and innovative heir of Stevenson and Haggard" (53).

83. A number of critics have discussed Morris's influence on Tolkien. Vaninskaya writes that Morris's late romances "brought together stylised archaism, pseudo-medieval settings, supernatural beings and artefacts and quest narratives: elements that have now come to characterise fantasy *tout court* in the popular imagination. Read avidly by Tolkien . . . they . . . furnished a storehouse of tropes that lasted Tolkien until the middle of the twentieth century" (9). On this topic see also Fitzpatrick.

84. On forests as "the shadow of civilization," see Robert Pogue Harrison.

85. This was not simply the sentence that began the novel; it was the sentence that began Tolkien's writing of the novel, for he jotted it down almost idly one day, and then concocted the story around it (Anderson 8). In appendix F to *The Lord of the Rings*, Tolkien explained that the origin for the word "hobbit" was "hole-builder" (510).

86. In the end Bilbo takes only a small portion of the treasure: "he would only take two small chests, one filled with silver, and the other with gold, such as one strong pony could carry. 'That will be quite as much as I can manage,' said he" (351).

87. Thorin says that the Lonely Mountain "had been discovered by my far ancestor, Thrain the Old" (54), but his grandfather Thror built the dwarf stronghold there. Jessica Seymour notes that "the dynastic nature of Dwarvish mining constructs the mine as a legacy to be cultivated and maintained" (33), a strangely agricultural conception, like the wood elves' underground palace with "pillars hewn out of the living stone" (223). This signals the dwarves' artisanal rather than industrial mining practices, for they use no engines or large machines, and it also signals the vast timescales of Middle-earth, where even resources that seem nonrenewable from a human perspective, such as stone, might be "living" in the eyes of the immortal elves.

88. A fair amount of criticism is devoted to the question of whether Tolkien's depiction of the dwarves is anti-Semitic. As characters the dwarves descend from the various little folk and chthonic folk of fairy tale and legend, but Tolkien openly acknowledged Semitic influences as well, such as in the dwarves' language. Although the dwarves in *The Hobbit* are not villainous and have many redeeming qualities, they are calculating and greedy and thus do express anti-Semitic stereotypes, despite the fact that Tolkien virulently rejected the "wholly pernicious and unscientific race doctrine" of the Nazis, as he put it in a 1938 letter (*Letters* 37), and despite the fact that the "dragon-sickness" of greed afflicts many of Tolkien's creatures (Vaninskaya 185). Several critics attribute Tolkien's softened depiction of the dwarves in *The Lord of the Rings* to his "firmer commitment to a cultural understanding of ethnic differences" in the wake of the horrors of the Holocaust (Saler 176). Seymour notes that within Tolkien's legendarium the dwarves are originally "carved from stone by their maker Aulë" who made them "stone-hard" and resilient so they could "withstand the ravages inflicted on Middle-earth by the dark powers which dwelt there" (29). On the broader question of race and fantasy literature, Jerng has made the persuasive argument that "racial worldmaking is not simply the deployment of stereotypes and racist language to build worlds. It is the instruction of the reader in specific modes of noticing race so as to make available certain knowledges about the world. It is the interaction of genre and race in the formation of a work's cognitive effects" (104). See Rebecca Brackmann on the dwarves and anti-Semitism; and see Robin Anne Reid for a review of scholarship on Tolkien and race (she specifically discusses the dwarves and Jewishness on pages 40–43).

89. On Haggard's influence on Tolkien, see Green, and William N. Rogers II and Michael R. Underwood.

90. Janet Brennan Croft connects the pastoral mode in Tolkien's work to Paul Fussell's argument that World War I writers made recourse to the pastoral to mark the calamitous antipastoralism of the war (9–12).

91. On mining, seismicity, and British coal mines, see Luckett. On earthquakes and fracking today, see Cook. I am grateful to George Hegarty for raising this connection.

92. In appendix E to *The Lord of the Rings*, Tolkien wrote that Middle-earth names with "au" should be pronounced "as in loud, how and not laud" (491). He also said in a 1938 letter that Smaug's name was from "the past tense of the primitive Germanic verb *Smugan*, to squeeze through a hole" (*Letters* 31). Neither of these points takes away from the name's evident pun on "smog," however, as several critics have noted. Fisher observes, too, that Smaug's name recalls the Old English verb *sméocan*, "to smoke" (108).

93. Tolkien objected to "the popular idea that the Ring of Power symbolized the Atomic bomb" (S. Kelly 127). As he wrote in a letter, "my story is not an allegory of Atomic power, but of *Power*." He followed this up, however, with a remark that seems to undercut his assertion: "If there is any contemporary reference in my story at all it is to what seems to me the most widespread assumption of our time: that if a thing can be done, it must be done" (*Letters* 246).

94. Canaday describes a series of 1908 lectures by Sir Frederick Soddy that did much to publicize the promise of newly discovered radioactive substances and their "apparently limitless use in the future" (228).

95. In a footnote to *The Hobbit*, Anderson discusses Tolkien's sources for the ring of invisibility, including two fairy stories by Andrew Lang published in the 1890s (Tolkien, *Hobbit* 133n31).

Conclusion

1. See Anna Kornbluh on the humanities' need for "thinking criticism constructively" (32).

2. See Kaiman and Lee and Wen for more on the toxicity of rare earth metals processing.

3. See Hsu for more on rare earth mining in the United States and Australia, and efforts to combat the environmental costs of this form of extraction.

4. Yusoff notes that Canada and Australia are now "the biggest extraterritorial mining countries" (48), which she links to their status as settler-colonial states; both were also part of the British Commonwealth and evolved within the shadow of Britain's nineteenth-century transition to extraction-based life.

5. My thanks to Colin Murphy for conversations on this point.

6. Here I am responding in very general terms to recent left debate about capitalism, the Green New Deal, and mineral resource extraction among such authors as Jasper Bernes, Kai Heron and Jodi Dean, and Thea Riofrancos ("Plan"). All these thought-provoking essays were published in 2019 or 2020, but I would note that this is a fast-moving conversation involving humanists, social scientists, engineers, and ecologists and one that concerns structural solutions as well as technological ones.

Achebe, Chinua. "An Image of Africa." *Massachusetts Review* 18.4 (Winter 1977): 782–94.

Ackermann, A.S.E. "The Utilisation of Solar Energy." *Journal of the Royal Society of Arts* 63.3258 (30 April 1915): 538–65.

Adams, William. *Solar Heat: A Substitute for Fuel in Tropical Countries for Heating Steam Boilers, and Other Purposes.* Bombay: Education Society's Press, 1878.

Adorno, Theodor. "Progress." 1964. Trans. Henry W. Pickford. *Can One Live after Auschwitz? A Philosophical Reader.* By Theodor W. Adorno. Ed. Rolf Tiedemann. Stanford, CA: Stanford University Press, 2003. 126–45.

Aguirre, Robert D. *Informal Empire: Mexico and Central America in Victorian Culture.* Minneapolis: University of Minnesota Press, 2004.

——. *Mobility and Modernity: Panama in the Nineteenth-Century Anglo-American Imagination.* Columbus: Ohio State University Press, 2017.

Ahuja, Neel. "Intimate Atmospheres: Queer Theory in a Time of Extinctions." *GLQ* 21:2–3 (2015): 365–85.

Alaimo, Stacy. *Bodily Natures: Science, Environment, and the Material Self.* Bloomington: Indiana University Press, 2010.

Albritton, Vicky, and Fredrik Albritton Jonsson. *Green Victorians: The Simple Life in John Ruskin's Lake District.* Chicago: University of Chicago Press, 2016.

Allewaert, Monique. "Electric Dialectics: Delany's Atlantic Materialism." Hensley and Steer, *Ecological Form.* 203–22.

Amatya, Alok, and Ashley Dawson. "Literature in an Age of Extraction: An Introduction." *Modern Fiction Studies* 66.1 (2020): 1–19.

Anderson, Douglas A. "Introduction." *The Annotated Hobbit.* New York: Houghton Mifflin, 2002. 1–25.

Andrews, W. S. "Fluorescence and Phosphorescence." 2-part series. *Engineering Review*, April 1911, 80; and May 1911, 78.

Andrews, William L. "Introduction." *Wonderful Adventures of Mrs. Seacole in Many Lands.* By Mary Seacole. Schomburg Library of Nineteenth-Century Black Women Writers. Oxford: Oxford University Press, 1988. xxvii–xxxiv.

Armstrong, Henry E. "Section L.—Educational Science." *Report of the Seventy-Second Meeting of the British Association for the Advancement of Science Held at Belfast in September 1902.* London: John Murray, 1903. 820–45.

Arrhenius, Svante. "On the Influence of Carbonic Acid in the Air upon the Temperature of the Ground." *London, Edinburgh, and Dublin Philosophical Magazine and Journal of Science* 41 (April 1896): 237–76.

Asselin, Steve. "Apocalypse, Inc.: Incorporating the Environment into the Boom/Bust Cycle in Fin-de-Siècle Science Fiction." *CR: The New Centennial Review* 19.1 (Spring 2019): 181–203.

Austen, Peter Townsend. "Harnessing the Sun." *North American Review*, June 1895, 160, 463.

Bakhtin, M. M. *The Dialogic Imagination: Four Essays.* Ed. Michael Holquist. Trans. Caryl Emerson and Michael Holquist. Austin: University of Texas Press, 1981.

Banerjee, Sukanya. "Drama, Ecology, and the Ground of Empire: The Play of Indigo." Hensley and Steer, *Ecological Form.* 21–41.

——. "Transimperial." *Victorian Literature and Culture* 46.3/4 (2018): 925–28.

Barrett, Ross, and Daniel Worden, eds. *Oil Culture*. Minneapolis: University of Minnesota Press, 2014.

Beards, Richard D. "*Sons and Lovers* as Bildungsroman." *College Literature* 1.3 (Fall 1974): 204–17.

Beaumont, Matthew. "The Party of Utopia: Utopian Fiction and the Politics of Readership, 1880–1900." *Exploring the Utopian Impulse: Essays on Utopian Thought and Practice*. Ed. Michael J. Griffin and Tom Moylan. Bern: Peter Lang, 2007. 163–82.

———. "Red Sphinx: Mechanics of the Uncanny in *The Time Machine*." *Science Fiction Studies* 33.2 (July 2006): 230–50.

———. *Utopia Ltd.: Ideologies of Social Dreaming in England, 1870–1900*. Leiden: Brill, 2005.

Beer, Gillian. *Darwin's Plots: Evolutionary Narrative in Darwin, George Eliot and Nineteenth-Century Fiction*. 1983. Cambridge: Cambridge University Press, 2000.

———. *Open Fields: Science in Cultural Encounter*. 1996. Oxford: Oxford University Press, 2006.

Benton, Ted. "Marxism and Natural Limits: An Ecological Critique and Reconstruction." *The Greening of Marxism*. Ed. Ted Benton. New York: Guilford, 1996. 157–83.

Bernes, Jasper. "Between the Devil and the Green New Deal." *Commune* 5 (Winter 2020). https://communemag.com/between-the-devil-and-the-green-new-deal/. Accessed 9 July 2020.

Bernstein, J. A. "'A Paradise of Snakes': Conrad's Ecological Ambivalence." Schneider-Rebozo, McCarthy, and Peters, *Conrad and Nature*. 196–210.

Birch, Dinah. "Introduction." *The Mill on the Floss*. By George Eliot. Oxford: Oxford University Press, 1998. vii–xxx.

[Biron, Chartres.] *King Solomon's Wives; or, The Phantom Mines*. By Hyder Ragged. London: Vizetelly, 1887.

Bishop, H. F. "A Suggestion for Solar Engine Inventors." *Scientific American* 98.19 (9 May 1908): 331.

Bolster, W. Jeffrey. *The Mortal Sea: Fishing the Atlantic in the Age of Sail*. Cambridge, MA: Harvard University Press, 2012.

Bonner, Frances. Introduction to "Sultana's Dream." *Inventing Women: Science, Technology, and Gender*. Ed. Gill Kirkup and Laurie Smith Keller. Cambridge, UK: Polity, 1992. 294–95.

Boone, Troy. "Dirty Weather." Schneider-Rebozo, McCarthy, and Peters, *Conrad and Nature*. 93–112.

———. "Early Dickens and Ecocriticism: The Social Novelist and the Nonhuman." Mazzeno and Morrison, *Victorian Writers and the Environment*. 97–113.

Boos, Florence S. "An Aesthetic Ecocommunist: Morris the Red and Morris the Green." *William Morris: Centenary Essays*. Ed. Peter Faulkner and Peter Preston. Exeter: University of Exeter Press, 1999. 21–46.

———. *History and Poetics in the Early Writings of William Morris, 1855–1870*. Columbus: Ohio State University Press, 2015.

Boos, Florence S., and Patrick O'Sullivan. "Morris and Devon Great Consols." *Journal of William Morris Studies* 19.4 (2012): 11–39.

Braber, Natalie. "Pit Talk in the East Midlands." *Sociolinguistics in England*. Ed. Natalie Braber and Sandra Jansen. London: Palgrave, 2018. 243–74.

Brackmann, Rebecca. "'Dwarves Are Not Heroes': Antisemitism and the Dwarves in J.R.R. Tolkien's Writing." *Mythlore* 28.3/4 (2010): 85–106.

Braun, Bruce. "Producing Vertical Territory: Geology and Governmentality in Late Victorian Canada." *Ecumene: A Journal of Cultural Geographies* 7.1 (2000): 7–46.

Brawley, Chris. "The Fading of the World: Tolkien's Ecology and Loss in 'The Lord of the Rings.'" *Journal of the Fantastic in the Arts* 18.3 (2007): 292–307.

Brians, Paul. *Nuclear Holocausts: Atomic War in Fiction, 1895–1984.* Kent, OH: Kent State University Press, 1987.

Brimblecombe, Peter. *The Big Smoke: A History of Air Pollution in London since Medieval Times.* London: Methuen, 1987.

Bristow, Joseph. *Empire Boys: Adventures in a Man's World.* London: HarperCollins, 1991.

Brown, Andrew. "Lytton, Edward George Earle Lytton Bulwer [formerly Edward George Earle Lytton Bulwer], first Baron Lytton." *Oxford Dictionary of National Biography*, 23 September 2004. https://doi.org/10.1093/ref:odnb/17314. Accessed 10 May 2019.

Brown, Jayna, and Alexis Lothian. "Speculative Life: An Introduction." *Social Text Periscope*, 4 January 2012. https://socialtextjournal.org/periscope_article/speculative _life_introduction/. Accessed 1 May 2020.

Bubandt, Nils. "Haunted Geologies: Spirits, Stones, and the Necropolitics of the Anthropocene." A. Tsing et al., *Arts of Living on a Damaged Planet.* G121–G141.

Buckland, Adelene. *Novel Science: Fiction and the Invention of Nineteenth-Century Geology.* Chicago: University of Chicago Press, 2013.

Bulmer-Thomas, Victor. *The Economic History of Latin America since Independence.* 3rd ed. Cambridge: Cambridge University Press, 2014.

[Bulwer Lytton, Edward]. *The Coming Race.* Edinburgh: William Blackwood and Sons, 1871.

Bulwer Lytton, Edward. *The Coming Race.* 1871. Ed. Peter W. Sinnema. Peterborough, ON: Broadview, 2008.

Bunster, Grosvenor. *Observations on Captain F. B. Head's "Reports Relative to the Failure of the Rio de la Plata Mining Association": With Additional Remarks, and an Appendix of Original Documents.* 2nd ed. London: E. Wilson, 1827.

Burkett, Paul. *Marx and Nature: A Red and Green Perspective.* New York: St. Martin's, 1999.

Burnett, D. Graham. *Masters of All They Surveyed: Exploration, Geography, and a British El Dorado.* Chicago: University of Chicago Press, 2000.

Burrow, Merrick. "The Imperial Souvenir: Things and Masculinities in H. Rider Haggard's *King Solomon's Mines* and *Allan Quatermain.*" *Journal of Victorian Culture* 18.1 (2013): 72–92.

Burton, Richard F. *The Highlands of the Brazil.* 2 vols. London: Tinsley Brothers, 1869.

Bushell, Sally. "Mapping Victorian Adventure Fiction: Silences, Doublings, and the Ur-Map in *Treasure Island* and *King Solomon's Mines.*" *Victorian Studies* 57.4 (2015): 611–37.

Butler, Robert. "Coal as an Actor in D. H. Lawrence's Early Plays: Challenging Anthropocentric Perspectives in the Cultural Representation of Human-Energy Relations." *Resilience: A Journal of the Environmental Humanities* 6.2–3 (2019): 53–71.

Cameron, D. G., T. Bide, S. F. Parry, A. S. Parker, and J. M. Mankelow. *Directory of Mines and Quarries, 2014.* 10th ed. Keyworth, Nottingham: British Geological Survey, 2014.

Canaday, John. *The Nuclear Muse: Literature, Physics, and the First Atomic Bombs.* Madison: University of Wisconsin Press, 2000.

Cantor, Paul A., and Peter Hufnagel. "The Empire of the Future: Imperialism and Modernism in H. G. Wells." *Studies in the Novel* 38.1 (Spring 2006): 36–56.

Carbon Brief and Duncan Clark. "How Long Do Greenhouse Gases Stay in the Air?" *Guardian*, 16 January 2012. https://www.theguardian.com/environment/2012/jan/16/greenhouse -gases-remain-air. Accessed 26 June 2020.

Carpenter, Edward. "The Smoke-Plague and Its Remedy." *Macmillan's Magazine* 62 (1890): 204–13.

Carpenter, Mary Wilson. *George Eliot and the Landscape of Time: Narrative Form and Protestant Apocalyptic History.* Chapel Hill: University of North Carolina Press, 1986.

Carr, Julie. *Surface Tension: Ruptural Time and the Poetics of Desire in Late Victorian Poetry.* Champaign, IL: Dalkey Archive, 2013.

Chakrabarty, Dipesh. "The Climate of History: Four Theses." *Critical Inquiry* 35.2 (Winter 2009): 197–222.

———. "The Planet: An Emergent Humanist Category." *Critical Inquiry* 46.1 (Autumn 2019): 1–31.

Chang, Elizabeth Hope. "Hollow Earth Fiction and Environmental Form in the Late Nineteenth Century." *Nineteenth Century Contexts* 38.5 (September 2016): 387–97.

———. *Novel Cultivations: Plants in British Literature of the Global Nineteenth Century.* Charlottesville: University of Virginia Press, 2019.

Chaudhuri, Maitrayee. "Ecology and Virtue in Rokeya Sakhawat Hussain, *Sultana's Dream.*" *Feminist Moments: Reading Feminist Texts.* Ed. Katherine Smits and Susan Bruce. London: Bloomsbury, 2016. 107–14.

Choi, Tina Young. "Forms of Closure: The First Law of Thermodynamics and Victorian Narrative." *ELH* 74.2 (2007): 301–22.

Choi, Tina Young, and Barbara Leckie. "Slow Causality: The Function of Narrative in an Age of Climate Change." *Victorian Studies* 60.4 (Summer 2018): 565–87.

Chrisman, Laura. *Rereading the Imperial Romance: British Imperialism and South African Resistance in Haggard, Schreiner, and Plaatje.* Oxford: Oxford University Press, 2000.

Churchill, Lord Randolph S., M.P. *Men, Mines and Animals in South Africa.* 2nd ed. London: Sampson Low, Marston, 1892.

Clark, Timothy. *Ecocriticism on the Edge: The Anthropocene as a Threshold Concept.* London: Bloomsbury, 2015.

———. "Nature, Post Nature." *The Cambridge Companion to Literature and the Environment.* Ed. Louise Westling. Cambridge: Cambridge University Press, 2013. 75–89.

Clarke, Bruce. *Energy Forms: Allegory and Science in the Era of Classical Thermodynamics.* Ann Arbor: University of Michigan Press, 2001.

Cleary, Thomas R., and Terry G. Sherwood. "Women in Conrad's Ironical Epic: Virgil, Dante, and *Heart of Darkness.*" *Conradiana* 16.3 (1984): 183–94.

Coatsworth, John H. "Economic and Institutional Trajectories in Nineteenth-Century Latin America." *Latin America and the World Economy since 1800.* Ed. John H. Coatsworth and Alan M. Taylor. Cambridge, MA: Harvard University Press, 1998. 23–44.

Cohen, Monica F. *Pirating Fictions: Ownership and Creativity in Nineteenth-Century Popular Culture.* Charlottesville: University of Virginia Press, 2017.

Cohen, William A. "Interiors: Sex and the Body in Dickens." *Critical Survey* 17.2 (2005): 5–19.

The Condition and Treatment of the Children Employed in the Mines and Collieries of the United Kingdom: Carefully Compiled from the Appendix to the First Report of the Commissioners Appointed to Inquire into This Subject. London: William Strange, 1842.

Conrad, Joseph. "Author's Note." 1917. Conrad, *Heart of Darkness.* 111–14.

———. *Heart of Darkness.* 1899. London: Penguin, 2007.

———. *Nostromo.* 1904. Oxford: Oxford University Press, 2009.

———. "Youth." 1898. *Selected Short Stories.* Ware: Wordsworth, 1997. 69–94.

"The Continuity of Nature." *English Leader,* 15 September 1866, 131–32.

Cook, Terri. "Ground-Shaking Research: How Humans Trigger Earthquakes." *Earth,* 15 April 2015. https://www.earthmagazine.org/article/ground-shaking-research-how -humans-trigger-earthquakes. Accessed 4 January 2021.

Coombs, David Sweeney, and Danielle Coriale. "V21 Forum on Strategic Presentism: Introduction." *Victorian Studies* 59.1 (Autumn 2016): 87–89.

Cork Industrial Exhibition, 1883: Report of Executive Committee, Awards of Jurors, and Statement of Accounts. Cork: Purcell, 1886.

"Cornish Mining in America." *Quarterly Review* 36 (1827): 81–106.

Cosgrove, Denis. *Apollo's Eye: A Cartographic Genealogy of the Earth in the Western Imagination.* Baltimore: Johns Hopkins University Press, 2001.

Courtemanche, Eleanor. "'Naked Truth Is the Best Eloquence': Martineau, Dickens, and the Moral Science of Realism." *ELH* 73 (2006): 383–407.

Cregan-Reid, Vybarr. "Ecologies of Labour: The Anthropocene Body as a Body of Work." *19: Interdisciplinary Studies in the Long Nineteenth Century* 26 (2018). https://19.bbk.ac.uk/article/id/1719/. Accessed 21 May 2020.

Croft, Janet Brennan. "The Great War and Tolkien's Memory: An Examination of World War I Themes in *The Hobbit* and *The Lord of the Rings.*" *Mythlore* 23.4 (2002): 4–21.

Crutzen, Paul J., and Eugene F. Stoermer. "The 'Anthropocene.'" *IGBP Newsletter* 41 (2000): 16–18.

Daggett, Cara New. *The Birth of Energy: Fossil Fuels, Thermodynamics, and the Politics of Work.* Durham, NC: Duke University Press, 2019.

Dahlgren, Charles B. *Historic Mines of Mexico: A Review of the Mines of That Republic for the Past Three Centuries.* New York: Charles B. Dahlgren, 1883.

Damian, Jessica. "A Novel Speculation: Mary Seacole's Ambitious Adventures in the New Granada Gold Mining Company." *Journal of West Indian Literature* 16.1 (November 2007): 15–36.

Daniszewski, John. "The Decision to Capitalize Black." *AP Style Blog*, 19 June 2020. https://blog.ap.org/_f07. Accessed 23 March 2021.

———. "Why We Will Lowercase White." *AP Style Blog*, 20 July 2020. https://blog.ap.org/_f1a. Accessed 23 March 2021.

Davies, Jeremy. *The Birth of the Anthropocene.* Berkeley: University of California Press, 2016.

Davis, Richard Harding. *Soldiers of Fortune.* 1897. Ed. Brady Harrison. Peterborough, ON: Broadview, 2006.

Debeir, Jean-Claude, Jean-Paul Deléage, and Daniel Hémery. *In the Servitude of Power: Energy and Civilization through the Ages.* 1986. Trans. John Barzman. London: Zed, 1991.

Deckard, Sharae. *Paradise Discourse, Imperialism, and Globalization: Exploiting Eden.* New York: Routledge, 2009.

Devas, Charles S. *Political Economy.* London: Longmans, 1892.

"Diabolical Plot against Human Life at Radstock.—Twelve Lives Lost." *Bristol Mercury*, 16 November 1839, 6.

Diamanti, Jeff, and Brent Ryan Bellamy, ed. "Energy Humanities." Special issue of *Reviews in Cultural Theory* 6.3 (2016).

Dickens, Charles. *Hard Times.* 1854. Ed. Jeff Nunokawa and Gage McWeeny. New York: Pearson, 2004.

Di Piazza, Elio. "Colonial Geographies in Two Novels by H. Rider Haggard." *Literary Landscapes, Landscape in Literature.* Ed. Michele Bottalico, Maria Teresa Chialant, and Eleonora Rao. Rome: Carocci, 2007. 88–98.

Disraeli, Benjamin. *Sybil; or, The Two Nations.* 1845. London: Penguin, 1980.

Dixon, Robert. *Writing the Colonial Adventure: Race, Gender and Nation in Anglo-Australian Popular Fiction, 1875–1914.* Cambridge: Cambridge University Press, 1995.

Doherty, Gerald. "D. H. Lawrence's *Sons and Lovers* and the Culture of Sacrifice." *D. H. Lawrence Review* 34/35 (2010): 5–24.

Dollar, David. "Chinese Investment in Latin America Continues to Expand." *Brookings*, 20 March 2018. https://www.brookings.edu/research/despite-slowdown-chinese-investment-in-latin-america-continues-to-expand/. Accessed 14 November 2019.

Do We Agree? A Debate between G. K. Chesterton and Bernard Shaw, with Hilaire Belloc in the Chair. London: Cecil Palmer, 1928.

"Dreadful Colliery Accident." *Leeds Intelligencer*, 25 January 1862, 5.

Dryden, Linda. *Joseph Conrad and the Imperial Romance.* Houndmills: Macmillan, 2000.

Duffield, A. J. *Peru in the Guano Age: Being a Short Account of a Recent Visit to the Guano Deposits.* London: Bentley, 1877.

Dugger, Julie M. "Editorial Interventions: *Hard Times*'s Industrial Imperative." *Dickens Studies Annual* 32 (2002): 151–77.

Duncan, Ian. "The Bildungsroman, the Romantic Nation, and the Marriage Plot." *Replotting Marriage in Nineteenth-Century British Literature*. Ed. Jill Galvan and Elsie Michie. Columbus: Ohio State University Press, 2018. 15–34.

——. *Human Forms: The Novel in the Age of Evolution*. Princeton, NJ: Princeton University Press, 2019.

——. "The Provincial or Regional Novel." *A Companion to the Victorian Novel*. Ed. Patrick Brantlinger and William B. Thesing. Oxford: Blackwell, 2002. 318–35.

Dunlop, Alexander. *Notes on the Isthmus of Panama with Remarks on Its Physical Geography and Its Prospects in Connection with the Gold Regions, Gold Mining, and Washing*. London: Joseph Thomas, 1852.

Dunn, Matthias. *An Historical, Geological, and Descriptive View of the Coal Trade of the North of England: Comprehending Its Rise, Progress, Present State, and Future Prospects*. Newcastle upon Tyne: Pattison and Ross, 1844.

Edelman, Lee. *No Future: Queer Theory and the Death Drive*. Durham, NC: Duke University Press, 2004.

"Egyptian and American Solar-Powered Plants." *Journal of the Royal Society of Arts*, 6 February 1914, 246.

E. H. "Subterranean Caverns, and Their Inhabitants." *Mining Journal*, 22 February 1868, 142.

Eliot, George. *Middlemarch*. 1871. Oxford: Oxford University Press, 2008.

——. *The Mill on the Floss*. 3 vols. Edinburgh: William Blackwood and Sons, 1860.

——. *The Mill on the Floss*. 1860. Ed. Oliver Lovesey. Peterborough, ON: Broadview, 2007.

——. *A Writer's Notebook, 1854–1879, and Uncollected Writings*. Ed. Joseph Wiesenfarth. Charlottesville: University of Virginia Press, 1981.

English, Henry. *A Compendium of Useful Information Relating to the Companies Formed for Working British Mines, Containing Copies of the Prospectuses, Amount of Capital, Number of Shares, Names of Directors, &c*. London: Boosey and Sons, 1826.

——. *A General Guide to the Companies Formed for Working Foreign Mines, with Their Prospectuses, Amount of Capital, Number of Shares, Names of Directors, &c*. London: Boosey and Sons, 1825.

"Equinoctial Storm, or Gale." *The New International Encyclopaedia*. Vol. 7. New York: Dodd, Mead, 1907. 166.

Esty, Jed. *Unseasonable Youth: Modernism, Colonialism, and the Fiction of Development*. Oxford: Oxford University Press, 2012.

Evans, Robert O. "Conrad's Underworld." *Modern Fiction Studies* 2.2 (1956): 56–62.

"extraction, n." *OED Online*. Oxford University Press, June 2017. www.oed.com/view/Entry/67087. Accessed 10 January 2018.

Field, Simon Quellen. *Boom! The Chemistry and History of Explosives*. Chicago: Chicago Review Press, 2017.

The Fiscal Problem. Cabinet Report, His Britannic Majesty's Government. 25 August 1903. National Archives (Kew, UK). Gale British Politics and Society Archive: British Cabinet Papers, 1880–1916. *Nineteenth Century Collections Online*. link.gale.com/apps/doc/ARV WEO053600554/NCCO?u=ucdavis&sid=NCCO&xid=8a16e7ea&pg=26. Accessed 28 May 2020.

Fisher, Jason. "Tolkien's Wraiths, Rings and Dragons: An Exercise in Literary Linguistics." Houghton et al., *Tolkien in the New Century*. 97–114.

Fitzpatrick, KellyAnn. "The Medievalism of William Morris: Teaching through Tolkien." *Teaching William Morris*. Ed. Jason D. Martinek and Elizabeth Carolyn Miller. Vancouver: Fairleigh Dickinson University Press, 2019. 65–78.

Flieger, Verlyn. "The Jewels, the Stone, the Ring, and the Making of Meaning." Houghton et al., *Tolkien in the New Century*. 65–77.

———. "Taking the Part of Trees: Eco-conflict in Middle-earth." *J.R.R. Tolkien and His Literary Resonances: Views of Middle-earth*. Ed. George Clark and Daniel Timmons. Westport, CT: Greenwood, 2000. 147–58.

Foster, John Bellamy. *Marx's Ecology: Materialism and Nature*. New York: Monthly Review Press, 2000.

Frederickson, Kathleen. "British Writers on Population, Infrastructure, and the Great Indian Famine of 1876–8." *BRANCH: Britain, Representation and Nineteenth-Century History*. https://www.branchcollective.org/?ps_articles=kathleen-frederickson-british-writers -on-population-infrastructure-and-the-great-indian-famine-of-1876-8. Accessed 7 October 2019.

Freedgood, Elaine. *The Ideas in Things: Fugitive Meaning in the Victorian Novel*. Chicago: University of Chicago Press, 2006.

Freeman, Elizabeth. *Time Binds: Queer Temporalities, Queer Histories*. Durham, NC: Duke University Press, 2010.

Fridell, Caleb. "The Extractive Logic of Fossil Capital in H. G. Wells's Scientific Prophecy." *Modern Fiction Studies* 66.1 (2020): 164–89.

"Frightful Colliery Accident." *Sheffield and Rotherham Independent*, 20 January 1862, 4.

Gagnier, Regenia. "Morris's Ethics, Cosmopolitanism, and Globalisation." *Journal of William Morris Studies* 16.2 (2005): 9–30.

Gagnier, Regenia, and Martin Delveaux. "Towards a Global Ecology of the *Fin de Siècle*." *Literature Compass* 3.3 (2006): 572–87.

Galeano, Eduardo. *Open Veins of Latin America: Five Centuries of the Pillage of a Continent*. 25th anniversary ed. Trans. Cedric Belfrage. New York: Monthly Review Press, 1997.

Gallagher, Catherine. *The Body Economic: Life, Death, and Sensation in Political Economy and the Victorian Novel*. Princeton, NJ: Princeton University Press, 2006.

Galvan, Jill, and Elsie Michie, ed. *Replotting Marriage in Nineteenth-Century British Literature*. Columbus: Ohio State University Press, 2018.

George Bagdanov, Kristin. *Fossils in the Making*. Boston: Black Ocean, 2019.

G. H. *The American Mines: Shewing Their Importance, in a National Point of View; With the Progress and Present Position of the Real del Monte Company; And Cursory Remarks on Other Similar Undertakings in South America*. London: Effingham Wilson, 1834.

Ghosh, Amitav. *The Great Derangement: Climate Change and the Unthinkable*. Chicago: University of Chicago Press, 2016.

Gibson, Mary Ellis. Introduction. *Science Fiction in Colonial India, 1835–1905: Five Tales of Speculation, Resistance and Rebellion*. Ed. Mary Ellis Gibson. London: Anthem, 2019. 1–28.

Gillis, Colin. "Lawrence's Bildungsroman and the Science of Sexual Development." *Twentieth-Century Literature* 60.3 (Fall 2014): 273–304.

"gin, n.1." *OED Online*. Oxford University Press, June 2020. www.oed.com/view/Entry/78357. Accessed 31 July 2020.

Gisborne, Thomas. *The Testimony of Natural Theology to Christianity*. London: Cadell and Davies, 1818.

Glasier, John Bruce. *William Morris and the Early Days of the Socialist Movement*. 1921. Bristol: Thommes, 1994.

Godfrey, Emelyne, ed. *Utopias and Dystopias in the Fiction of H. G. Wells and William Morris: Landscape and Space*. London: Palgrave Macmillan, 2016.

"Gold." *True Briton* 2, n.s., no. 2 (20 October 1853): 184.

Gold, Barri J. *ThermoPoetics: Energy in Victorian Literature and Science*. Cambridge, MA: MIT Press, 2010.

"The Gold Hunt." *True Briton.* Part 1 (no. 29, 20 January 1853): 469–70. Part 2 (no. 31, 3 February 1853): 501–3. Part 3 (no. 42, 21 April 1853): 674–76.

Goldmann, Charles Sydney, with the co-operation of Joseph Kitchin. *South African Mines: Their Position, Results, and Developments; Together with an Account of Diamond, Land, Finance, and Kindred Concerns.* Vol. 1. *Rand Mining Companies.* London: E. Wilson; Johannesburg: Argus, 1895–96.

Gomel, Elana. "Shapes of the Past and the Future: Darwin and the Narratology of Time Travel." *Narrative* 17.3 (October 2009): 334–52.

Gómez-Barris, Macarena. *The Extractive Zone: Social Ecologies and Decolonial Perspectives.* Durham, NC: Duke University Press, 2017.

Goodbody, Axel, and Bradon Smith. "Stories of Energy: Narrative in the Energy Humanities." Introduction to special issue of *Resilience: A Journal of the Environmental Humanities* 6.2–3 (Spring–Fall 2019): 1–25.

Goonan, Thomas G. "Rare Earth Elements—End Use and Recyclability: U.S. Geological Survey Scientific Investigations Report 2011-5094." Reston, VA: US Geological Survey, 2011. http://pubs.usgs.gov/sir/2011/5094. Accessed 9 July 2020.

Gore, Clare Walker. *Plotting Disability in the Nineteenth-Century Novel.* Edinburgh: Edinburgh University Press, 2020.

Gorelik, Boris. "'Africa . . . Always Moves Me Deeply': Tolkien in Bloemfontein." *Mallorn* 55 (Winter 2014): 4–10.

Gould, Peter C. *Early Green Politics: Back to Nature, Back to the Land, and Socialism in Britain, 1880–1900.* Sussex: Harvester, 1988.

Granofsky, Ronald. "'His Father's Dirty Digging': Recuperating the Masculine in D. H. Lawrence's *Sons and Lovers.*" *Modern Fiction Studies* 55.2 (Summer 2009): 242–64.

Graver, Suzanne. *George Eliot and Community: A Study in Social Theory and Fictional Form.* Berkeley: University of California Press, 1984.

Green, William H. "King Thorin's Mines: *The Hobbit* as Victorian Adventure Novel." *Extrapolation* 42.1 (2001): 53–64.

"Green Peace Report on Air Pollution in Mpumalanga: DEA, Eskom and Sasol Responses." *Parliamentary Monitoring Group,* 21 November 2018. https://pmg.org.za/committee -meeting/27603/. Accessed 11 May 2020.

Griffiths, Devin. *The Age of Analogy: Science and Literature between the Darwins.* Baltimore: Johns Hopkins University Press, 2016.

——. "Petrodrama: Melodrama and Energetic Modernity." *Victorian Studies* 60.4 (Summer 2018): 611–38.

Grimm, Jacob, and Wilhelm. "Hansel and Gretel." 1812. Trans. Ralph Manheim. *Folk and Fairy Tales.* 4th ed. Ed. Martin Hallett and Barbara Karasek. Peterborough, ON: Broadview, 2009. 142–47.

Grundy, James. *Report of the Inspection of Mines in India, for the Year Ending the 30th of June 1894.* Calcutta: Office of the Superintendent of Government Printing, 1894.

Gubar, Marah. *Artful Dodgers: Reconceiving the Golden Age of Children's Literature.* Oxford: Oxford University Press, 2009.

Haggard, H. Rider. "A Farmer's Year." Part 13. *Longman's Magazine,* September 1899, 421–50.

——. *King Solomon's Mines.* London: Cassell, 1885.

——. *King Solomon's Mines.* 1885. Ed. Gerald Monsman. Peterborough, ON: Broadview, 2002.

——. *Montezuma's Daughter.* 1893. London: Longmans, 1898.

Halliday, Stephen. *Underground to Everywhere: London's Underground Railway in the Life of the Capital.* London: Sutton, 2001.

Hanley, Anne G. *Native Capital: Financial Institutions and Economic Development in São Paulo, Brazil, 1850–1920.* Palo Alto, CA: Stanford University Press, 2005.

Hannam, Peter. "What's Next for Adani, the Coal Project That Helped to Return Morrison to Power?" *Sydney Morning Herald*, 22 May 2019. https://www.smh.com.au/environment /conservation/what-s-next-for-the-coal-mine-that-helped-return-morrison-to-power -20190520-p51p7j.html. Accessed 11 May 2020.

Haraway, Donna. *Staying with the Trouble: Making Kin in the Chthulucene.* Durham, NC: Duke University Press, 2016.

Harrison, Robert Pogue. *Forests: The Shadow of Civilization.* Chicago: University of Chicago Press, 1993.

"The Hartley Colliery Disaster." *Leeds Mercury*, 24 January 1862, 3.

Harvey, Charles, and Jon Press. *Art, Enterprise and Ethics: The Life and Works of William Morris.* London: Frank Cass, 1996.

Harvey, David. "Cartographic Identities: Geographical Knowledges under Globalization." 2000. *Spaces of Capital: Towards a Critical Geography.* Edinburgh: Edinburgh University Press, 2001. 208–33.

———. "The Geopolitics of Capitalism." 1985. *Spaces of Capital: Towards a Critical Geography.* Edinburgh: Edinburgh University Press, 2001. 312–44.

Hasan, Md. Mahmudul. "Commemorating Rokeya Sakhawat Hossain and Contextualising Her Work in South Asian Muslim Feminism." *Asiatic* 7.2 (2013): 39–59.

———. "Muslim Bengal Writes Back: A Study of Rokeya's Encounter with and Representation of Europe." *Journal of Postcolonial Writing* 52.6 (2016): 739–51.

Hasanat, Fayeza. "Sultana's Utopian Awakening: An Ecocritical Reading of Rokeya Sakhawat Hossain's *Sultana's Dream*." *Asiatic* 7.2 (December 2013): 114–25.

Hashimoto, Keizaburo. "Oil Consumption for Power Use in Japan." *Far-Eastern Review* 26.2 (February 1930): 15–22.

Hatch, Frederick H., and J. A. Chalmers. *The Gold Mines of the Rand: Being a Description of the Mining Industry of Witwatersrand South African Republic.* London: Macmillan, 1895.

Hausfather, Zeke. "Common Climate Misconceptions: Atmospheric Carbon Dioxide." *Yale Climate Connections*, 16 December 2010. https://www.yaleclimateconnections .org/2010/12/common-climate-misconceptions-atmospheric-carbon-dioxide/. Accessed 11 May 2020.

Hay, Eloise Knapp. "Nostromo." *The Cambridge Companion to Joseph Conrad.* Ed. J. H. Stape. Cambridge: Cambridge University Press, 2004. 81–99.

Heady, Emily. "Reading to Mourn: Narrative Gaps and Ethical Burdens in Dickens's *Hard Times*." *Victorians Institute Journal* 44 (2016): 130–56.

Heath, Hilarie J. "British Merchant Houses in Mexico, 1821–1860: Conforming Business Practices and Ethics." *Hispanic American Historical Review* 73.2 (May 1993): 261–90.

Heise, Ursula K. *Imagining Extinction: The Cultural Meanings of Endangered Species.* Chicago: University of Chicago Press, 2016.

Hejnol, Andreas. "Ladders, Trees, Complexity, and Other Metaphors in Evolutionary Thinking." A. Tsing et al., *Arts of Living on a Damaged Planet.* G87–G102.

Hensley, Nathan K. *Forms of Empire: The Poetics of Victorian Sovereignty.* Oxford: Oxford University Press, 2016.

Hensley, Nathan K., and Philip Steer. "Introduction: Ecological Formalism; or, Love among the Ruins." Hensley and Steer, *Ecological Form.* 1–17.

———, eds. *Ecological Form: System and Aesthetics in the Age of Empire.* New York: Fordham University Press, 2019.

———. "Signatures of the Carboniferous: The Literary Forms of Coal." Hensley and Steer, *Ecological Form.* 63–82.

Heron, Kai, and Jodi Dean. "Revolution or Ruin." *e-flux* 110 (June 2020). https://www.e-flux .com/journal/110/335242/revolution-or-ruin/. Accessed 9 July 2020.

Hewlett, Maurice. "A Materialist's Paradise." *National Review* 102 (August 1891): 818–27.

Higgins, David M., and Hugh C. O'Connell. "Introduction: Speculative Finance / Speculative Fiction." *CR: The New Centennial Review* 19.1 (Spring 2019): 1–9.

Higney, Robert. "'Law, Good Faith, Order, Security': Joseph Conrad's Institutions." *Novel* 48.1 (2015): 85–102.

Holderness, Graham. "Miners and the Novel: From Bourgeois to Proletarian Fiction." *The British Working-Class Novel in the Twentieth Century.* Ed. Jeremy Hawthorn. London: Edward Arnold, 1984. 19–32.

[Holland, John.] *The History and Description of Fossil Fuel, the Collieries, and Coal Trade of Great Britain.* 2nd ed. London: Whittaker, 1841.

[Hossain, Rokeya Sakhawat.] "Sultana's Dream." By Mrs. R. S. Hossein. *The Indian Ladies' Magazine* 5.3 (September 1905): 82–86.

——. *Sultana's Dream: A Feminist Utopia; and Selections from "The Secluded Ones."* Ed. and trans. Roushan Jahan. New York: Feminist Press, 1988.

Hossain, Yasmin. "The Begum's Dream: Rokeya Sakhawat Hossain and the Broadening of Muslim Women's Aspirations in Bengal." *South Asia Research* 12.1 (May 1992): 1–19.

Houghton, John Wm., Janet Brennan Croft, Nancy Martsch, John D. Rateliff, and Robin Anne Reid, eds. *Tolkien in the New Century: Essays in Honor of Tom Shippey.* Jefferson, NC: McFarland, 2014.

Hovanec, Caroline. "Rereading H. G. Wells's *The Time Machine*: Empiricism, Aestheticism, Modernism." *English Literature in Transition* 58.4 (2015): 459–85.

Howell, Jessica. *Exploring Victorian Travel Literature: Disease, Race and Climate.* Edinburgh: Edinburgh University Press, 2014.

Howitt, William. *Land, Labour, and Gold; or, Two Years in Victoria with Visits to Sydney and Van Diemen's Land.* 1855. Sydney: Sydney University Press, 1972.

Hsu, Jeremy. "Don't Panic about Rare Earth Elements." *Scientific American*, 31 May 2019. https://www.scientificamerican.com/article/dont-panic-about-rare-earth-elements/. Accessed 21 May 2020.

Hudson, William Henry. *Green Mansions: A Romance of the Tropical Forest.* 1904. Mineola, NY: Dover, 2014.

Huggan, Graham, and Helen Tiffin. *Postcolonial Ecocriticism: Literature, Animals, Environment.* 2nd ed. Abingdon: Routledge, 2015.

Hughes, David McDermott. *Energy without Conscience: Oil, Climate Change, and Complicity.* Durham, NC: Duke University Press, 2017.

Hull, Edward. *The Coal Fields of Great Britain: Their History, Structure, and Resources; With Descriptions of the Coal-Fields of Our Indian and Colonial Empire, and of Other Parts of the World.* 5th ed., rev., embodying the reports of the Royal Coal-Commission of 1904. London: H. Rees, 1905.

Hultgren, Neil. "Haggard Criticism since 1980: Imperial Romance before and after the Postcolonial Turn." *Literature Compass* 8/9 (2011): 645–59.

Humboldt, Alexander de. *Political Essay on the Kingdom of New Spain.* Trans. John Black. Vol. 1. London: Longman, 1811.

——. *Selections from the Works of the Baron de Humboldt, Relating to the Climate, Inhabitants, Productions, and Mines of Mexico.* Ed. and adapted by John Taylor. London: Longman, 1824.

Humpherys, Anne. "Louisa Gradgrind's Secret: Marriage and Divorce in *Hard Times*." *Dickens Studies Annual* 25 (1996): 177–95.

Humphreys, R. A. "British Merchants and South American Independence." *Proceedings of the British Academy* 51 (London: Oxford University Press, 1965): 151–74.

Hurley, Jessica. "Impossible Futures: Fictions of Risk in the *Longue Durée*." *American Literature* 89.4 (2017): 761–89.

[Hutton, Richard Holt.] "In A.D. 802,701." *Spectator*, 13 July 1895, 41–43. Rpt. in H. G. Wells, *The Time Machine*. Ed. Nicholas Ruddick. Peterborough, ON: Broadview, 2001. 263–67.

"Hydrogen Fuel Technology: A Cleaner and More Secure Energy Future." *The White House, President George W. Bush*. US Government. https://georgewbush-whitehouse.archives. gov/infocus/technology/economic_policy200404/chap2.html. Accessed 19 June 2019.

Industrial Prospects in the Union of South Africa: A Country of Growing Possibilities. Compiled by the Department of Mines and Industries in collaboration with the South African Railways and Harbours Administration. Pretoria: Government Printing and Stationery Office, 1923.

Jackson, Roland. "Eunice Foote, John Tyndall and a Question of Priority." *Notes and Records: The Royal Society Journal of the History of Science* 74 (2020): 105–18.

Jameson, Fredric. *Archaeologies of the Future: The Desire Called Utopia and Other Science Fictions*. London: Verso, 2005.

Jeffers, Susan. *Arda Inhabited: Environmental Relationships in "The Lord of the Rings."* Kent, OH: Kent State University Press, 2014.

Jennings, Humphrey. *Pandæmonium 1660–1886: The Coming of the Machine as Seen by Contemporary Observers*. Ed. Mary-Lou Jennings and Charles Madge. New York: Free Press, 1985.

Jerng, Mark C. *Racial Worldmaking: The Power of Popular Fiction*. New York: Fordham University Press, 2018.

Jevons, W. Stanley. *The Coal Question: An Inquiry concerning the Progress of the Nation, and the Probable Exhaustion of Our Coal-Mines*. London: Macmillan, 1865.

Johnson, Bob. *Mineral Rites: An Archaeology of the Fossil Economy*. Baltimore: Johns Hopkins University Press, 2019.

Johnson, Patricia E. "'Hard Times' and the Structure of Industrialism: The Novel as Factory." *Studies in the Novel* 21.2 (Summer 1989): 128–37.

Johnson, Walter. "The Pedestal and the Veil: Rethinking the Capitalism/Slavery Question." *Journal of the Early Republic* 24.2 (Summer 2004): 299–308.

Jonsson, Fredrik Albritton. *Enlightenment's Frontier: The Scottish Highlands and the Origins of Environmentalism*. New Haven, CT: Yale University Press, 2013.

Joshi, Priti. "An Old Dog Enters the Fray; or, Reading *Hard Times* as an Industrial Novel." *Dickens Studies Annual* 44 (2013): 221–41.

Judge, Jennifer. "The 'Seamy Side' of Human Perfectibility: Satire on Habit in Edward Bulwer-Lytton's *The Coming Race*." *Journal of Narrative Theory* 39.2 (2009): 137–58.

Kaiman, Jonathan. "Rare Earth Mining in China: The Bleak Social and Environmental Costs." *Guardian*, 20 March 2014. https://www.theguardian.com/sustainable-business /rare-earth-mining-china-social-environmental-costs. Accessed 21 May 2020.

Kains-Jackson, H. [Henry]. "The Earth Losing Primal Fertility." *Times*, 3 February 1890, 13.

Kaufman, Heidi. "'King Solomon's Mines'? African Jewry, British Imperialism, and H. Rider Haggard's Diamonds." *Victorian Literature and Culture* 33.2 (2005): 517–39.

Kavanagh, Brendan. "The 'Breaking-Up' of the Monsoon and *Lord Jim*'s Atmospherics." Schneider-Rebozo, McCarthy, and Peters, *Conrad and Nature*. 113–45.

Kearns, Katherine. "A Tropology of Realism in *Hard Times*." *ELH* 59.4 (Winter 1992): 857–81.

Kelly, Annie. "Apple and Google Named in US Lawsuit over Congolese Child Cobalt Mining Deaths." *Guardian*, 16 December 2019. https://www.theguardian.com/global -development/2019/dec/16/apple-and-google-named-in-us-lawsuit-over-congolese -child-cobalt-mining-deaths/. Accessed 30 June 2020.

Kelly, Steven. "Breaking the Dragon's Gaze: Commodity Fetishism in Tolkien's Middle-earth." *Mythlore* 34.2 (2016): 113–32.

Kent, Eddy. "William Morris's Green Cosmopolitanism." *Journal of William Morris Studies* 19.3 (2011): 64–78.

Kestner, Joseph A. "Fanny N. Mayne's 'Jane Rutherford' and the Tradition of the Social-Protest Novel in England." *Studies in the Novel* 19.3 (1987): 368–80.

Ketabgian, Tamara. *The Lives of Machines: The Industrial Imaginary in Victorian Literature and Culture.* Ann Arbor: University of Michigan Press, 2011.

Kinane, Ian. "Less Noise, More Green: Cultural Materialism and the Reverse Discourse of the Wild in Tolkien's *The Hobbit.*" *Tolkien: The Forest and the City.* Ed. Helen Conrad-O'Briain and Gerard Hynes. Dublin: Four Courts, 2013. 144–53.

Kirby, Peter. *Child Workers and Industrial Health in Britain, 1780–1850.* Woodbridge: Boydell and Brewer, 2013.

Klein, Naomi. "Dancing the World into Being: A Conversation with Idle No More's Leanne Simpson." *Yes! Magazine,* 5 March 2013. http://www.yesmagazine.org/peace-justice/dancing-the-world-into-being-a-conversation-with-idle-no-more-leanne-simpson. Accessed 19 January 2018.

———. *This Changes Everything: Capitalism vs. the Climate.* New York: Simon and Schuster, 2014.

Kohlmann, Benjamin. "'The End of Laissez-Faire': Literature, Economics, and the Idea of the Welfare State." *Late Victorian into Modern.* Ed. Laura Marcus, Michèle Mendelssohn, and Kirsten E. Shepherd-Barr. Oxford: Oxford University Press, 2016. 448–62.

Komsta, Marta. "On Utopian (Im)Perfection and Solidarity in Edward Bulwer-Lytton's *The Coming Race.*" *Utopian Studies* 29.2 (2018): 159–75.

Kornbluh, Anna. *The Order of Forms: Realism, Formalism, and Social Space.* Chicago: University of Chicago Press, 2019.

Kreisel, Deanna K. "'Form against Force': Sustainability and Organicism in the Work of John Ruskin." Hensley and Steer, *Ecological Form.* 101–20.

———. "Superfluity and Suction: The Problem with Saving in *The Mill on the Floss.*" *Novel* 35.1 (Fall 2001): 69–103.

———. "Sustainability." *Victorian Literature and Culture* 46.3/4 (Fall/Winter 2018): 895–900.

———. "Teaching Morris the Utopian." *Teaching William Morris.* Ed. Jason Martinek and Elizabeth Carolyn Miller. Vancouver: Fairleigh Dickinson University Press, 2019. 161–74.

Lane, Christopher. "Bulwer's Misanthropes and the Limits of Victorian Sympathy." *Victorian Studies* 44.4 (2002): 597–624.

Lang, Andrew. *Essays in Little.* New York: Scribner's, 1907.

Larabee, Mark D. "Guano, Globalization and Ecosystem Change in *Lord Jim.*" Schneider-Rebozo, McCarthy, and Peters, *Conrad and Nature.* 268–85.

Laslett, John H. M. "America before and after Emigration: Scottish Miners' Views of the U.S. through the Columns of the *Glasgow Sentinel,* 1850–1876." *The Press of Labor Migrants in Europe and North America, 1880s to 1930s.* Ed. Christiane Harzig and Dirk Hoerder. Bremen: Publications of the Labor Newspaper Preservation Project, 1985. 521–45.

Laughlin, Tom. "Karl Marx and the Lancashire 'Cotton Famine.'" Interdisciplinary Nineteenth Century Studies Conference. Sheraton Grand, Los Angeles. 7 March 2020. Precirculated conference paper.

Law, Jules. *The Social Life of Fluids: Blood, Milk, and Water in the Victorian Novel.* Ithaca, NY: Cornell University Press, 2010.

Lawrence, D. H. "Odour of Chrysanthemums." 1911. *Selected Short Stories.* Ed. Brian Finney. London: Penguin, 1982. 88–105.

———. *Sons and Lovers.* 1913. Penguin, 2006.

Leader, Alexandra, Gabrielle Gaustad, and Callie Babbitt. "The Effect of Critical Material Prices on the Competitiveness of Clean Energy Technologies." *Materials for Renewable and Sustainable Energy* 8.8 (2019): 1–17.

Lee, Jason C. K., and Zongguo Wen. "Rare Earths from Mines to Metals: Comparing Environmental Impacts from China's Main Production Pathways." *Journal of Industrial Ecology* 21.5 (2016): 1277–90.

Le Guin, Ursula K. "The Critics, the Monsters, and the Fantasists." *Wordsworth Circle* 38.1–2 (2007): 83–87.

Leitch, Thomas M. "Closure and Teleology in Dickens." *Studies in the Novel* 18.2 (Summer 1986): 143–56.

LeMenager, Stephanie. *Living Oil: Petroleum Culture in the American Century.* Oxford: Oxford University Press, 2014.

———. "Sediment." *Veer Ecology: A Companion for Environmental Thinking.* Ed. Jeffrey Jerome Cohen and Lowell Duckert. Minneapolis: University of Minnesota Press, 2017. 168–82.

Lesjak, Carolyn. *The Afterlife of Enclosure: British Realism, Character, and the Commons.* Palo Alto, CA: Stanford University Press, 2021.

Levine, Caroline. *Forms: Whole, Rhythm, Hierarchy, Network.* Princeton, NJ: Princeton University Press, 2015.

Levine-Clark, Marjorie. "'The Entombment of Thomas Shaw': Mining Accidents and the Politics of Workers' Bodies." *Victorian Review* 40.2 (Fall 2014): 22–26.

Logan, Deborah Anna. *The Indian Ladies' Magazine, 1901–1938: From Raj to Swaraj.* Bethlehem, PA: Lehigh University Press, 2017.

Loman, Andrew. "The Sea Cook's Wife: Evocations of Slavery in *Treasure Island.*" *Children's Literature* 38 (2010): 1–26.

Lootens, Tricia. *The Political Poetess: Victorian Femininity, Race, and the Legacy of Separate Spheres.* Princeton, NJ: Princeton University Press, 2017.

Lothian, Alexis. *Old Futures: Speculative Fiction and Queer Possibility.* New York: New York University Press, 2018.

Loughlin, Marie H. "Tolkien's Treasures: Marvellous Objects in *The Hobbit* and *The Lord of the Rings.*" *Tolkien Studies* 16 (2019): 21–58.

Lovesey, Oliver. Introduction. *The Mill on the Floss.* By George Eliot. Peterborough, ON: Broadview, 2007. 8–43.

Lowe, Lisa. *The Intimacies of Four Continents.* Durham, NC: Duke University Press, 2015.

Luciano, Dana. *Arranging Grief: Sacred Time and the Body in Nineteenth-Century America.* New York: New York University Press, 2007.

———. "Romancing the Trace: Edward Hitchcock's Speculative Ichnology." Menely and Taylor, *Anthropocene Reading.* 96–116.

Luckett, R. "East Midlands Coal Mining Seismicity." British Geological Survey Commercial Report. CR/18/115. 2018.

Lupton, Christina. "Walking on Flowers: The Kantian Aesthetics of *Hard Times.*" *ELH* 70.1 (Spring 2003): 151–69.

Lyell, Charles. *Principles of Geology; or, The Modern Changes of the Earth and Its Inhabitants, Considered as Illustrative of Geology.* Vol. 1. 6th ed. Boston: Hilliard, 1842.

MacDuffie, Allen. "Charles Darwin and the Victorian Pre-history of Climate Denial." *Victorian Studies* 60.4 (Summer 2018): 543–64.

———. "*The Jungle Books*: Rudyard Kipling's Lamarckian Fantasy." *PMLA* 129.1 (2014): 18–34.

———. *Victorian Literature, Energy, and the Ecological Imagination.* Cambridge: Cambridge University Press, 2014.

Malm, Andreas. *Fossil Capital: The Rise of Steam Power and the Roots of Global Warming.* London: Verso, 2016.

Marcus, Laura. *Dreams of Modernity: Psychoanalysis, Literature, Cinema.* Cambridge: Cambridge University Press, 2014.

Marder, Michael. *Energy Dreams: Of Actuality*. New York: Columbia University Press, 2017.

Martel, Michael. "Radioactive Forms: Radium, the State, and the End of Victorian Narrative." *Genre* 52.3 (December 2019): 151–77.

Martin, Regina. "Absentee Capitalism and the Politics of Conrad's Imperial Novels." *PMLA* 130.3 (2015): 584–98.

Marx, Karl. *Capital: A Critique of Political Economy*. Vol. 1. 1867. Trans. Ben Fowkes. London: Penguin, 1990.

Marzec, Robert P. "The Monstrous and the Secure: Reading Conrad in the Anthropocene." Schneider-Rebozo, McCarthy, and Peters, *Conrad and Nature*. 68–89.

Mason, Courtney W. *Spirits of the Rockies: Reasserting an Indigenous Presence in Banff National Park*. Toronto: University of Toronto Press, 2014.

Maxwell, J. Byers. *A Passion for Gold: The Story of a South African Mine*. London: Anthony Treherne, 1902.

[Mayne, Fanny.] *Jane Rutherford: or, The Miners' Strike*. By "A Friend of the People." London: Clarke, Beeton, 1854.

Mazovick, David. "The Hydrogen Fuel Cell Scam—from George W. Bush and 'The Big 3' to Toyota, Honda, and Japan." *CleanTechnica*, Sustainable Enterprises Media, 2 January 2019. https://cleantechnica.com/2019/01/02/the-hydrogen-fuel-cell-scam-from -george-w-bush-the-big-3-to-toyota-honda-japan/. Accessed 19 June 2019.

Mazzeno, Laurence W., and Ronald D. Morrison, eds. *Victorian Writers and the Environment: Ecocritical Perspectives*. Abingdon: Routledge, 2017.

McCarthy, Jeffrey Mathes. *Green Modernism: Nature and the English Novel, 1900 to 1930*. London: Palgrave Macmillan, 2015.

McClintock, Anne. *Imperial Leather: Race, Gender and Sexuality in the Colonial Contest*. New York: Routledge, 1995.

McCutcheon, John Elliott. *The Hartley Colliery Disaster, 1862*. Seaham: McCutcheon, 1963.

Meade, Richard. *The Coal and Iron Industries of the United Kingdom*. London: Crosby Lockwood, 1882.

Meharg, Andy. "The Arsenic Green." *Nature* 423 (12 June 2003): 688.

Menely, Tobias, and Jesse Oak Taylor, ed. *Anthropocene Reading: Literary History in Geologic Times*. University Park: Pennsylvania State University Press, 2017.

——. "Introduction." Menely and Taylor, ed., *Anthropocene Reading*. 1–24.

Merchant, Carolyn. *Reinventing Eden: The Fate of Nature in Western Culture*. New York: Routledge, 2003.

Meth, Oliver. "New Satellite Data Reveals the World's Largest Air Pollution Hotspot Is Mpumalanga—South Africa." *Greenpeace*, 29 October 2018. https://www.greenpeace .org/africa/en/press/4202/new-satellite-data-reveals-the-worlds-largest-air-pollution -hotspot-is-mpumalanga-south-africa/. Accessed 11 May 2020.

Mezzadra, Sandro, and Brett Neilson. "Operations of Capital." *South Atlantic Quarterly* 114.1 (January 2015): 1–9.

Miah, Mohammad Moniruzzaman. "A Feminist Critical Evaluation of How Rokeya Sakhawat Hossain's Language of Protest Deplored Patriarchy and Social Anachronism in the British Bengal." *Journal of Arts and Humanities* 4.10 (2014): 41–51.

Miéville, China. "Symposium: Marxism and Fantasy; Editorial Introduction." *Historical Materialism* 10.4 (2002): 39–49.

Miller, Elizabeth Carolyn. "Ecology." *Victorian Literature and Culture* 46, no. 3/4 (2018): 653–56.

——. "Fixed Capital and the Flow: Water Power, Steam Power, and *The Mill on the Floss*." Hensley and Steer, *Ecological Form*. 85–100.

——. *Slow Print: Literary Radicalism and Late Victorian Print Culture*. Palo Alto, CA: Stanford University Press, 2013.

Miller, J. Hillis. *Communities in Fiction*. New York: Fordham University Press, 2015.

Miller, John. "The Environmental Politics and Aesthetics of Rider Haggard's *King Solomon's Mines*: Capital, Mourning, and Desire." Mazzeno and Morrison, *Victorian Writers and the Environment*. 157–73.

———. "Fiction, Fashion, and the Victorian Fur Seal Hunt." *Reading Literary Animals*. Ed. Karen L. Edwards, Derek Ryan, and Jane Spencer. New York: Routledge, 2019. 212–26.

Miller, John MacNeill. "The Ecological Plot: A Brief History of Multispecies Storytelling, from Malthus to *Middlemarch*." *Victorian Literature and Culture* 48.1 (2020): 155–85.

———. "Mischaracterizing the Environment: Hardy, Darwin, and the Art of Ecological Storytelling." *Texas Studies in Literature and Language* 62.2 (2020): 149–77.

———. "Weird beyond Description: Weird Fiction and the Suspicion of Scenery." *Victorian Studies* 62.2 (Winter 2020): 244–52.

"Mine Inspection in Australia." *Mining Journal*, 15 July 1882, 851.

"Mineral Exhibits at the Colonial and Indian Exhibition: Empire of India and Asiatic Possessions." *Mining Journal*, 14 August 1886, 949.

Missemer, Antoine. "Fossil Fuels in Economic Theory: Back to the 19th Century British Debates." *French Journal of British Studies* 23.3 (2018). https://doi.org/10.4000/rfcb.2685. Accessed 8 April 2020.

Mitchell, Timothy. *Carbon Democracy: Political Power in the Age of Oil*. London: Verso, 2011.

Monsman, Gerald. "Of Diamonds and Deities: Social Anthropology in H. Rider Haggard's *King Solomon's Mines*." *English Literature in Transition, 1880–1920* 43.3 (2000): 280–97.

Moore, Jason W. *Capitalism in the Web of Life: Ecology and the Accumulation of Capital*. London: Verso, 2015.

Moore, Lisa. "Greenhouse Gases: How Long Will They Last?" *Climate 411*, 26 February 2008. http://blogs.edf.org/climate411/2008/02/26/ghg_lifetimes/. Accessed 11 May 2020.

Morgan, Benjamin. "After the Arctic Sublime." *New Literary History* 47.1 (2016): 1–26.

———. "How We Might Live: Utopian Ecology in William Morris and Samuel Butler." Hensley and Steer, *Ecological Form*. 139–60.

———. *The Outward Mind: Materialist Aesthetics in Victorian Science and Literature*. Chicago: University of Chicago Press, 2017.

———. "Scale as Form: Thomas Hardy's Rocks and Stars." Menely and Taylor, *Anthropocene Reading*. 132–49.

Morris, William. "Art and Socialism." 1884. Rpt. in Morris, *Political Writings*. 109–33.

———. "Art and the Beauty of the Earth." Lecture delivered at Burslem Town Hall on 13 October 1881. London: Chiswick, 1898.

———. "Art under Plutocracy." 1883. Rpt. in Morris, *Political Writings*. 57–85.

———. "The Lesser Arts." 1877. Rpt. in Morris, *Political Writings*. 31–56.

———. *News from Nowhere; or, An Epoch of Rest, Being Some Chapters from a Utopian Romance*. 1890. Ed. Stephen Arata. Peterborough, ON: Broadview, 2003.

———. *News from Nowhere; or, An Epoch of Rest*. 1890. London: Kelmscott, 1892.

———. *Political Writings of William Morris*. Ed. A. L. Morton. New York: International, 1973.

Morton, Timothy. "The Mesh." *Environmental Criticism for the Twenty-First Century*. Ed. Stephanie LeMenager, Teresa Shewry, and Ken Hiltner. New York: Routledge, 2011. 19–30.

Moten, Fred. "Resistance of the Object: Aunt Hester's Scream." *In the Break: The Aesthetics of the Black Radical Tradition*. Minneapolis: University of Minnesota Press, 2003. 1–24.

Mufti, Nasser. *Civilizing War: Imperial Politics and the Poetics of National Rupture*. Evanston, IL: Northwestern University Press, 2018.

Munich, Adrienne. *Empire of Diamonds: Victorian Gems in Imperial Settings.* Charlottesville: University of Virginia Press, 2020.

Murphy, Benjamin J. "Not So New Materialism: Homeostasis Revisited." *Configurations* 27.1 (Winter 2019): 1–36.

Murphy, Ryan Francis. "Exterminating the Elephant in 'Heart of Darkness.'" *Conradian* 38.2 (Autumn 2013): 1–17.

Murray, Cara. "Catastrophe and Development in the Adventure Romance." *English Literature in Transition, 1880–1920* 53.2 (2010): 150–69.

Neuman, Justin. "Anthropocene Interruptions: Energy Recognition Scenes and the Global Cooling Myth." Menely and Taylor, *Anthropocene Reading.* 150–66.

Newell, Peter. "Race and the Politics of Energy Transitions." *Energy Research and Social Science* 71 (2021): 1–5.

Newman, Ian. "Property, History, and Identity in Defoe's *Captain Singleton.*" *SEL* 51.3 (Summer 2011): 565–83.

Niedecker, Lorine. *Lake Superior.* 1968. Seattle: Wave Books, 2013.

Nijhawan, Shobna. *Women and Girls in the Hindi Public Sphere: Periodical Literature in Colonial North India.* Oxford: Oxford University Press, 2012.

Nixon, Rob. *Slow Violence and the Environmentalism of the Poor.* Cambridge, MA: Harvard University Press, 2011.

Observations on Foreign Mining in Mexico. By "A Resident." London: Pelham Richardson, 1838.

Oglethorpe, Miles K. "Mines, Quarries, and Mineral Works." *Scotland's Buildings.* Ed. Geoffrey Stell, John Shaw, and Susan Storrier. East Lothian: Tuckwell, 2003. 551–70.

"Old Pit Shafts." *Mining Journal,* 4 January 1868, 11.

O'Malley, Seamus. *Making History New: Modernism and Historical Narrative.* Oxford: Oxford University Press, 2015.

Orwell, George. *The Road to Wigan Pier.* 1937. London: Penguin, 2001.

O'Sullivan, Patrick. "'Morris the Red, Morris the Green,'—A Partial Review." *Journal of William Morris Studies* 19.3 (Winter 2011): 22–38.

———. "William Morris and Arsenic—Guilty or Not Proven?" *William Morris Society Newsletter* (Spring 2011): 10–17.

Otjen, Nathaniel. "Energy Anxiety and Fossil Fuel Modernity in H. G. Wells's *The War of the Worlds.*" *Journal of Modern Literature* 43:2 (2020): 118–33.

Padmini. "An Answer to Sultana's Dream." *Indian Ladies' Magazine,* October 1905, 115–17.

Parham, John. "Bleak Intra-Actions: Dickens, Turbulence, Material Ecology." Mazzeno and Morrison, *Victorian Writers and the Environment.* 114–29.

———. "Sustenance from the Past: Precedents to Sustainability in Nineteenth-Century Literature and Culture." *Literature and Sustainability: Concept, Text and Culture.* Ed. Adeline Johns-Putra, John Parham, and Louise Squire. Manchester: Manchester University Press, 2017. 33–51.

Parkes, Christopher. *Children's Literature and Capitalism: Fictions of Social Mobility in Britain, 1850–1914.* Houndmills: Palgrave, 2012.

Parrinder, Patrick. "*News from Nowhere, The Time Machine* and the Break-Up of Classical Realism." *Science Fiction Studies* 3.3 (November 1976): 265–74.

Pauley, Daniel. "Anecdotes and the Shifting Baseline Syndrome of Fisheries." *Trends in Ecology and Evolution* 10.10 (10 October 1995): 430.

Pearson, Richard. "Personal and National Trauma in H. Rider Haggard's *Montezuma's Daughter.*" *English Literature in Transition* 58.1 (2015): 30–53.

Perelman, Michael. "Marx and Resource Scarcity." *The Greening of Marxism.* Ed. Ted Benton. New York: Guilford, 1996. 64–80.

Perkins, Frank C. "A New Solar Power Plant." *Scientific American* 98.6 (8 February 1908): 97.

Perks, Samuel. "'He Can't Throw Any of His Coal-Dust in My Eyes': Adventurers and Entrepreneurs in *Victory*'s Coal Empire." Schneider-Rebozo, McCarthy, and Peters, *Conrad and Nature*. 252–68.

Perlin, John. *Let It Shine: The 6,000-Year Story of Solar Energy*. Novato, CA: New World Library, 2013.

Phillips, John. *Descriptive Notice of the Silver Mines and Amalgamation Process of Mexico*. London: Pelham Richardson, 1846.

Plotz, John. *Portable Property: Victorian Culture on the Move*. Princeton, NJ: Princeton University Press, 2008.

——. "The Provincial Novel." *A Companion to the English Novel*. Ed. Stephen Arata, Madigan Haley, J. Paul Hunter, and Jennifer Wicke. Chichester: Wiley Blackwell, 2015. 360–72.

——. *Semi-detached: The Aesthetics of Virtual Experience since Dickens*. Princeton, NJ: Princeton University Press, 2018.

Pocock, Tom. *Rider Haggard and the Lost Empire*. London: Weidenfeld and Nicolson, 1993.

Pomeranz, Kenneth. *The Great Divergence: China, Europe, and the Making of the Modern World Economy*. Princeton, NJ: Princeton University Press, 2000.

Pope, Charles Henry. *Solar Heat: Its Practical Applications*. Boston: Charles H. Pope, 1903.

Povinelli, Elizabeth A. *Geontologies: A Requiem to Late Liberalism*. Durham, NC: Duke University Press, 2016.

"power, n.1." *OED Online*. Oxford University Press, June 2020. www.oed.com/view /Entry/149167. Accessed 31 July 2020.

Pratt, Mary Louise. *Imperial Eyes: Travel Writing and Transculturation*. London: Routledge, 1992.

Quiggin, John. "Explaining Adani: Why Would a Billionaire Persist with a Mine That Will Probably Lose Money?" *Conversation*, 2 June 2019. https://theconversation.com /explaining-adani-why-would-a-billionaire-persist-with-a-mine-that-will-probably-lose-money-117682. Accessed 11 May 2020.

Rabinbach, Anson. *The Human Motor: Energy, Fatigue, and the Origins of Modernity*. New York: Basic, 1990.

Ramirez, Luz Elena. *British Representations of Latin America*. Gainesville: University Press of Florida, 2007.

Rappaport, Erika. *A Thirst for Empire: How Tea Shaped the Modern World*. Princeton, NJ: Princeton University Press, 2017.

Ray, Bharati. "A Voice of Protest: The Writings of Rokeya Sakhawat Hossain (1880–1932)." *Women of India: Colonial and Post-colonial Periods*. Ed. Bharati Ray. New Delhi: Sage, 2005. 427–53.

Redford, Catherine. "'Great Safe Places Down Deep': Subterranean Spaces in the Early Novels of H. G. Wells." *Utopias and Dystopias in the Fiction of H. G. Wells and William Morris*. Ed. Emelyne Godfrey. London: Palgrave Macmillan, 2016. 123–38.

Reid, Robin Anne. "Race in Tolkien Studies: A Bibliographic Essay." *Tolkien and Alterity*. Ed. Chris Vaccaro and Yvette Kisor. Cham: Palgrave Macmillan, 2017. 33–74.

The Report of the South Shields Committee, Appointed to Investigate the Causes of Accidents in Coal Mines. London: Longman, 1843.

Reunert's Diamond Mines of South Africa. Kimberley South African and International Exhibition. London: Sampson Low, Marston; Cape Town: J. C. Juta, 1892.

Review of *Jane Rutherford*. *Tait's Edinburgh Magazine* 21.245 (1854): 313.

Richards, Robert J. *The Tragic Sense of Life: Ernst Haeckel and the Struggle over Evolutionary Thought*. Chicago: University of Chicago Press, 2008.

Riofrancos, Thea. "*Extractivismo* Unearthed: A Geneology of a Radical Discourse." *Cultural Studies* 31.2–3 (2017): 277–306.

———. "Plan, Mood, Battlefield—Reflections on the Green New Deal." *Viewpoint Magazine*, 16 May 2019. https://www.viewpointmag.com/2019/05/16/plan-mood-battlefield-reflections-on-the-green-new-deal/. Accessed 9 July 2020.

Rippy, J. Fred. *British Investments in Latin America, 1822–1949: A Case Study in the Operations of Private Enterprise in Retarded Regions*. Minneapolis: University of Minnesota Press, 1959.

Robertson, Michael. *The Last Utopians: Four Late Nineteenth-Century Visionaries and Their Legacy*. Princeton, NJ: Princeton University Press, 2018.

Robinson, Kim Stanley. *2312*. New York: Orbit, 2013.

Rogers, William N., II, and Michael R. Underwood. "Gagool and Gollum: Exemplars of Degeneration in *King Solomon's Mines* and *The Hobbit.*" *J.R.R. Tolkien and His Literary Resonances: Views of Middle-earth*. Ed. George Clark and Daniel Timmons. Westport, CT: Greenwood, 2000. 121–31.

Rolston, Jessica Smith. *Mining Coal and Undermining Gender: Rhythms of Work and Family in the American West*. New Brunswick, NJ: Rutgers University Press, 2014.

Romero, Matías. "British Investors and the Real del Monte Mine: Mexico, 1824–48." *Foreign Investment in Latin America: Cases and Attitudes*. Ed. Marvin D. Bernstein. New York: Knopf, 1966. 71–75.

Ronda, Margaret. "Nature." *Understanding Marx, Understanding Modernism*. Ed. Mark Steven. New York: Bloomsbury, 2021. 226–28.

———. *Remainders: American Poetry at Nature's End*. Palo Alto, CA: Stanford University Press, 2018.

———. "The Social Forms of Speculative Poetics." *Post45*, 26 April 2019. http://post45.org/2019/04/the-social-forms-of-speculative-poetics/. Accessed 1 May 2020.

Rosen, Michael. "Charms for Grime's Graves." Walls of Visitor's Center and handbill, Grime's Graves, England (https://www.english-heritage.org.uk/visit/places/grimes-graves-prehistoric-flint-mine/). English Heritage, 2009.

Rosenberg, Aaron. "Romancing the Anthropocene: H. G. Wells and the Genre of the Future." *Novel* 51.1 (May 2018): 79–100.

Rosenthal, Jesse. *Good Form: The Ethical Experience of the Victorian Novel*. Princeton, NJ: Princeton University Press, 2017.

Ruskin, John. *Sesame and Lilies*. 1865. New Haven, CT: Yale University Press, 2002.

———. "The Storm-Cloud of the Nineteenth Century: Two Lectures Delivered at the London Institution, February 4th and 11th, 1884." *The Works of John Ruskin*. Ed. E. T. Cook and Alexander Wedderburn. Vol. 34. London: George Allen, 1908. 1–80.

Saler, Michael. *As If: Modern Enchantment and the Literary Prehistory of Virtual Reality*. Oxford: Oxford University Press, 2012.

Santiago, Myrna. *The Ecology of Oil: Environment, Labor, and the Mexican Revolution, 1900–1938*. Cambridge: Cambridge University Press, 2006.

Schaffer, Talia. *Romance's Rival: Familiar Marriage in Victorian Fiction*. Oxford: Oxford University Press, 2016.

Schmitt, Cannon. *Darwin and the Memory of the Human: Evolution, Savages, and South America*. Cambridge: Cambridge University Press, 2009.

———. "Rumor, Shares, and Novelistic Form: Joseph Conrad's *Nostromo.*" *Victorian Investments: New Perspectives on Finance and Culture*. Ed. Nancy Henry and Cannon Schmitt. Bloomington: Indiana University Press, 2008. 182–201.

Schneider-Rebozo, Lissa, Jeffrey Mathes McCarthy, and John G. Peters. *Conrad and Nature: Essays*. New York: Routledge, 2019.

Scott, Heidi C. M. *Chaos and Cosmos: Literary Roots of Modern Ecology in the British Nineteenth Century*. University Park: Pennsylvania State University Press, 2014.

——. *Fuel: An Ecocritical History*. London: Bloomsbury, 2018.

——. "Industrial Souls: Climate Change, Immorality, and Victorian Anticipations of the Good Anthropocene." *Victorian Studies* 60.4 (Summer 2018): 588–610.

Scott, Heidi V. "Colonialism, Landscape and the Subterranean." *Geography Compass* 2.6 (2008): 1853–69.

Seacole, Mary. *Wonderful Adventures of Mrs Seacole in Many Lands*. London: James Blackwood. 1857.

——. *Wonderful Adventures of Mrs Seacole in Many Lands*. 1857. London: Penguin, 2005.

Seccombe, Wally. *Weathering the Storm: Working-Class Families from the Industrial Revolution to the Fertility Decline*. London: Verso, 1993.

Seymour, Jessica. "'As We Draw Near Mountains': Nature and Beauty in the Hearts of Dwarves." *Representations of Nature in Middle-earth*. Ed. Martin Simonson. Zollikofen: Walking Tree, 2015. 29–47.

Shackleton, David. "H. G. Wells, Geology, and the Ruins of Time." *Victorian Literature and Culture* 45 (2017): 839–55.

Shaw, George Bernard. "Common Sense about the War." 1914. *What I Really Wrote about the War*. London: Constable, 1931. 22–110.

——. *Fabianism and the Empire: A Manifesto by the Fabian Society*. London: Grant Richards, 1900.

Shea, Daniel P. "'Abortions of the Market': Production and Reproduction in *News from Nowhere*." *Nineteenth-Century Contexts* 32.2 (2010): 153–72.

Sheafer, P. W. *The Anthracite Coal Fields of Pennsylvania, and Their Exhaustion*. Read before the American Association for the Advancement of Science, at Saratoga, August 1880. Harrisburg, PA: Lane S. Hart, 1881.

Sheehan, Paul. *Modernism, Narrative and Humanism*. Cambridge: Cambridge University Press, 2004.

Sheldon, Rebekah. *The Child to Come: Life after the Human Catastrophe*. Minneapolis: University of Minnesota Press, 2016.

Shiel, M. P. *The Purple Cloud*. 1901. London: Penguin, 2012.

Shuttleworth, Sally. *George Eliot and Nineteenth-Century Science: The Make-Believe of a Beginning*. Cambridge: Cambridge University Press, 1984.

Sieferle, Rolf Peter. *The Subterranean Forest: Energy Systems and the Industrial Revolution*. 1982. Trans. Michael P. Osman. Cambridge: White Horse, 2001.

Simonin, Louis. *Mines and Miners; or, Underground Life*. Trans., adapted, and ed. by H. W. Bristow. London: Mackenzie, [1868].

Simpson, Eleanor R. "The Evolution of J.R.R. Tolkien's Portrayal of Nature: Foreshadowing Anti-speciesism." *Tolkien Studies* 14 (2017): 71–89.

Sinnema, Peter W. "Introduction." *The Coming Race*. By Edward Bulwer Lytton. Peterborough, ON: Broadview, 2008. 8–25.

Smee, Ben. "'Adani Mine Would Be 'Unviable' without $4.4bn in Subsidies, Report Finds." *Guardian*, 28 August 2019. https://www.theguardian.com/environment/2019/aug/29/adani-mine-would-be-unviable-without-44bn-in-subsidies-report-finds. Accessed 11 May 2020.

Smith, Bradon. "Imagined Energy Futures in Contemporary Speculative Fictions." *Resilience: A Journal of the Environmental Humanities*. 6.2–3 (Spring–Fall 2019): 136–54.

Smith, Jonathan. *Fact and Feeling: Baconian Science and the Nineteenth-Century Literary Imagination*. Madison: University of Wisconsin Press, 1994.

Solnit, Rebecca. *Savage Dreams: A Journey into the Hidden Wars of the American West*. 20th-anniversary ed. Berkeley: University of California Press, 2014.

Spence, Mark David. *Dispossessing the Wilderness: Indian Removal and the Making of the National Parks*. Oxford: Oxford University Press, 1999.

Spicer, Arwen. "Toward Sustainable Change: The Legacy of William Morris, George Bernard Shaw, and H. G. Wells in the Ecological Discourse of Contemporary Science Fiction." PhD diss., University of Oregon, 2005. http://hdl.handle.net/1794/3932. Accessed 1 May 2020.

Stableford, Brian. *Scientific Romance in Britain 1890–1950*. London: Fourth Estate, 1985.

Starr, Elizabeth. "Manufacturing Novels: Charles Dickens on the Hearth in Coketown." *Texas Studies in Literature and Language* 51.3 (Fall 2009): 317–40.

Stauffer, Robert C. "Haeckel, Darwin, and Ecology." *Quarterly Review of Biology* 32.2 (June 1957): 138–44.

Steer, Philip. "Gold and Greater Britain: Jevons, Trollope, and Settler Colonialism." *Victorian Studies* 58.3 (2016): 436–63.

——. *Settler Colonialism in Victorian Literature: Economics and Political Identity in the Networks of Empire*. Cambridge: Cambridge University Press, 2020.

Stevenson, Robert Louis. "My First Book—'Treasure Island.'" *The Works of Robert Louis Stevenson*. Vol. 4, *Miscellanies*. London: Longmans, 1896. 285–97.

——. *Silverado Squatters*. 1883. *From Scotland to Silverado*. Ed. James D. Hart. Cambridge, MA: Harvard University Press, 1966. 189–287.

——. *Treasure Island*. London: Cassell, 1883.

——. *Treasure Island*. 1883. Oxford: Oxford University Press, 2011.

Stiebel, Lindy. "Creating a Landscape of Africa: Baines, Haggard and Great Zimbabwe." *English in Africa* 28.2 (October 2001): 123–33.

Stone-Blackburn, Susan. "Consciousness Evolution and Early Telepathic Tales." *Science Fiction Studies* 20.2 (July 1993): 241–50.

Stott, Rebecca. "'Scaping the Body: Of Cannibal Mothers and Colonial Landscapes." *The New Woman in Fiction and Fact: Fin-de-Siècle Feminisms*. Ed. Angelique Richardson and Chris Willis. New York: Palgrave, 2001. 150–66.

Suvin, Darko. "The Extraordinary Voyage, the Future War, and Bulwer's *The Coming Race*: Three Sub-Genres in British Science Fiction, 1871–1885." *Literature and History* 10.2 (1984): 231–48.

——. *Metamorphoses of Science Fiction: On the Poetics and History of a Literary Genre*. New Haven, CT: Yale University Press, 1979.

——. "Victorian Science Fiction, 1871–85: The Rise of the Alternative History Sub-genre." *Science Fiction Studies* 10.2 (1983): 148–69.

Svampa, Maristella. "Commodities Consensus: Neoextractivism and Enclosure of the Commons in Latin America." *South Atlantic Quarterly* 114.1 (2015): 65–82.

Swanson, Heather, Anna Tsing, Nils Bubandt, and Elaine Gan. "Introduction: Bodies Tumbled into Bodies." A. Tsing et al., *Arts of Living on a Damaged Planet*. M1–M12.

Sympson, Edward Mansel. *Lincolnshire*. Cambridge: Cambridge University Press, 1913.

Szeman, Imre. "Conjectures on World Energy Literature; or, What Is Petroculture?" *Journal of Postcolonial Writing* 53.3 (2017): 277–88.

——. "Literature and Energy Futures." *PMLA* 126.2 (March 2011): 323–25.

Tally, Robert T., Jr. "Places Where the Stars Are Strange: Fantasy and Utopia in Tolkien's Middle-earth." Houghton et al., *Tolkien in the New Century*. 41–56.

Taylor, Jesse Oak. *The Sky of Our Manufacture: The London Fog in British Fiction from Dickens to Woolf*. Charlottesville: University of Virginia Press, 2016.

——. "Storm-Clouds on the Horizon: John Ruskin and the Emergence of Anthropogenic Climate Change." *19: Interdisciplinary Studies in the Long Nineteenth Century* 26 (2018). https://doi.org/10.16995/ntn.802. Accessed 8 April 2020.

——. "While the World Burns: Joseph Conrad and the Delayed Decoding of Catastrophe." *19: Interdisciplinary Studies in the Long Nineteenth Century* 25 (2017). https://doi .org/10.16995/ntn.798. Accessed 25 November 2019.

——. "Wilderness after Nature: Conrad, Empire, and the Anthropocene." Schneider-Rebozo, McCarthy, and Peters, *Conrad and Nature.* 21–42.

Taylor, John. Introduction. Humboldt, *Selections from the Works.* i–xxii.

Taylor, Philip. "Description of a Horizontal Pumping Engine Erected on the Mine of Moran in Mexico." *Philosophical Magazine and Annals of Philosophy* 1.4 (April 1827): 241–45.

Tenenbaum, Barbara A., and James N. McElveen, "From Speculative to Substantive Boom: The British in Mexico, 1821–1911." *English-Speaking Communities in Latin America.* Ed. Oliver Marshall. New York: Palgrave Macmillan, 2000. 51–79.

Tennyson, Alfred. "Ulysses." 1842. *Tennyson: A Selected Edition.* Ed. Christopher Ricks. Berkeley: University of California Press, 1989. 141–45.

Thesing, William B., and Ted Wojtasik. "Poetry, Politics, and Coal Mines in Victorian England: Elizabeth Barrett Browning, Joseph Skipsey, and Thomas Llewelyn Thomas." *Caverns of Night: Coal Mines in Art, Literature, and Film.* Ed. William B. Thesing. Columbia: University of South Carolina Press, 2000. 32–49.

Thiessen, David. "A Baggins Back Yard: Environmentalism, Authorship and the Elves in Tolkien's Legendarium." *The Hobbit and Tolkien's Mythology: Essays on Revisions and Influences.* Ed. Bradford Lee Eden. Jefferson, NC: McFarland, 2014. 195–207.

Thompson, E. P. *William Morris: Romantic to Revolutionary.* 1955. New York: Pantheon, 1976.

Tolkien, J.R.R. *The Hobbit.* 1937. *The Annotated Hobbit.* Ed. Douglas A. Anderson. New York: Houghton Mifflin, 2002. 27–364.

——. *The Letters of J.R.R. Tolkien.* Ed. Humphrey Carpenter. Boston: Houghton Mifflin, 2000.

——. Appendices A through F. In *The Return of the King: Being the Third Part of the Lord of the Rings.* Vol. 3 of *The Lord of the Rings.* New York: Ballantine, 1985. 387–520.

——. "On Fairy-Stories." 1939, 1947. *The Monsters and the Critics, and Other Essays.* London: Allen and Unwin, 1983. 109–62.

Tondre, Michael. "Conrad's Carbon Imaginary: Oil, Imperialism, and the Victorian Petroarchive." *Victorian Literature and Culture* 48.1 (2020): 57–90.

Trollope, Anthony. *John Caldigate.* 1879. London: Zodiac, 1978.

Trout, Teresa. "Scaling Down: H. G. Wells's *The Time Machine* (1895), *Tono-Bungay* (1909), and the Uppark Dolls House." *Journal of Victorian Culture* 24.1 (2019): 88–105.

Tsing, Anna, Heather Swanson, Elaine Gan, and Nils Bubandt, eds. *Arts of Living on a Damaged Planet.* Minneapolis: University of Minnesota Press, 2017.

Tsing, Anna Lowenhaupt. *The Mushroom at the End of the World: On the Possibility of Life in Capitalist Ruins.* Princeton, NJ: Princeton University Press, 2015.

Uncertain Commons. *Speculate This!* Durham, NC: Duke University Press, 2013.

Underwood, Ted. *The Work of the Sun: Literature, Science, and Political Economy, 1760–1860.* New York: Palgrave Macmillan, 2005.

United Mexican Mining Association. *Report of the Court of Directors, Addressed to the Proprietors at a General Meeting on the 7th March, 1827.* London: Philanthropic Society, 1827.

Usher, Phillip John. *Exterranean: Extraction in the Humanist Anthropocene.* New York: Fordham University Press, 2019.

Valeur, Bernard, and Mário N. Berberan-Santos. "A Brief History of Fluorescence and Phosphorescence before the Emergence of Quantum Theory." *Journal of Chemical Education* 88.6 (2011): 731–38.

Van Gosen, Bradley S., Philip L. Verplanck, and Poul Emsbo. "Rare Earth Element Mineral Deposits in the United States (Ver 1.1, 15 April 2019): U.S. Geological Survey Circular 1454." Reston, VA: US Geological Survey, 2019. https://doi.org/10.3133/cir1454. 9 July 2020.

Vaninskaya, Anna. *Fantasies of Time and Death: Dunsany, Eddison, Tolkien*. London: Palgrave Macmillan, 2020.

Verne, Jules. *Journey to the Centre of the Earth*. 1864. Oxford: Oxford University Press, 2008.

Vicinus, Martha. *The Industrial Muse: A Study of Nineteenth Century British Working-Class Literature*. London: Croom Helm, 1974.

[Vizetelly, Henry, published under pseudonym J. Tyrwhitt Brooks.] *Four Months among the Gold-Finders in California: Being the Diary of an Expedition from San Francisco to the Gold Districts*. New York: D. Appleton, 1849.

Wallace-Wells, David. *The Uninhabitable Earth: Life After Warming*. New York: Tim Duggan, 2019.

Wang, Huei-Ju. "Haunting and the Other Story in Joseph Conrad's *Nostromo*: Global Capital and Indigenous Labor." *Conradiana* 44.1 (2012): 1–28.

Ward, H. G. "Mexico in 1827." *Westminster Review*, April 1828, 480–500.

Warde, Paul. "The Invention of Sustainability." *Modern Intellectual History* 8.1 (2011): 153–70.

Watt, Ian. *Conrad in the Nineteenth Century*. Berkeley: University of California Press, 1979.

Webb, Sidney. *The Story of the Durham Miners (1662–1921)*. London: Fabian Society, 1921.

Weber, Max. *The Protestant Ethic and the Spirit of Capitalism*. 1905. Trans. Talcott Parsons. London: Routledge, 1992.

Wells, H. G. *The Time Machine*. 1895. Ed. Nicholas Ruddick. Peterborough, ON: Broadview, 2001.

———. *Tono-Bungay*. 1909. London: Penguin, 2005.

———. *The War of the Worlds*. 1898. London: Penguin, 2005.

———. *The World Set Free*. London: Macmillan, 1914.

"Wellsway Colliery Shaft Accident—Midsomer Norton—1839." Northern Mine Research Society. https://www.nmrs.org.uk/mines-map/accidents-disasters/somerset/wellsway-colliery-shaft-accident-midsomer-norton-1839/. Accessed 11 June 2020.

Wenzel, Jennifer. *The Disposition of Nature: Environmental Crisis and World Literature*. New York: Fordham University Press, 2020.

White, Michael. *Tolkien: A Biography*. 2001. New York: New American Library, 2003.

Whitehead, Tony. "Haswell Colliery and the Disaster of September 1844." *Durham Records Online*, 30 July 2013. https://durhamrecordsonline.com/library/haswell-colliery/. Accessed 9 June 2020.

Whyte, Nicola, and Axel Goodbody. "Pandaemonium: Narratives of Energy-System Change in Historical and Literary Perspective." *Resilience: A Journal of the Environmental Humanities* 6.2–3 (Spring–Fall 2019): 26–52.

Wiedenfeld, Logan. "Excess and Economy in *Sons and Lovers*." *Studies in the Novel* 48.3 (Fall 2016): 301–17.

Williams, Daniel. "The Clouds and the Poor: Ruskin, Mayhew, and Ecology." *Nineteenth-Century Contexts* 38.5 (2016): 319–31.

Williams, John. *The Natural History of the Mineral Kingdom*. Vol. 1. Edinburgh: Williams, 1789.

Williams, M., J. A. Zalasiewicz, C. N. Waters, and E. Landing. "Is the Fossil Record of Complex Animal Behaviour a Stratigraphical Analogue for the Anthropocene?" *A Stratigraphical Basis for the Anthropocene*. Ed. C. N. Waters et al. London: Geological Society, 2013. 143–48.

Williams, Raymond. *The Country and the City*. Oxford: Oxford University Press, 1973.

——. "Socialism and Ecology." *Capitalism, Nature, Socialism* 1 (1995): 41–57.

——. "Working-Class, Proletarian, Socialist: Problems in Some Welsh Novels." *The Socialist Novel in Britain: Towards the Recovery of a Tradition*. Ed. H. Gustav Klaus. Brighton: Harvester, 1982. 110–21.

——. *Writing in Society*. London: Verso, 1983.

Williams, Rosalind. *Notes on the Underground: An Essay on Technology, Society, and the Imagination*. Cambridge, MA: MIT Press, 1990.

Williams, W. "Report of the Inspector of Cotton Cloth Factories." *Annual Report of the Chief Inspector of Factories and Workshops for the Year 1901: Part I.—Reports*. London: Darling and Son, 1902. 194–203.

Winter, James. *Secure from Rash Assault: Sustaining the Victorian Environment*. Berkeley: University of California Press, 1999.

Wisnicki, Adrian S. *Fieldwork of Empire, 1840–1900: Intercultural Dynamics in the Production of British Expeditionary Literature*. New York: Routledge, 2019.

Woloch, Alex. *Or Orwell: Writing and Democratic Socialism*. Cambridge, MA: Harvard University Press, 2016.

Wood, Naomi J. "Gold Standards and Silver Subversions: *Treasure Island* and the Romance of Money." *Children's Literature* 26 (1998): 61–85.

Worthen, John. *D. H. Lawrence: The Early Years 1885–1912*. Cambridge: Cambridge University Press, 1992.

Wrigley, E. A. *Energy and the English Industrial Revolution*. Cambridge: Cambridge University Press, 2010.

——. *The Path to Sustained Growth: England's Transition from an Organic Economy to an Industrial Revolution*. Cambridge: Cambridge University Press, 2016.

Wulf, Andrea. *The Invention of Nature: Alexander von Humboldt's New World*. New York: Knopf, 2015.

Wybergh, W. J. *The Coal Resources of Union of South Africa*. 3 vols. Pretoria: Government Printing and Stationery Office, 1922.

Wynter, Sylvia. "On How We Mistook the Map for the Territory, and Re-imprisoned Ourselves in Our Unbearable Wrongness of Being, of *Désêtre*: Black Studies toward the Human Project." *Not Only the Master's Tools: African-American Studies in Theory and Practice*. Ed. Lewis R. Gordon and Jane Anna Gordon. Boulder, CO: Paradigm, 2006. 107–69.

——. "Unsettling the Coloniality of Being/Power/Truth/Freedom: Towards the Human, after Man, Its Overrepresentation—An Argument." *CR: The New Centennial Review* 3.3 (2003): 257–337.

Yaeger, Patricia. Introduction. "Editor's Column: Literature in the Ages of Wood, Tallow, Coal, Whale Oil, Gasoline, Atomic Power, and Other Energy Sources." *PMLA* 126.2 (March 2011): 305–10.

Yusoff, Kathryn. *A Billion Black Anthropocenes or None*. Minneapolis: University of Minnesota Press, 2018.

Yuval-Naeh, Naomi. "Cultivating the Carboniferous: Coal as a Botanical Curiosity in Victorian Culture." *Victorian Studies* 61.3 (Spring 2019): 419–45.

Zalasiewicz, Jan, Colin N. Waters, and Mark Williams. "Human Bioturbation, and the Subterranean Landscape of the Anthropocene." *Anthropocene* 6 (2014): 3–9.

Zemka, Sue. *Time and the Moment in Victorian Literature and Society*. Cambridge: Cambridge University Press, 2012.

Zola, Émile. *Germinal*. 1885. Trans. L. W. Tancock. Harmondsworth: Penguin, 1973.

INDEX

Page numbers in *italics* indicate illustrations.

Abarodhbasini (Hossain), 233n43

Achebe, Chinua, 133

Ackermann, A.S.E., "The Utilisation of Solar Energy," 163

Adams, William, *Solar Heat: A Substitute for Fuel in Tropical Countries*, 165, 166

Adorno, Theodor, 172

adventure literature, 82–139; centrifugal extractivism of, 85–88; colonial frontiers of, 82–85, 86–88, 91, 93–97, 139; concealments and unknowns in, 87, 94–95, 139; as genre of extraction-based life, 85, 139; and geognosy, 95–96; Global South settings of, 88–93; and hollow earth fiction, 136, 223n40; limited perspective in, 96–97, 123–24, 139; *vs.* provincial realism, 82–83, 86, 99, 123–24; and treasure hunting, 84, 88, 94, 96–97, 139; unintended consequences in, 94, 96; vertical frontiers of, 85–88

Africa: coal exported to, 168; ivory trade in, 134, 227n81; as setting in adventure literature, 88–93, 221n19; and solar power, 163–164. *See also* central Africa; South Africa

Aguirre, Robert D., 88–89, 101, 212n24, 222n32, 224n49

Ahuja, Neel, 28

Alaimo, Stacy, 74, 217n69

Algeria, and solar power, 164

Allewaert, Monique, 226n78

alloy, 66

aluminum, 178, 184, 235n63

Amatya, Alok, 210n38

American Mines, The (G. H.), 36

Anderson, Douglas A., 238n95

Andrews, W. S., 218n85

Andrews, William L., 100

animacy, 133–35

"Answer to Sultana's Dream, An" (Padmini), 232n36

Anthropocene, the, 14–17, 29, 199, 205n1, 206n11; and adventure literature, 104–105, 130–31; and *Hard Times* (Dickens), 70; literature of, 14–17; and provincial realism, 212n20; and speculative fiction, 150; temporality of, 61–62; uses of term, 209n32, 209n35

anthroturbation, 1, 84, 136

Antipodes, the, 93, 221n19

Archimedes, and solar power, 164, 232n39

Armstrong, Henry E., 10–11, 80

Arrhenius, Svante, 217n66

arsenic: and Morris, 173–74, 234n59

"Art and the Beauty of the Earth" (Morris), 176

"Art and Socialism" (Morris), 172–73, 203

"Art under Plutocracy" (Morris), 172, 174–75

asbestos, 125

Association Internationale Africaine, 226n76

Atlantic slave trade, 209n35, 220n17. *See also* slavery

atmospheric carbon dioxide, 17, 35, 47, 123, 146, 210n36, 214n38, 217n66, 232n43

atomic power. *See* nuclear power

Austen, Peter Townsend, 164

Australia: coal mining in, 148; exhaustion of mines in, 225n65; extraterritorial mining of, 238n4; gold rushes in, 20, 33, 83–84, 93, 221n19, 221n20; in maps within *Heart of Darkness* (Conrad), 107; mining accidents in, 216n58; rare earth mining in, 238n3; as setting in *John Caldigate* (Trollope), 83

Babbage, Charles, 15, 209n33

Bacon, Francis, 233n48

Baines, Thomas, 225n65

Bakhtin, M. M., 30

Ballantine, 229n6

A NOTE ON THE TYPE

THIS BOOK has been composed in Miller, a Scotch Roman typeface designed by Matthew Carter and first released by Font Bureau in 1997. It resembles Monticello, the typeface developed for The Papers of Thomas Jefferson in the 1940s by C. H. Griffith and P. J. Conkwright and reinterpreted in digital form by Carter in 2003.

Pleasant Jefferson ("P. J.") Conkwright (1905–1986) was Typographer at Princeton University Press from 1939 to 1970. He was an acclaimed book designer and AIGA Medalist.

The ornament used throughout this book was designed by Pierre Simon Fournier (1712–1768) and was a favorite of Conkwright's, used in his design of the *Princeton University Library Chronicle*.

CPSIA information can be obtained
at www.ICGtesting.com
Printed in the USA
LVHW102332011122
732152LV00013B/257